VOLUME SEVEN

Advances in
CHEMICAL POLLUTION, ENVIRONMENTAL MANAGEMENT AND PROTECTION

Biochar: Fundamentals and Applications in Environmental Science and Remediation Technologies

VOLUME SEVEN

ADVANCES IN
CHEMICAL POLLUTION, ENVIRONMENTAL MANAGEMENT AND PROTECTION

Biochar: Fundamentals and Applications in Environmental Science and Remediation Technologies

Edited by

AJIT K. SARMAH
Department of Civil and Environmental Engineering,
The Faculty of Engineering, The University of Auckland,
Auckland, New Zealand

Academic Press is an imprint of Elsevier
50 Hampshire Street, 5th Floor, Cambridge, MA 02139, United States
525 B Street, Suite 1650, San Diego, CA 92101, United States
The Boulevard, Langford Lane, Kidlington, Oxford OX5 1GB, United Kingdom
125 London Wall, London, EC2Y 5AS, United Kingdom

First edition 2021

Copyright © 2021 Elsevier Inc. All rights reserved.

No part of this publication may be reproduced or transmitted in any form or by any means, electronic or mechanical, including photocopying, recording, or any information storage and retrieval system, without permission in writing from the publisher. Details on how to seek permission, further information about the Publisher's permissions policies and our arrangements with organizations such as the Copyright Clearance Center and the Copyright Licensing Agency, can be found at our website: www.elsevier.com/permissions.

This book and the individual contributions contained in it are protected under copyright by the Publisher (other than as may be noted herein).

Notices
Knowledge and best practice in this field are constantly changing. As new research and experience broaden our understanding, changes in research methods, professional practices, or medical treatment may become necessary.

Practitioners and researchers must always rely on their own experience and knowledge in evaluating and using any information, methods, compounds, or experiments described herein. In using such information or methods they should be mindful of their own safety and the safety of others, including parties for whom they have a professional responsibility.

To the fullest extent of the law, neither the Publisher nor the authors, contributors, or editors, assume any liability for any injury and/or damage to persons or property as a matter of products liability, negligence or otherwise, or from any use or operation of any methods, products, instructions, or ideas contained in the material herein.

ISBN: 978-0-12-820178-7
ISSN: 2468-9289

For information on all Academic Press publications
visit our website at https://www.elsevier.com/books-and-journals

Publisher: Zoe Kruze
Acquisitions Editor: Jason Mitchell
Developmental Editor: Jhon Michael Peñano
Production Project Manager: Denny Mansingh
Cover Designer: Victoria Pearson

Typeset by STRAIVE, India

Contents

Contributors	*ix*
Series editor's preface	*xiii*
Preface	*xv*

1. Biochar amalgamation with clay: Enhanced performance for environmental remediation   1

Thilakshani Atugoda, Ahmed Ashiq, S. Keerthanan, Prabuddhi Wijekoon, Sammani Ramanayaka, and Meththika Vithanage

1. Introduction	2
2. Preparation methods of clay-biochar composites	3
3. Properties of clay-biochar composites	10
4. Adsorptive removal of organic contaminants using clay-biochar composite	22
5. Clay biochar composites for inorganic contaminant adsorption	25
6. Mechanisms governing adsorptive removal by clay-biochar composites	29
7. Future perspectives	32
References	34

2. Remediation of heavy metal contaminated soil: Role of biochar   39

Lina Gogoi, Rumi Narzari, Rahul S. Chutia, Bikram Borkotoki, Nirmali Gogoi, and Rupam Kataki

1. Introduction	40
2. Inorganic contaminants in soil	41
3. Biochar as a soil amendment	45
4. Mechanisms of heavy metal remediation by biochar	48
5. Biochar applications in heavy metal remediation	51
6. Conclusion and recommendations for further research	51
References	57

3. Application of biochar for emerging contaminant mitigation   65

Elsa Antunes, Arun K. Vuppaladadiyam, Ajit K. Sarmah, S.S.V. Varsha, Kamal Kishore Pant, Bhagyashree Tiwari, and Ashish Pandey

1. Introduction	66
2. Emerging contaminants classes	68
3. Adsorption technology for the removal of ECs	74

v

4. Adsorption of ECs	79
5. Technoeconomic feasibility of Biochar's use as adsorbent	83
6. Conclusions and future prospects	84
References	86

4. Role of biochar as a cover material in landfill waste disposal system: Perspective on unsaturated hydraulic properties 93

Sanandam Bordoloi, Janarul Shaikh, Ján Horák, Ankit Garg, S. Sreedeep, and Ajit K. Sarmah

1. Introduction	94
2. Materials and methods	96
3. Results and discussion	101
4. Conclusions	104
References	105
Further reading	106

5. Effects of modified biochar on As-contaminated water and soil: A recent update 107

Jingzi Beiyuan, Yiyin Qin, Qiqi Huang, Hailong Wang, Daniel C.W. Tsang, and Jörg Rinklebe

1. Introduction	108
2. Biochar for As contamination: Limitation and challenge	112
3. Effect of Fe/Mn-based modified biochar	116
4. Effects of other modified biochar in As-contaminated soil	124
5. The way forward	129
References	130

6. Biochar for modification of manure properties 137

Sören Thiele-Bruhn and Anastasiah N. Ngigi

1. Introduction	138
2. Biochar as a feed supplement and amendment to bedding material	140
3. Biochar as a bulking agent for manure handling, storage and further use	142
4. Biochar for improving manure composting	144
5. Biochar effects on anaerobic digestion and biogas formation	148
6. Nutrient availability and delivery in soil of biochar amended manure	150
7. Impact of biochar on the fate and effects of micropollutants in manure	157
8. Summary and conclusion	161
References	162

Contents

vii

7. Biochar role in improving pathogens removal capacity of stormwater biofilters
175

Renan Valenca, Annesh Borthakur, Huong Le, and Sanjay K. Mohanty

1. Introduction — 176
2. Testing methods to evaluate biochar capacity to remove pathogens in stormwater — 179
3. Pathogen removal processes in biochar-amended filter media — 184
4. Challenges — 189
5. Opportunities — 193
6. Summary — 195
References — 195

8. A relationship paradigm between Biochar amendment and greenhouse gas emissions
203

Mohd Ahsaan, Pratibha Tripathi, Anupama, and Puja Khare

1. Introduction — 204
2. Strategies adopted for mitigation of GHG emission — 206
3. What is biochar? — 207
4. Role of biochar in altering the factor responsible for GHGs emissions — 209
5. Effect of biochar addition on major soil GHGs emission — 212
6. Limitations — 216
Acknowledgments — 217
References — 217

9. Biochar for sustainable agriculture: Prospects and implications
221

Kumar Raja Vanapalli, Biswajit Samal, Brajesh Kumar Dubey, and Jayanta Bhattacharya

1. Introduction — 222
2. Agronomic properties of biochar — 225
3. Biochar induced soil quality improvement — 229
4. Biochar-induced crop growth and production — 237
5. Economics of biochar application for agriculture — 248
6. Concluding remarks and future research needs — 250
References — 251

10. Structure and function of biochar in remediation and as carrier of microbes
263

Kim Yrjälä and Eglantina Lopez-Echartea

1. Introduction — 264
2. Structures of biochar — 265

3. Application of biochar in remediation	274
4. Microbes and biochar	283
References	288

11. Influence of process parameters for production of biochar: A potential tool for an energy transition 295

Biswajit Samal, Kumar Raja Vanapalli, Brajesh Kumar Dubey, and Jayanta Bhattacharya

1. Introduction	296
2. Source of biomass	298
3. Conversion techniques	299
4. Implications of process parameters on physicochemical and thermal properties of biochar	304
5. Application of biochar	307
6. Summary and future prospects	308
References	309

Contributors

Mohd Ahsaan
Crop Production and Protection Division, CSIR-Central Institute of Medicinal and Aromatic Plants, Lucknow; Academy of Scientific and Innovative Research (AcSIR), Ghaziabad, India

Elsa Antunes
College of Science and Engineering, James Cook University, Townsville, QLD, Australia

Anupama
Crop Production and Protection Division, CSIR-Central Institute of Medicinal and Aromatic Plants, Lucknow; Academy of Scientific and Innovative Research (AcSIR), Ghaziabad, India

Ahmed Ashiq
Ecosphere Resilience Research Centre, Faculty of Applied Sciences, University of Sri Jayewardenepura, Nugegoda, Sri Lanka

Thilakshani Atugoda
Ecosphere Resilience Research Centre, Faculty of Applied Sciences, University of Sri Jayewardenepura, Nugegoda, Sri Lanka

Jingzi Beiyuan
School of Environmental and Chemical Engineering, Foshan University, Foshan, China

Jayanta Bhattacharya
School of Environmental Science and Engineering; Department of Mining Engineering, Indian Institute of Technology, Kharagpur, West Bengal, India

Sanandam Bordoloi
Department of Civil Engineering, Indian Institute of Technology Guwahati, Assam, India

Bikram Borkotoki
Biswanath College of Agriculture, Assam Agricultural University, Sonitpur, India

Annesh Borthakur
Department of Civil and Environmental Engineering, University of California Los Angeles, California, United States

Rahul S. Chutia
Department of Energy Engineering, North Eastern Hill University, Shillong, India

Brajesh Kumar Dubey
School of Environmental Science and Engineering; Department of Civil Engineering, Indian Institute of Technology, Kharagpur, West Bengal, India

Ankit Garg
Department of Civil and Environmental Engineering, Shantou University, Shantou, China

Lina Gogoi
Department of Energy; Department of Environmental Sciences, Tezpur University, Tezpur, India

Nirmali Gogoi
Department of Environmental Sciences, Tezpur University, Tezpur, India

Ján Horák
Department of Biometeorology and Hydrology, Faculty of Horticulture and Landscape Engineering, Slovak University of Agriculture in Nitra, Nitra, Slovakia

Qiqi Huang
School of Environmental and Chemical Engineering, Foshan University, Foshan, China

Rupam Kataki
Department of Energy, Tezpur University, Tezpur, India

S. Keerthanan
Ecosphere Resilience Research Centre, Faculty of Applied Sciences, University of Sri Jayewardenepura, Nugegoda, Sri Lanka

Puja Khare
Crop Production and Protection Division, CSIR-Central Institute of Medicinal and Aromatic Plants, Lucknow; Academy of Scientific and Innovative Research (AcSIR), Ghaziabad, India

Huong Le
Department of Civil and Environmental Engineering, University of California Los Angeles, California, United States

Eglantina Lopez-Echartea
Department of Biochemistry and Microbiology, University of Chemistry and Technology, Prague, Czech Republic

Sanjay K. Mohanty
Department of Civil and Environmental Engineering, University of California Los Angeles, California, United States

Rumi Narzari
Department of Energy, Tezpur University, Tezpur, India

Anastasiah N. Ngigi
Department of Chemistry, Multimedia University of Kenya, Nairobi, Kenya

Ashish Pandey
Catalytic Reaction Engineering Laboratory, Department of Chemical Engineering, IIT Delhi, New Delhi, India

Kamal Kishore Pant
Catalytic Reaction Engineering Laboratory, Department of Chemical Engineering, IIT Delhi, New Delhi, India

Yiyin Qin
School of Environmental and Chemical Engineering, Foshan University, Foshan, China

Sammani Ramanayaka
Ecosphere Resilience Research Centre, Faculty of Applied Sciences, University of Sri Jayewardenepura, Nugegoda, Sri Lanka

Jörg Rinklebe
University of Wuppertal, Institute of Foundation Engineering, Waste and Water Management, School of Architecture and Civil Engineering, Soil and Groundwater Management, Wuppertal, Germany

Biswajit Samal
School of Environmental Science and Engineering, Indian Institute of Technology, Kharagpur, West Bengal, India

Ajit K. Sarmah
Department of Civil and Environmental Engineering, The Faculty of Engineering, The University of Auckland, Auckland, New Zealand

Janarul Shaikh
Department of Biometeorology and Hydrology, Faculty of Horticulture and Landscape Engineering, Slovak University of Agriculture in Nitra, Nitra, Slovakia; Department of Civil Engineering, C.V. Raman Global University, Bhubaneswar, Odisha, India

S. Sreedeep
Department of Civil Engineering, Indian Institute of Technology Guwahati, Assam, India

Sören Thiele–Bruhn
Soil Science, University of Trier, Trier, Germany

Bhagyashree Tiwari
INRS-Eau, Terre et Environment, Quebec City, QC, Canada

Pratibha Tripathi
Crop Production and Protection Division, CSIR-Central Institute of Medicinal and Aromatic Plants, Lucknow, India

Daniel C.W. Tsang
Department of Civil and Environmental Engineering, Hong Kong Polytechnic University, Hong Kong, China

Renan Valenca
Department of Civil and Environmental Engineering, University of California Los Angeles, California, United States

Kumar Raja Vanapalli
School of Environmental Science and Engineering, Indian Institute of Technology, Kharagpur, West Bengal, India

S.S.V. Varsha
Department of Civil Engineering, AVN institute of Engineering & Technology, Hyderabad, Telengana, India

Meththika Vithanage
Ecosphere Resilience Research Centre, Faculty of Applied Sciences, University of Sri Jayewardenepura, Nugegoda, Sri Lanka

Arun K. Vuppaladadiyam
College of Science and Engineering, James Cook University, Townsville, QLD, Australia; Catalytic Reaction Engineering Laboratory, Department of Chemical Engineering, IIT Delhi, New Delhi, India

Hailong Wang
School of Environmental and Chemical Engineering, Foshan University, Foshan, China

Prabuddhi Wijekoon
Ecosphere Resilience Research Centre, Faculty of Applied Sciences, University of Sri Jayewardenepura, Nugegoda, Sri Lanka

Kim Yrjälä
Zhejiang A & F University, State Key Laboratory of Subtropical Silviculture, Hangzhou, Zhejiang, China; Department of Forest Sciences, University of Helsinki, Helsinki, Finland

Series editor's preface

In October 2018 during my visit at the University of Auckland, in New Zealand, I was able to convince my old friend Ajit Sarmah to edit a biochar book for this series. He accepted immediately and today the final product is ready for the scientific community as well as for the general public. The thing is that during the last few years, biochar research has been very popular worldwide with the immediate consequences that many papers are being published every year. This is the reason why this book on fundamentals and applications of biochar in environmental sciences was urgently needed. Needless to say that I am very happy about its publication that is also a good follow-up of Volume 2 from this series on Soil Degradation, Restoration and Management edited by Paulo Pereira, who is another good friend of mine. With this new volume on biochar, we added another brick to soil-related issues in this series.

Biochar has many properties and applications; perhaps one of the key ones is to help sustainable development. In addition to cleaning up wastes, biochar also plays a key role in a variety of human activities in the realization of a circular economy and sustainable development. Driven by its charged surface and multiple functional groups, biochar is emerging as an effective and safe natural adsorbent that can remove diverse organic contaminants (e.g., antibiotics, agrochemicals, polychlorinated biphenyls, and polycyclic aromatic hydrocarbons) and inorganic contaminants (e.g., phosphate, ammonia, arsenic, sulfide, and heavy metals) from solid, aqueous, and/or gaseous media. As a soil amendment, it can improve plant productivity and photosynthesis rate by enhancing the physical, chemical, and biological properties of the soil. Biochar addition to agricultural soils has improved soil water availability, water-holding capacity, and nutrient availability; increased soil microbial biomass and activity; and reduced risk of crust formation and soil erosion. Biochar together with nutrients and microorganisms can be used as a carrier material for agricultural inputs, thus increasing the nutrient use efficiency of the inoculated microorganisms in the soil. In addition, biochar can also reduce the emission of CH_4, N_2O, and other air pollutants during the degradation of biomass in the soil, mainly by adsorbing free C and N compounds to its surface and changing the properties of the systems. For example, biochar used as a soil amendment can reduce soil CH_4 and N_2O emissions between 30% and 40%.

Furthermore, biochar has been shown to mitigate the emission of GHGs (CH_4, N_2O, and CO_2) during composting. The high porosity and variable charged sites of biochar make it a good sink for GHGs. Therefore, the conversion of agricultural waste into biochar with the aim of soil fertilization is regarded as a promising strategy for storing soil nutrients and reducing GHG emissions.

All these aspects cited above and few more are being discussed in this book that contains 11 chapters covering fundamental issues like unsaturated soil properties as well as many remediation and useful applications like removal strategies for arsenic, metals, organic pollutants, and pathogens among other examples. Reduction of GHGs using biochar deserved a whole chapter as well as the last two chapters on how biochar can be used as a carrier of microbes for decontamination purposes and the last one on the optimization of the different parameters controlling physicochemical and thermal characteristics of pyrolyzed biochar.

In short, this book will be very useful for soil scientists, environmental and chemical engineers, microbiologists, and waste managers. It is a multi-purpose book that can be used for undergraduates, PhD students, stake-holders, and government officers working in agricultural, environmental, and waste departments.

Lastly, I would like to thank especially Ajit for the great effort as editor as well as to all the well-known coauthors who contributed to this unique and timing volume on biochar. Thanks to all of you and hopefully most of you, authors and readers of this book, are already vaccinated. Please stay safe and healthy!

D. Barceló

IDAEA-CSIC, Barcelona and ICRA-CERCA, Girona

October 12, 2021

Editor-in-Chief of the *Advances in Chemical Pollution, Environmental Management and Protection* Series

Preface

A large body of scientific literature in the form of journal articles and book chapters on the fundamentals of biochar and its applications in diverse sectors such as soil amendment, greenhouse gas mitigation, contaminant mitigation, construction and building, composite development, catalysts, and energy storage is currently available. Given the advances in biochar production, characterization, and its applications in the aforementioned areas over the past few decades, a plethora of new knowledge/information has been generated through research conducted in laboratories and at the field scale, and it is now time to present some of these new findings in a more comprehensive manner in a book volume. Therefore, the purpose of this volume is not to repeat what has already been discussed but to add new knowledge and fundamentals of science and technologies pertinent to biochar application in soil and water remediation, as landfill cover material, in altering manure properties, removing pathogens in stormwater filters, greenhouse gas mitigation, sustainable agriculture, and as microbial carriers for remediation, as well as to discuss how process parameters of biochar production can play a role in energy transition.

Chapter 1 focuses on the development of clay–biochar as a novel composite material for remediation of environmental contaminants. The overarching objective is to provide an overview of various preparation techniques for the preparation of a clay–biochar composite, its properties, use in adsorptive removal of both organic and inorganic contaminants from water, and the underlying mechanisms responsible for the removal of contaminants from the concerned media.

Chapter 2 focuses on the role of biochar on heavy metal remediation with particular emphasis on the different types and sources of heavy metal contamination in soil to understand the biochar-induced underlying mechanisms of heavy metal remediation in soil.

Chapter 3 focuses on the application of biochar as an effective adsorbent for emerging contaminant mitigation focusing on recent updates. The authors provide a holistic assessment on the source and categories of emerging contaminants, adsorption as a potential technology for their removal, and the mechanisms of adsorption. The chapter also includes critical analyses of adsorption, the mechanisms involved during the removal of contaminants from water, and a perspective on the use and importance of artificial neural

xv

networks in adsorption technology. Finally, technoeconomic feasibility assessment of using biochar as an effective sorbent is discussed.

The overarching objective of Chapter 4 is to investigate the unsaturated soil properties (SWRC, gas permeability, and infiltration rate), which play important roles in landfill cover functioning. The authors emphasize how various hydraulic properties can influence the functioning of the cover material in a landfill waste disposal system by providing findings from a series of laboratory experiments.

Chapter 5 reviews the latest findings on how modified biochar can be utilized for remediating arsenic (a highly toxic and nonmetallic element)-contaminated soils. New modification methods for arsenic removal/immobilization in both water and soil are reviewed, such as Fe-modified biochar, Mn-modified biochar, Zn-modified biochar, Si-modified biochar, rare earth element-modified biochar, and acid–base-modified biochar.

Chapter 6 focuses on how the properties of manure can be altered by biochar addition. The overarching aim of the chapter is to review the recent literature on the use of biochar as an additive to manure or similar organic waste materials. The authors specifically discuss the use of biochar as a bulking agent for manure handling, storage and further uses, effects on anaerobic digestion and biogas formation, nutrient availability, and delivery in soil of biochar-amended manure and the impact of biochar on the fate and effects of micropollutants in manure.

Chapter 7 focuses on how biochar addition in stormwater treatment systems such as biofilters affects the pathogen-removal capacity of the biofilters. The authors discuss the available literature data on the flow-through column studies examining the potential of biochar in removing pathogens from stormwater. In addition, the removal processes of pathogens from biochar-amended filter media are discussed with special attention on attachment and straining as well as die-off and inactivation of virus.

Chapter 8 focuses on how biochar can be used as a promising and cost-effective mitigation strategy/solution to reduce greenhouse gas (GHG) emissions without compromising farm productivity in terms of quality or quantity of farm produce. The authors specifically discuss the role and impact of biochar on indigenous soil organic carbon, methanotrophs, and the underlying mechanisms responsible for soil C sequestration and mitigation of GHG emissions.

Chapter 9 mainly focuses on the agronomic potential of biochar for improved crop growth and productivity. The chapter is specifically aimed at how biochar influences the soil nutrient cycle along with its stability and durability in the long term. The economics of integrating biochar application into the current agricultural practices and the future research needs for achieving economic and environmental sustainability of the process are also reviewed.

Chapter 10 focuses on the current knowledge in sustainable remediation and discusses the possibilities of using biochar as a carrier of remedial microbes. The specific aim of the chapter is to provide an in-depth review of the carbon structure effects on remediation, seeking suitable designer biochar for use in environmental remediation, considering different waste streams as a source of feedstock and how microbes can improve decontamination.

The overarching objective of Chapter 11 is to discuss the effects of various process parameters such as flow rate, pressure, heating rate, residence time, pyrolysis temperature, and the biomass composition on the physico-chemical and thermal characteristics of pyrolyzed biochar.

I acknowledge all the authors of this volume for their contribution and timely submission of the chapters. I thank Shellie Bryant and Jason Mitchell for the initial communication. A special thanks goes to Jhon Michael Peñano who has been extremely helpful in the valuable and friendly collaboration right from the first contact through to the completion of this editorial venture. Last but not the least, I express my sincere thanks to the series editor, Damià Barceló, for inviting me to act as the volume editor of this book series.

Ajit K. Sarmah, PhD, MEng, MS, BSc AgEng (Distn)
Department of Civil and Environmental Engineering
The Faculty of Engineering
The University of Auckland
Auckland, New Zealand

CHAPTER ONE

Biochar amalgamation with clay: Enhanced performance for environmental remediation

Thilakshani Atugoda, Ahmed Ashiq, S. Keerthanan,
Prabuddhi Wijekoon, Sammani Ramanayaka,
and Meththika Vithanage*

Ecosphere Resilience Research Centre, Faculty of Applied Sciences, University of Sri Jayewardenepura, Nugegoda, Sri Lanka
*Corresponding author: e-mail address: meththika@sjp.ac.lk

Contents

1. Introduction	2
1.1 Composite with biochar and clay	3
2. Preparation methods of clay-biochar composites	3
2.1 Slurry method	4
2.2 Biomass-clay pyrolysis method	9
3. Properties of clay-biochar composites	10
3.1 Physical properties	10
3.2 Chemical properties	14
4. Adsorptive removal of organic contaminants using clay-biochar composite	22
4.1 Removal of pharmaceuticals and personal care products	22
4.2 Removal of dyes and pigments	23
5. Clay biochar composites for inorganic contaminant adsorption	25
5.1 Removal of inorganic ions	25
5.2 Removal of heavy metals	27
6. Mechanisms governing adsorptive removal by clay-biochar composites	29
7. Future perspectives	32
References	34

Abstract

The scope of environmental remediation governed by biochar is limited to that of activated carbon due to its lower functionality and lower surface area. Incorporation of clay to biochar could possibly impart excellent sportive properties to treat a diverse range of pollutants effectively through adsorption. Clays are considered as natural pollutant scavengers for their unique characteristics such as high specific surface area, surface chemistry, and structural properties. Clay-biochar composites are generally prepared through pyrolyzing a mixture of biomass and clay or through mixing biochar and clay.

Montmorillonite, bentonite, kaolinite and attapulgite clays are widely used in the synthesis of clay-biochar composites. This chapter assesses the physicochemical characteristics of the composite based biomass and clay type, clay biomass ratio, and pyrolysis condition. Pore diameter, pore-volume, surface area, and induced functional groups on the adsorbent material are the major alterations made on the biochar surface. Predominant mechanism for organic pollutant adsorption on to the composite occurs mainly through electrostatic and hydrophobic interactions, while the binding of heavy metal and nutrient ions on to the composite occurs through cationic exchange mechanism and strong covalent bonding.

Keywords: Biochar composites, Clay, Environmental remediation, Physicochemical properties, Adsorption

1. Introduction

Extensive studies have been undertaken on the use of biochar for pollution abatement in diverse environmental matrices. Biochar possess has shown excellent adsorption potential due to its large specific surface area, porous structure, and multi-functional groups as compared with other adsorbents. Biochar is also known to be economically more viable and environment-friendly, as it is derived from organic materials.[1] Modifications to biochar, are quite common, to enhance the adsorption capacity of biochar toward organic and inorganic pollutants. In order to improve performance of biochar, feedstock pretreatment and biochar post-treatment strategies such as physical (grinding, sieve, wash, etc.) chemical methods (acid treatment, loading of precursors, functional agents, etc.) and thermochemical treatment methods (pyrolysis temperature and duration) have been conducted.[2,3]

Biochar-based composites are sought to have improved properties such as high surface area, and porosity than regular biochar including alterations to the functional groups on the biochar surfaces.[4] Thus, developing a biochar-composite to compensate any drawback that pristine biochar withholds, would be a viable option that potentially leads to remediate a wider range of pollutants in environmental matrices. A composite of biochar and clay is advantageous over other strategies for environmental remediation since both materials are cost-effective and possess good adsorption performances. Hence, the overarching objective of this chapter is to review how tailoring biochar with clays as a multi-functional adsorbent can help enhanced biochar's properties and its effectiveness in environmental remediation.[5–8]

1.1 Composite with biochar and clay

Incorporating clay with biochar, enhances biochar to be a better adsorbent material for environmental remediation that cannot be achieved through the use of pristine biochar and clay, individually.[9,10] For example, attapulgite-biochar composite showed a great affinity for quinolone antibiotics with an adsorption capacity of $5.24\,\mathrm{mgg^{-1}}$, which is two-fold higher than the pristine biochar.[11] Adsorption performance of montmorillonite (MMT) and biochar composite[12] improved heavy metal remediation wherein, Zn^{2+} was removed at optimum adsorption capacity of $8.163\,\mathrm{mgg^{-1}}$ for MMT-biochar composite prepared at $350\,°C$. A dramatic improvement was noticed in the presence of clay materials during the pre-treatment process of biochar, to remediate methylene blue. Adsorption experiments indicated a 5-fold increase in adsorption capacity for MMT-bagasse biochar compared to the pristine biochar for the removal of methylene blue.[13] Higher surface area, surface activity, and increased inter-lamellar spaces of the clay in the composite was also found to improve the adsorption of tetracycline by MMT- biochar composite.[14] Increased surface area of MMT-biochar composite, enhanced adsorption capacities for NH_4^+ and $PO_4{}^{3-}$ compared to the pristine biochar.[15] Through the studies reviewed here,[12–14] the techniques such as infrared and X-ray spectral analysis along with batch adsorption studies, led to the conclusion that due to the relatively larger surface area and pore channels of biochar, clay minerals get easily steeped into the pores of the biochar and further improves the adsorption capacity. An exemplification of these characterization techniques and batch adsorption studies are detailed in this chapter. Clay-biochar composites represent a straightforward but effective way to combine the best properties of both materials as an individual, obtaining acceptable morphological and adsorptive characteristics, which cannot be fulfilled with standalone pristine biochar.

2. Preparation methods of clay-biochar composites

Synthesis of clay-based carbonaceous composites is challenging as the surface area, and pore volumes of the so-produced composites have been improved while retaining the clay structure (e.g., 2:1 for bentonite/MMT or 1:1 for kaolinite).[12,16] Thus, the thermal treatment is a crucial step for the synthesis of the final composite that can counter-balance with the water molecules, by holding the clay structure and the cations. This can tremendously influence the physicochemical characteristics of the composite, which should be taken into account during composite preparation.[17] Researchers have also focused on the final yield of the composite that is mostly influenced by the selection of biomass utilized for the clay-biochar synthesis. Moreover, the cellulose, lignin, and hemicellulose content of the biomass control the final yield of the clay-biochar composite, which are further elaborated in the later parts of the sections.[13,18]

The preparation of clay-biochar composites is followed by two main preparation strategies. Both methods involve pyrolyzing of the feedstock; however, based on the feeds provided to the pyrolyzer, distinct methods are followed: slurry method and biomass-clay pyrolysis method. Flowchart, given below (Fig. 1), outlines the preparation strategies for clay-biochar composite preparation. The predominant method followed for the preparation of clay-biochar composite is the biomass-clay pyrolysis method. Table 1 summarizes the preparation methods of clay-biochar composites with distinct attributes detailed out, upon the incorporation of clay.

2.1 Slurry method

In the slurry method, composites are prepared by mixing a fixed ratio of clay and biochar (ratio by w/w) in a slurry suspension. Prepared biochar is used directly as the primary feedstock in this method for the synthesis of the composite.[28] A clay suspension is then made by adding a certain amount of clay to deionized water, followed by sonication. Then biochar, is added to the clay suspension and stirred in a closed system for an elongated period of time for the purpose of hydration of the biochar in the clay suspension. Solid-liquid phases are then separated through centrifugation wherein the solid is collected and oven-dried.[12,14]

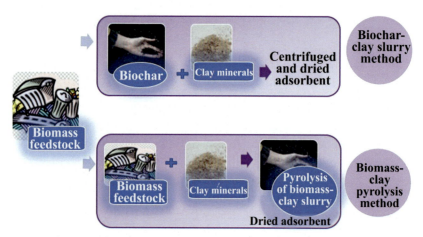

Fig. 1 Brief outline of preparation strategies for clay-biochar composites.

Table 1 Summary of clay-biochar synthesis under different conditions and attributes to incorporation of clays.

Biochar feedstock	Clay	Clay:biochar (w/w)	Preparation method	Pollutant remediated/soil health	Characteristic feature attributed to clay modification	Maximum adsorption capacity (mg g^{-1})	Reference
Corn straw	MMT	1:5	Slurry (biochar at temperature) 350 °C / 500 °C / 700 °C	Zn^{2+}	Optimum at 350 °C and retained the original clay structure	8.163	Song et al. [12]
Cauliflower leaves	ATP	1:5	Biomass–clay pyrolysis at 500 °C for 6 h	OTC	Increment in specific surface area and mesopore structure and decreased dramatically after adsorption of OTC	33.31	Wang et al. [19]
Saccharina japonica (kelp, macroalga)	Bentonite	1:4	Slurry at (biochar) 600 °C for 4 h	Crystal violet	Crystal violet strongly adhered to slurry method (leading to adsorption sites blockage) at higher biochar content	1480	Sewu et al. [20]
		1:2 / 1:1	Biomass–clay pyrolysis at 600 °C for 4 h	Congo red		1275	
Cauliflower (*Brassica oleracea L.*) leaves	MMT	1:9 / 1:1	Biomass–clay pyrolysis at 500 °C for 6 h	OTC	Specific surface area decreased attributed to coating or blockage of pores on biochar, however, act as supporting materials to disperse MMT	55.77	Liang et al. [21]

Continued

Table 1 Summary of clay-biochar synthesis under different conditions and attributes to incorporation of clays.—cont'd

Biochar feedstock	Clay	Clay: biochar (w/w)	Preparation method	Pollutant remediated/soil health	Characteristic feature attributed to clay modification	Maximum adsorption capacity (mg g^{-1})	Reference
Municipal solid waste	MMT	1:5	Biomass–clay pyrolysis at 500 °C for 30 min	OTC	Adsorption of OTC attributed to intercalation between inter-layer of MMT and enhanced available binding sites for MMT deposition on biochar	78	Premarathna et al.[14]
Municipal solid waste	Bentonite	1:1	Slurry at (biochar) 450 °C for 4 h	CPX	Low dependency on the specific surface area, but a strong dependency on the functional group's interactions between adsorbate and adsorbent	167.36	Ashiq et al.[8]
Municipal solid waste	MMT	1:5	Biomass–clay pyrolysis at 450 °C for 1 h	CPX	More active sites and functional groups generated	286.60	Ashiq et al.[22]
Agriculture bamboo waste	MMT	1:5	Biomass–clay pyrolysis at 460 °C for 2 h	Nitrate	Composite has a higher tendency for adsorption of anionic species due to the positive charge	9	Viglašová et al.[23]
Wheat straw	MMT	1:4	Biomass–clay pyrolysis at 400 °C for 6 h	Norfloxacin	Newer pores and active sites on the biochar surface	25.53	Zhang et al.[24]

Yak dung	ATP	1:1	Biomass–clay pyrolysis at 400/500 °C for 2 h	Improving soil nutrition	The improved abundance of microorganisms enhanced nutrient cycling	–	Dou and Goldfarb.[25]
Residual wood and paper	MMT	1:4	Slurry at 600 °C for 4 h	Cu^{2+}	Organic-inorganic nature with acid/basic functional groups with enhanced surface area and mesoporous structure to efficiently immobilize the heavy metals	25.42	Arabyarmohammadi et al.[7]
				Pb^{2+}		76.68	
				Zn^{2+}		24.96	
Peanut shell	KAOLINITE	1:5	Slurry at (biochar) 500 °C and retained for 30–40 min	Soil nutrient enhancement	Micropore distributed on the surface of biochar by clay. Water retention enhanced with the application of clay in biochar	–	Wong et al.[26]
Bamboo powder (*Phyllostachys pubescens*)	MMT	1:1	Biomass–clay pyrolysis at 300, 350, 400, 450 and 500 °C for 1 h	NH_4^+	Adsorption controlled by surface adsorption through surface texture and active binding sites. Adsorption partially controlled by CEC property of the composite	12.52	Chen et al.[15]
				PO_4^{3-}		105.28	
Potato stem	ATP	1:5	Biomass–clay pyrolysis at and 500 °C for 6 h	Norfloxacin	Attapulgite introduced ultrafine SiO_2 particles on biochar matrix and increased oxygen-containing functional groups	5.24	Li et al.[11]

Continued

Table 1 Summary of clay-biochar synthesis under different conditions and attributes to incorporation of clays.—cont'd

Biochar feedstock	Clay	Clay: biochar (w/w)	Preparation method	Pollutant remediated/soil health	Characteristic feature attributed to clay modification	Maximum adsorption capacity (mg g^{-1})	Reference
Cassava peal for hydrochar	Bentonite	1:5	Biomass-clay pyrolysis at and 500 °C for 1 h	NH_4^+	Enhanced specific surface area and developed a porous structure. During the thermal process, CO_2 enters and oxidizes, leading to larger and newer micropores in the composite structure	23.67	Ismadji et al.[27]
Bamboo, Bagasse (selected), Hickory chips	MMT Kaolinite	1:5	Biomass-clay pyrolysis at and 600 °C for 1 h	Methylene blue	Bagasse-MMT composite showed the optimum and thus studied further for adsorption mechanisms. Higher recyclability with enhanced binding sites and pore channels	60.17	Yao et al.[13]

MMT, montmorillonite; ATP, attapuligite; OTC, oxytetracycline; CPX, ciprofloxacin.

Yao et al.[13] have used three types of feedstock, viz., bamboo, bagasse, and hickory chips to prepare the biochar and as clay materials, both MMT and kaolin powders were used to synthesis the composite (Table 1). Biochar of each kind, 10 g was added to the 2 g of MMT/kaolinite suspension, and the slurry was stirred for overnight. Finally, filtered solids were centrifuged and separated through successive filtration, which was then oven-dried at 80 °C for 24 h. before their intended use for the experiments.[13,15]

Furthermore, several authors studied clay-biomass composites, which have been pyrolyzed after the preparation of the mixture.[8,14,20] Dried municipal solid waste (250 g), dipped in a suspended bentonite clay (50 g), was allowed to stir for few hours. The suspension was centrifuged, and the precipitated biomass-clay composite was oven-dried at 80 °C for overnight. The dried composite mixture was then pyrolyzed at 450 °C under 15 °C/min rate of increasing temperature in the muffle furnace. After pyrolysis, the composite was quenched by transferring the composites to a cold-water bath to activate the pores in biochar. Finally, the solid particles were separated and oven-dried for overnight to obtain the composite.[22]

2.2 Biomass-clay pyrolysis method

In this method, clay-biochar composite is prepared by pyrolyzing both biomass and clay, which are mixed in a closed system in their respective ratios. Biomass, used as a primary feedstock in this method, is initially washed, dried and dipped in a clay suspension over a period of time. The solid fraction of the clay-biomass mixture is separated and allowed to dry and then fed to the pyrolizer (Table 1).

Rawal et al.[29] used lignocellulose-based feedstock such as bamboo-sourced for the synthesis of kaolinite/bentonite-biochar composite. Clay suspensions were prepared with bentonite and kaolinite in separate systems, wherein bamboo cubes were soaked and stirred in both the clay slurries. The maximum infiltration of the clay into the biomass pores, was achieved through prolonging the contact time of stirring. As a pre-pyrolysis step, the biomass-clay mixtures were dried and solid particles were separated and fed into the muffle furnace for pyrolysis. The temperature ramp rate was 3 °C/min until the pyrolysis temperature was reached. Four sets of pyrolysis temperatures (250, 350, 450 and 550 °C) were employed for all the slurries individually. This method is known to suppress gas diffusion through the biochar matrix, thereby reducing gas permeability in the pore channels of the overall clay-biochar composite.[29,30]

3. Properties of clay-biochar composites

Physicochemical properties of clay-biochar composites mainly depend on the pyrolysis conditions of the biochar prepared, biomass used and clay minerals incorporated in the composite preparations. In this section, we cover the physical and chemical properties of clay-biochar composites with a summary of studies on different types of characterizations performed on clay-biochar composites.

3.1 Physical properties

3.1.1 Pore volume/diameter and surface area

Surface area and porosity are the major physical characteristics of biochar influencing the adsorptive removal of contaminants. The incorporation of clay to biochar has been proved to increase the surface area, pore volume and the average pore volume of the composites by several folds with respect to the pristine biochar (Table 2). Viglašová et al.[23] observed a five-fold increase of specific surface area for the bamboo biochar when treated with MMT due to the well-developed porosity in the composite. The deposition of clay particles on the biochar surface provides additional surface area for the composite.[14] The attapulgite (ATP) clay consists of mineral elements $(Mg_5Si_8O_{20}(OH)_2(OH_2)_4 \cdot 4H_2O)$, which have small surface areas and rich in transitional pores (2–50 nm).[11] The intrusion of ATP into rice straw biochar[32] and yak dung biochar[33] has been found to improve the surface area, total pore volume and the average pore width with respect to the pristine biochar.

The high surface area of the clay, does not merely contribute to the improved surface area of the composite. In fact, intrinsic properties of clay can catalyze the volatilization and destruction of the cellulose and hemicellulose structure, creating an additional surface for more adsorption sites on the surface. Montmorillonite can act as a solid acid catalyst improving dehydration of biomass and intramolecular polymerization of the carbon structure during pyrolysis.[34] In the case of the activation process, biomass precursor is treated with acid (e.g., HCl, H_2SO_4, H_3PO_4 and HNO_3) or alkali solution (e.g., NaOH, KOH) or chemical agents (e.g., $ZnCl_2$) and carbonized to produce activated biochar with higher surface area and pore volume.[32,35] Moreover, certain components in the clay mineral can act as an alkali catalyst. For instance, in the study of Chen et al.,[15] alkali and alkaline earth metal present in MMT facilitated the activation of the MMT- bamboo

Table 2 Physical properties of clay-biochar composites.

Clay-BC composite	Pyrolysis temp (°C)	Surface area ($m^2\,g^{-1}$)	Pore volume ($cm^3\,g^{-1}$)	Avg pore width (nm)	Reference
Bamboo BC	400	2.272	0.00246	6.994	Chen et al.[15]
Bamboo BC-MMT		19.928	0.0622	11.467	
Bamboo BC	400	28	0.0395	–	Viglašová et al.[23]
Bamboo BC-MMT		156	0.1382		
Wheat straw BC	600	20.1	0.138	9.85	Zhang et al.[24]
Wheat straw BC-MMT		112.6	0.604	21.7	
BC	500	53.31	0.0434	3.26	Liang et al.[21]
BC-MMT		37.44	0.969	10.35	
Corn straw 350 BCE	350	–	–	–	Song et al.[12]
Corn straw BC-MMT 350		1.57	0.010	33.76	
Corn straw 500 BCE	500	–	–	–	
Corn straw BC-MMT 500		2.53	0.009	38.34	
Corn straw 700 BCE	700	–	–	–	
Corn straw BC-MMT 700		4.01	00016	34.68	
Bamboo BC	600	375.5	–	–	Yao et al.[13]
Bamboo BC-MMT		408.1			
Bamboo BC-kaolinite		239.8			
Bagasse BC		388.3			
Bagasse BC-MMT		407.0			
Bagasse BC-kaolinite		328.6			
Hickory chips BC		401.0			
Hickory chips BC-MMT		376.1			
Hickory chips BC-kaolinite		224.5			
Municipal solid waste BC	500	4.33	–	–	Premarathna et al.[14]
Municipal solid waste BC-MMT		8.72			
Municipal solid waste BC-RE		8.44			

Continued

Table 2 Physical properties of clay-biochar composites.—cont'd

Clay-BC composite	Pyrolysis temp (°C)	Surface area ($m^2 g^{-1}$)	Pore volume ($cm^3 g^{-1}$)	Avg pore width (nm)	Reference
Municipal solid waste BC	450	4.33	–	–	Ashiq et al.[8]
Municipal solid waste BC–MMT		6.51			
Bagasse BC	700	219.49	127.0	2.31	Chen et al.[31]
Bagasse BC-ATP		178.51	134.3	3.01	
Rice straw BC	450	11.13	0.01	5.63	Yin et al.[32]
Rice straw BC-ATP		203.38	0.09	1.94	
Potato stem BC	500	99.43	0.00075	3.12	Li et al.[11]
Potato stem BC-ATP		90.40	0.1225	5.42	
Macro algae BC	600	203	0.0904	1.78	Sewu et al.[20]
Macro algae BC-bentonite 5%		29.2	0.0195	2.67	
Macro algae BC-bentonite 10%		2.30	0.0043	7.43	
Macro algae BC-bentonite 20%		6.62	0.0211	12.7	
Mango pit 280 BCE	280	7.39	–	–	Sewu et al.[25]
Mango pit BC-FE 280		12.15			
Mango pit 350 BCE	350	6.01	–	–	
Mango pit BC-FE 350		105.57			
Pineapple plant 280 BCE	280	12.67	–	–	
Pineapple plant BC-FE 280		61.08			
Pineapple plant 350 BCE	350	4.2	–	–	
Pineapple plants BC-FE 350		65.22			

BC, biochar; MMT, montmorillonite; ATP, attapulgite; RE, red earth's clay; FE, fuller's earth clay.

composite (MMT:bamboo = 1:1). Hence, the clay-biochar composite has been found to increase the surface area, pore volume and pore diameter with respect to original biochar. Similar effects were shown by MMT mixed with biochar derived from a wide variety of feedstock; i.e. bamboo, municipal solid waste, bagasse and hickory chips.[13,15,23] Such activation catalyzes the carbonization process, thereby widening the existing pores as well as favoring the development of new micropores and mesopore in the biochar.[36] Additionally, the physical properties of the clay and biochar itself play a vital role in the composition preparation. For instance, the high heat transfer capacity of bentonite promoted the devolatilization and improved the surface area of the biochar (mango and pineapple biomass) while the oxygenated groups of the clay partially oxidized the biochar creating porous voids.[25]

Incorporation of clay can cause pore blockage resulting in low surface area as opposed to the original biochar (Table 2). Composites made with kaolinite and MMT had shown a reduction in the surface area due to the coating or pore blockage of the biochar caused by clay particles.[13,21] The pore blockage is also likely to occur when the clay content increases during composite preparation.[13] The ATP-bagasse biochar composites had a larger surface area and pore volume for 1:5 by mass and reduced when the clay content increased to 1:3 by mass. However, elsewhere it was found that in the case of ATP and yak dung biochar, the increase of clay content (1:9 to 1:1) increased the surface area of the composite, which enhanced further at elevated temperature.[33] The increase of pyrolysis temperature obviously promotes the degradation of the biomass. In contrast, at high temperature (>500 °C), kaolinite converts to meta-kaolinite with low surface area, causing blockage of the mesoporous structure in the biochar, as discussed previously in the preparation section. Although, bentonite did not undergo such structural changes, a reduction in pore size was observed for the composite.[29] This was attributed to the re-condensation of volatile organic matter on the composite pores at high temperature to block the pores that led to surface area reduction.[15]

3.1.2 Morphological characteristics and crystallinity

Pyrolysis of the clay-biochar mixture at high-temperature triggers intrusion of clay minerals to the porous interior structure of the biochar or either attach the external surface of the biochar.[13,37] The scanning electron microscope (SEM) provides clear visualization of the morphological structures of the surface-modified biochar. Generally, the composite can be observed with adhered fine particles on the surface, which could not be found in the pristine biochar.[9] A rough surface on biochar is generated as a resulted of the deposition of clay particles and their agglomerates on the surface. The infiltered mineral phases can be observed, especially in macropores, as crystalline and amorphous nano deposits.[29] In bentonite-macroalgae composite, the bubblelike structures/blisters were observed on the composite due to the trapped volatile compounds in the clay coating on the biochar surface.[20] The increase of clay ratio results in greater coverage of the biochar, which occurred to have a higher deposition of fibrous structures in ATP: biochar 1:3 than 1:5 ratio. At micron level, mineral depositions could be observed as acicular structures or plate-like structures on the composites surface and the unmodified areas appear to be as highly porous biomass structure.[25,33] In the composite cross-sectional view of the solid waste biochar, the carbon bound MMT was distinguished as irregular flaky structures embedded in the voids of biochar counterpart.[8,22]

The changes in the crystallographic structure of the composite and the presence of clay crystals can be distinguished by the XRD analysis. During pyrolysis, dehydration occurs within the interlayered space of the clay leading to shrinkage of the interlayer spaces. Studies have shown that as temperature was elevated from 300 to 500 °C, the interlayer space of MMT decreased from 1.55 to 0.95 nm as a result of enhanced dehydration of the clay.[8,9,14,22] High temperature accelerates the decomposition of the crystalline structure of the biochar to produce amorphous biochar, which is attributed to the disappearance of crystalline peaks in the XRD spectrum.[15] The absence of mineral phases (quartz; SiO_2) and reduced crystalline intensities of municipal solid waste biochar-MMT indicated amorphous characteristics imparted on the composite.[14] However, the increase of interlayer spacing of MMT was also observed during pyrolysis, which was likely to the intercalation of the biochar components in the clay structure.[8] The peak indication for quarts (SiO_2) and dolomite ($CaMg(CO_3)_2$) in the XRD pattern displayed successful loading of ATP on the potato stem biochar and rice straw biochar.[11,32] Similarly, during the study of Ashiq et al.[22] the presence of dolomite ($CaMg(CO_3)_2$) and feldspar peaks in the composite indicated the modification by bentonite on the biochar surface.

3.2 Chemical properties

3.2.1 Aromaticity

The modification with clay can impart changes in the atomic ratios, with respect to the biomass type, clay type and temperature. Although the majority of the carbon input is from the biomass, clay incorporation can either alter the C% through activation of the carbonization process or leave it unaltered.[23] The carbon content of the clay-biochar composite was found to increase by 16% due to the accelerated dehydration caused by the catalytic effect of MMT.[15] Increasing of the pyrolysis time and temperature also elevates the carbon content (C%) and decrease the surface functionalities (i.e. carboxylic, hydroxyl, amino group) leading to low H/C, O/C and N/C ratios.[38] The low H/C ratio reflects the high aromaticity and the degree of carbonization and the low O/C derived to represent the loss of functional groups and the chemical stability of the composite.[39] However, the levels of O and N of the composites can be influenced by the clay, whereas in kaolinite and MMT modified biochar, the increase of O was relatable to the elemental addition from clay (Table 3). Pinewood biochar and MMT composites had a lower H/C atomic ratio (0.24) than the pristine biochar (2.36), indicating higher aromaticity of the composite. The polarity index $[(O+N)/C]$ found to increase to 0.40 for the composites from 0.37 of the biochar, which was due to the number of functional groups and hydrophilicity that has been imparted by the clay on the biochar.[24] The modification of bagasse biochar with ATP was found to elevate O/C ratio and high polarity index also supporting the evidence of polar functional groups on the composite surface.[31]

3.2.2 Functional groups

The functionalization of the clay-biochar composites can be clearly observed through infrared (FTIR) analysis. At elevated temperatures, the presence of clay catalyzes the cross-linking of aliphatic carbon condensing them to aromatic structures. Reduced peak intensities were observed for aliphatic species and/or increased intensities or emergence of the new peak for alkyl and alkene aromatic stretching ($-CH$ and $C{=}C$) were observed with the introduction of clay.[29] Aliphatic $-CH_2$ and $-CH$ stretching of wheat straw biochar was found to reduce with the addition of MMT and a strong peak for aromatic $C{=}C$ stretching was observed for the composites.[24] Similarly, the presence of $-CH$ stretching peak in municipal solid waste biochar disappeared in the MMT composite due to the aromatic ring formation.[8] The intrusion of clay could be deduced by the bands appearing below $1100\,cm^{-1,}$ indicating the presence of siloxane functional groups arising from the clay elements.[14] The presence of MMT in the composites was attributed to Si—O stretching, Si–O–Al and Si–O–Mg bending vibrations and $-OH$ stretching with lower intensities in comparison to the pristine MMT.[23] All peaks assigned for Si–O–Si, Fe—O, Al—O, Al–O–Si, Si—O in the composites indicated successful modification of biochar in

Table 3 Chemical properties of clay-biochar composites.

Clay-BC composite	Pyrolysis temp (°C)	pH	CEC Meq/ 100g	Ultimate analysis (wt%)											Ash	Polarity [(N+O)/C]	Reference
				C	H	O	N	Si	Fe	Al	Ca	Na	K	Mg			
Bamboo BC	400	5.3	–	76.05	3.47	19.04	0.37	–	–	–	–	–	–	–	0.70	0.2548	Chen et al.[15]
Bamboo BC-MMT		7.1		64.01	3.81	10.03	0.41								21.30	0.1626	
Bamboo BC	400	–	–	80.97	3.89	–	0.52	–	–	–	–	–	–	–	–		Viglašová et al.[23]
Bamboo BC-MMT				46.22	1.69		0.35										
Wheat straw BC	600	–	–	52.4	10.3	25.2	2.03	–	–	–	–	–	–	–	–	0.39	Zhang et al.[24]
Wheat straw BC -MMT				58.8	1.16	27.0	1.85									0.37	
BC	500	–	–	83.12	–	14.03	–	0.09	0.07	0.09	1.76	–	–	0.84	–	–	Liang et al.[21]
BC-MMT				24.93		38.16		22.36	2.32	4.54	3.43	0.20		1.05			
Bamboo BC	600	–	–	80.89	2.43	14.86	0.15	–	0.00	0.04	0.34	0.01	0.52	0.23	–	0.186	Yao et al.[13]
Bamboo BC-MMT				83.27	2.26	12.41	0.25		0.23	0.68	0.21	0.14	0.33	0.14		0.152	
Bamboo BC-kaolinite				81.02	2.15	15.85	0.25		0.08	0.30	0.19	–	0.07	0.05		0.199	

Feedstock														Reference
Bagasse BC			76.45	2.93	18.32	0.79	0.05	0.11	0.91	–	0.15	0.21	0.250	Premarathna et al.[14]
Bagasse BC–MMT			75.31	2.25	18.87	0.75	0.47	0.75	0.85	0.13	0.32	0.22	0.261	
Bagasse BC–kaolinite			7.20	2.44	24.44	0.74	0.46	0.53	0.88	–	0.06	0.16	3.497	
Hickory chips BC			81.81	2.17	14.02	0.73	0.01	0.06	0.82	–	0.24	0.13	0.180	
Hickory chips BC–MMT			80.93	2.21	15.14	0.28	0.15	0.32	0.57	0.04	0.11	0.19	0.191	
Hickory chips BC–kaolinite			78.08	2.11	18.12	0.33	0.07	0.51	0.52	–	0.05	0.18	0.236	
Municipal solid waste BC		9.55	49.58	3.30	12.99	9.52	9.12	–	–	–	–	–	–	Premarathna et al.[14]
Municipal solid waste–MMT	500	9.51	46.4	8.84	8.66	20.13	–	–						
Municipal solid waste–RE		8.99	44.26	19.51	12.76	12.29	7.39							
Corn straw 350 BCE	350	–	72.69	4.71	20.44	1.87	–	–	–	–			0.307	Song et al.[12]
Corn straw BC–MMT 350			36.13	3.06	8.11	0.75							0.245	

Continued

Table 3 Chemical properties of clay-biochar composites.—cont'd

Clay-BC composite	Pyrolysis temp (°C)	pH	CEC Meq/ 100 g	Ultimate analysis (wt%)												Polarity [(N+O)/C]	Reference
				C	H	O	N	Si	Fe	Al	Ca	Na	K	Mg	Ash		
Corn straw 500 BCE	500	–	–	77.74	2.27	10.37	1.28	–	–	–	–	–	–	–	–	0.150	
Corn straw BC-MMT 500				32.70	2.34	3.82	0.60									0.135	
Corn straw 700 BCE	700	–	–	83.57	1.94	3.78	0.96	–	–	–	–	–	–	–	–	0.057	
Corn straw BC-MMT 700				41.97	1.70	0.92	0.59									0.036	
Municipal solid waste BC	450	9.58	–	–	–	–	–		–	–	–	–	–	–	–	–	Ashiq et al.[8]
Municipal solid waste BC-MMT		9.46															
Bagasse BC	700	–	–	78.27	0.98	5.13	0.46	6.28	0.75	0.66	2.05	3.25	1.37	0.71	–	0.0885	Chen et al.[31]
Bagasse BC-ATP				54.76	0.62	8.53	0.18	16.31	1.77	2.58	3.68	5.58	2.92	1.0		0.1074	
Rice straw BC	450	–	–	40.07	5.30	0.28	0.38	-	–	-	–	–	–	-	–	0.016	Yin et al.[32]
Rice straw BC-ATP				51.10	2.19	0.22	0.19	13.05		1.10				0.07		0.008	

Sample																Reference
Potato stem BC	500	10.38	–	75.81	–	19.30	–	0.09	–	0.39	1.51	–	–	1.49	24.67 –	Li et al.[11]
Potato stem BC–ATP		9.93		59.75		30.92		3.19		1.15	2.13			2.05	5.300	
Yak dung BC		10.6	66.5	43.16	1.73	–	1.72	–	–	–	–	–	1.82	–	– –	Rafiq et al.[33]
Yak dung BC–ATP 10%		10.6	109.4	33.12	1.48		1.42						1.59			
Yak dung BC–ATP 20%	500	10.3	93.7	26.86	1.03		0.89						1.65			
Yak dung BC–ATP 30%		10.1	84.0	23.75	1.09		0.99						1.61			
Yak dung BC–ATP 40%		9.8	76.5	14.21	0.75		0.54						1.51			
Yak dung BC–ATP 50%		9.5	64.7	11.39	0.63		0.38						1.41			
Yak dung BC	400	10.1	45.2	46.90	3.07	–	1.76	–	–	–	–	–	1.42	–	– –	
Yak dung BC–ATP 10%		10.1	59.3	36.25	2.45		1.53						1.07			
Yak dung BC–ATP 20%		9.8	51.9	27.35	1.91		1.15						1.14			
Yak dung BC–ATP 30%		9.7	48.8	24.63	1.82		1.02						1.20			

Continued

Table 3 Chemical properties of clay-biochar composites.—cont'd

Clay-BC composite	Pyrolysis temp (°C)	pH	CEC Meq/ 100 g	Ultimate analysis (wt%)											Ash	Polarity [(N+O)/C]	Reference
				C	H	O	N	Si	Fe	Al	Ca	Na	K	Mg			
Yak dung BC–ATP 40%		9.6	44.1	18.32	1.41		0.74						1.25				
Yak dung BC–ATP 50%		9.2	42.3	13.74	1.12		0.56						1.15				
Macro algae BC	600	10.2	–	24.5	0.47	11.7	1.21	–	–	–	–	–	–	–	66.67	0.527	Sewu et al.[20]
Macro algae BC-bentonite 5%		10.1		26.2	0.42	8.37	1.54								66.2	0.378	
Macro algae BC-bentonite 10%		10.1		23.3	0.25	5.40	1.33								70.80	0.289	
Macro algae BC-bentonite 20%		10.3		20.2	0.21	3.03	0.68								75.40	0.184	

BC, biochar; MMT, montmorillonite; ATP, attapulgite; RE, red earth clay; FE, Fuller's earth clay.

the case of ATP clay modified biochar.[31,33] Indication of C$=$O stretch and O$-$H deformations of the carboxylic group in the composite of MMT with wheat straw biochar and bamboo biochar was evident in the increased of polar functional groups induced by the clay on the biochar surface.[15,24]

3.2.3 Elemental composition

The elemental distribution provides insight into the clay induced modification of the composite. The common elements of clays are Ca, Na, Al, Mg, Fe, Zn, and Si and the presence or enrichment of these minerals in the composite indicates successful modification of the biochar.[9] Modification carried out with MMT ($(Na,Ca)_{0.33}(Al,Mg)_2(Si_4O_{10})(OH)_2 \cdot nH_2O$), kaolinite ($Al_2Si_2O_5(OH)_4$) significantly increased the Al and Fe content of bagasse, hickory chips and bamboo with regard to the untreated biochar.[13] Aluminium is the main element of clays corresponding to 20% of kaolinite weight and 10% of the MMT weight. Similarly, in the presence of bentonite and kaolinite, the content of Al, Si and Fe was several orders of magnitude greater than the pristine bamboo biochar. At high temperature, the volatilization process and biomass degradation enhance the levels of clay minerals in the biochar composites.[29] Modification of bagasse biochar with ATP ($Si_8O_{20}(OH)_2(OH_2)_4 \cdot 4H_2O$), had elevated levels of Na, Ca, Mg, Al, Fe and Si with regard to the unmodified biochar.[31] Correspondingly, high content of Si and Mg, Al elements were found in the yak dung and potatoes stem biochar after being modified by ATP.[11,32]

3.2.4 pH and surface charge

The pH of the clay-biochar composite can vary depending on the acidic and basic properties of the modifying material, the biochar feedstock and the temperature. Biochar can exhibit different pH's depending on the feedstock composition varying in the range of pH 4–11. Generally, most biochars are alkaline, and the pH tends to increase with the rise of pyrolysis temperature since the ash production and decomposition of acidic functional groups (–COOH) are promoted at high temperature.[38,40] Clay can either increase or decrease the original pH value of biochar according to the acidity or alkalinity of the particular clay used for the composite preparation (Table 3). The pH drop of solid waste biochar modified by Fuller's earth was attributed to the acidic property of the clay.[9] Moreover, clay can impart additional oxygen-containing functional groups (carboxylic and phenolic groups) on the composite surface. The presence of carboxylic groups increased the acidity of the ATP modified potato stem biochar composite with respect to the pristine biochar, reducing them from pH 10.38 to 9.93.[11] As opposed to this behavior MMT increased the pH of the composite to 7.1 from 5.4 of the pristine bamboo biochar. This was attributed to the catalytic effect of MMT, which intensified the decomposition of acidic groups (deoxygenation) and formation of more alkali salts that lead to inclined pH values.[15] The point of zero charges (pHpzc) or the pH at which the biochar's net surface charge becomes zero could also be altered by the clay. The surface becomes more negative as more anionic functional groups (-COO⁻ and OH⁻) are imparted on the biochar surface by the clay, shifting the pHpzc further to lower pH end.[40] The recorded pHpzc of the composite was found to reduce to 7.75 from the pHpzc 9.55 of the pristine biochar due to the carboxylic groups promoted by ATP on the surface.[11] This is particularly important as the adsorptive removal of inorganic or organic contaminants is thoroughly dependent on the pHpzc of the composite. For instance, maximum adsorption capacities of reactive brilliant red X-3B dye on the ATP-bagasse composites varied as pH 5, 7 and 9 for the pristine biochar, 1:5 and 1:3 ratios respectively which was entirely governed by the surface charge of the composite.[31]

4. Adsorptive removal of organic contaminants using clay-biochar composite

4.1 Removal of pharmaceuticals and personal care products

The clay-biochar composites demonstrated high adsorption capacity toward the removal of different pharmaceuticals and personal care products (PPCPs) with improved affinity for adsorption of PPCPs than pristine biochars.[11,22] The application of composite of potato stem biochar and natural ATP (BC-ATP) in the adsorptive removal of norfloxacin was studied by Li et al.[11] The removal of norfloxacin by BC-ATP was pH-independent and observed high affinity to norfloxacin than pristine potato stem biochar. This may be due to the introduction of ultrafine SiO_2 particles and oxygen contained functional groups on the surface of clay-biochar composites.[11]

In a study by Liang et al.,[21] the authors successfully prepared MMT-biochar (MMT-BC) and magnetic-MMT-biochar (M-MMT-BC) composites from the feedstock of cauliflower leaves, MMT, and $FeCl_3$, for the removal of oxytetracycline from aqueous solution. Results indicated a strong dependency on the pH of the solution. The M-MMT-BC and MMT-BC showed 98.9% and 91.8% of removal affinity toward oxytetracycline, whereas the pristine biochar showed 44.5% of removal. This may occur due to the accumulation of Fe_3O_4 nanoparticles, which provided extra active sites for the adsorption of oxytetracycline as well as functional groups such as Si—O, Si–O–Al, Si–O–Si imparted by the clay itself.[21] Similarly, Ashiq et al.[8] and Sarkar et al.,[41] observed the enhancement of active adsorptive sites on the clay-biochar derived from municipal solid waste biochar and MMT. The engineered composite has also been utilized to remediate the ciprofloxacin. The highest adsorption of ciprofloxacin was recorded by composite prepared from the MMT and municipal solid waste biochar with a 40% increment of removal efficiency in comparison to the pristine biochar.[8]

The adsorption capacity of an adsorbent is mainly influenced by physicochemical properties such as surface area, porosity, surface functional group, etc.[40] Moreover, the clay-biochar composite exhibited minimization and maximization of surface area when compared to their respective pristine materials.[14,20,42] Although the clay-biochar composite exhibited low surface area, it exhibited a higher adsorption affinity toward the PPCPs as reported in the literature (Table 4). This may be due to the intercalation of PPCPs within the clay counterparts of the composite, the involvement of oxygen containing functional groups and the extra active sites for the adsorption.

4.2 Removal of dyes and pigments

Clay-biochar composites have shown outstanding performance in the adsorption of some dyes in comparison to pristine forms. Removal ability of reactive brilliant red (RBR) dye using clay-biochar composites from bagasse and natural attapulgite have been studied in detail using different ratios of biochar and clay ratios.[31] The reported maximum adsorption capacities were 65.1 and $72.2 \, mg \, g^{-1}$ respectively for 1:5 and 1:3 clay: biochar ratios. The study revealed different mechanisms of dye removal with an important role of π–π interactions and hydrogen bonding followed by electrostatic interactions, hydrophobic interactions and surface participation. The characterization of clay-biochar composite has revealed that it comprises more Al_2O_3 and SiO_2 particles, and oxygen-containing functional groups on the surface, in comparison to pure biochar. Thus, adsorption capacity has been greatly improved via the clay modification.[31]

Clay-biochar composites synthesized using the from three different feedstock materials were modified by MMT and kaolinite in their respective composition and ratio. These were then assessed for the removal of

Table 4 Efficiency of biochar-clay composites on the removal of pharmaceutical and personal care products.

PPCPs	Adsorbent	Adsorption efficiency (mg g^{-1})	Experiment conditions			Fitted isotherm models	Reference
			pH	CT (h)	T (°C)		
Norfloxacin	BC- ATP	5.02	2	24	25	Langmuir	Li et al.[11]
	Pristine biochar	2.72					
Ciprofloxacin	MSBC- MMT	167.36	5–6	12	25	Hill	Ashiq et al.[8]
	Pristine biochar	122.16					
Tetracycline	BBC-S	17.51	Neutral	48	25	Freundlich	Zhao et al.[42]
	Pristine biochar	2.45					
Tetracycline	MSBC- MMT	77.96	7–8	6	–	Freundlich	Premarathna et al.[14]
						Langmuir	
Ciprofloxacin	MSBC-bentonite	286.60	7–8	12	25	Hill	Ashiq et al.[22]
	Pristine biochar	167.61					

MSB-bentonite, municipal solid waste biochar-bentonite composite; BC-ATP, potato stem biochar-natural attapulgite composite; MSBC-MMT, municipal solid waste biochar-montmorillonite composite; BBC-S, bamboo-silica biochar composite; CT, contact time; T, temperature.

methylene blue dye by Yao et al.[13] The authors reported that the biochar derived from bagasse exhibited a higher adsorption capacity in comparison to the other two feedstock materials (bamboo and hickory chips).[13] Moreover, MMT-bagasse biochar composite enhanced methylene blue adsorption approximately by 5 folds than pristine biochar. Regeneration study revealed that MMT was the main factor for adsorption, and predominated by cation exchange mechanism, while electrostatic interactions can govern methylene blue adsorption into biochar, the subsidiary factor. However, biochar modification with clay does not have the ability to change the native ion exchange capacity resulting in relatively low methylene blue adsorption capacity in comparison to other modifications. The ability of regenerating the adsorbent by using KCl solution is concluded, stressing the fact that the composite can be recycled and reused for dye adsorption.[13] Regeneration ability of the adsorbent is considered as a vital economic benefit for their industrial application. It is noteworthy that composites present 70% dye removal after six regeneration cycles, which is a higher rate compared to activated carbon.[43] Adsorption properties can be altered due to the type of clay and properties of feedstock used to derive biochar. However, there is no direct correlation that has been identified between dye adsorption and surface area.[6]

5. Clay biochar composites for inorganic contaminant adsorption

5.1 Removal of inorganic ions

Clay-biochar composites are found to be promising for a wide variety of contaminant removal, including inorganic ions. The concept of clay incorporation to biochar has been developed to improve functionalities in order to restrict nutrient release. However, compared to organic contaminant remediation, the utilization of clay-biochar for inorganic ion removal is limited to a handful of studies. Viglašová et al.[23] examined the effectiveness of the MMT-bamboo biochar (1:5) composites for the removal of nitrate (NO_3^-) from aqueous media. The composite achieved a maximum sorption capacity of $9\,mg\,g^{-1}$ at pH 4, while pristine biochar could achieve only an adsorption capacity of $5\,mg\,g^{-1}$. Since NO_3^- does not show any speciation behavior, the pH influenced the surface charge of the composite solely affected the sorption. The modification by MMT imparted a high positive charge on the surface through adsorption of excess H^+ and thereby enhancing the NO_3^- sorption capability of the composite.[23] Also, the removal of NO_3^- is influenced by outer-sphere complexation and cationic exchange to some extent. However, the biochar modification by clay should be able to impart positive charge on the matrix to remove negatively charged ions effectively.[44]

The same composite prepared of biochar: MMT with a ratio of 1:1, was assessed by Chen et al.[15] for the adsorption of NH_4^+ and PO_4^{3-}. The maximum adsorption capacity for NH_4^+ removal derived from the Langmuir isotherm was 2.44 and $12.52\,mg\,g^{-1}$ for the pristine biochar and the composite, respectively. The maximum adsorption for PO_4^{3-} was $105.28\,mg\,g^{-1}$ for the composite. Incorporation of clay extended the surface area of the biochar, providing more binding sites for the attachment of NH_4^+ and PO_4^{3-} ions. In addition, ions could attach to the interlayer spaces of the MMT, through attractive forces generated in between the cations (Ca^{2+}, Mg^{2+}, Al^{3+} and Fe^{3+}) present in the interlayer spaces of the clay. Although the exchangeability of cations increases as in the order of $Na^+ < K^+ < Ca^{2+} < Mg^{2+} < NH_4^+$.[45] due to the shrinkage of the MMT layers the intercalation of NH_4^+ ions becomes unfavorable, whereas the removal was assisted with physical sorption on the surface and partially by cationic exchange. However, PO_4^{3-} displayed higher affinity to the composites ($105.28\,mg\,g^{-1}$) than NH_4^+, due to the formation of strong ionic bonds and electrostatic attractions between PO_4^{3-} and cations in the composite. The desorption studies of the composite exhibited excellent reduction of NH_4^+ and PO_4^{3-} leaching by 4.92 and 1.06% respectively after 88h, while the pristine biochar achieved 95.21 and 5.09% for NH_4^+ and PO_4^{3-} respectively. Hence, the MMT-biochar composites showed effective adsorptive properties as well as slow-release behavior for NH_4^+ and PO_4^{3-} ions, enhanced by the introduction of the clay.

The removal of aqueous NH_4^+ produced from Koi fish excretion was assessed by another clay hydrochar composite produced by cassava peel hydrochar and bentonite (2:1). The composite achieved maximum adsorption of $23.67\,mg\,g^{-1}$ as opposed to the $9.49\,mg\,g^{-1}$ displayed by the cassava peel hydrochar. When $pH < pHpzc$, the bentonite biochar composite displayed anionic exchange properties and when the $pH > pHpzc$, displays cationic exchange properties. Since the pH of the medium is higher than the pHpzc (5.4), the removal of NH_4^+ was promoted through the cationic exchange.[27] Sorption of NH_4^+ through ion exchange is an effective mechanism than bonding interaction.[44] This was also clearly observed using pig manure biochar, which had low acidic functional groups and high ash content due to the presence of metal elements (Mg, K, Ca).[46] Similarly, more metal elements were introduced to the matrix, which in turn enhances the cationic exchanging property.

5.2 Removal of heavy metals

Clay-biochar composites have been demonstrated to enhance the properties and capacities of adsorbents, improving their application for the removal of heavy metals.[47] Vithanage et al.[48] have evaluated the efficacy of clay-biochar composite, derived from red earth clay and municipal solid waste (MSW) to remove As^{3+}. The maximum adsorption capacity was recorded as $0.012\,mgg^{-1}$ and the authors concluded that the clay-biochar composite showed a great removal of As^{3+} from water at environmental pH values.[48]

Song et al.[12] have assessed the effect of temperature on modified corn straw biochar with MMT clay for the removal of Zn^{2+}. The composite was prepared in four temperatures of 200, 350, 500 and 700 °C and characterization and adsorption experiment data revealed that 350 °C as the optimum temperature at which the best adsorption performances were achieved. Generally, adsorption of Zn^{2+} has been described using Langmuir isotherm model which was the best-fitted curve with a maximum adsorption capacity of $8.163\,mgg^{-1}$ for biochar-clay composite prepared at 350 °C.[12] Table 5 summarizes the experimental data for the studies conducted in the context to remove heavy metals using clay-biochar composites.

Extensive studies were undertaken for heavy metal removal by Vithanage et al.[50] and Arabyarmohammadi et al.[7] using clay biochar composites synthesized under varied pyrolysis temperatures and then composites with other materials, followed by acid treatment. Three organic acids and three inorganic acids have been used together with two types of clay-biochar composites to assess the As^{3+} desorption capacity. Clay-biochar were derived using MMT and NRE with biochar. The MMT-biochar composite had shown 90% of desorption capacity, while biochar-red earth composite exhibited 95% with citric acid.[50] Arabyarmohammadi et al.[7] have combined the organic material, chitosan into nano clay, and biochar in order to

Table 5 Experimental data of the studies for heavy metal removal using clay-biochar composites.

Clay-biochar composite	Clay:BC ratio	Heavy metal contaminant	Maximum adsorption $(mg\,g^{-1})$	Isotherm model	Kinetic model	Reference
MSW BC-NRE	1:5	As^{3+}	0.012	—	Pseudo second order	Vithanage et al.[48]
Corn straw BC-MMT	1:5	Zn^{2+}	8.163	Freundlich	Pseudo second order	Song et al.[12]
Peanut shell BC-MMT		Cr^{4+}	12.18	Langmuir	Pseudo second order	Wang et al.[49]
Residual plant bark chips BC-kaolinite	1:10	Cu^{2+}	121.5	Freundlich	Pseudo second order	Arabyarmohammadi et al.[7]
		Pb^{2+}	336	Temkin		
		Zn^{2+}	134.6	Freundlich		

BC, biochar; NRE, natural red earth; MMT, montmorillonite.

develop a bio-nano composite for the removal of Cu^{2+}, Pb^{2+}, and Zn^{2+} metal ions from contaminated soil media Batch adsorption experiments data showed maximum adsorption values of 121.5, 336, and 134.6 mg g^{-1}, respectively, which was found to be higher than the pristine biochar. The characterization study has revealed that the immobilization of metal ions can be predominantly driven by the mechanism of binding with –NH$_2$ groups which could be attributed to chemisorption.

6. Mechanisms governing adsorptive removal by clay-biochar composites

Potential adsorption mechanism/s for a particular material depends on the surface functional groups of the adsorbent, speciation of the adsorbate, pH of the media and pHpzc of the adsorbent. Clay-biochar composites can be either positively or negatively charged, contingent on the medium pH. Almost all individual materials consist of a specific pHpzc, which varies with pH. The adsorbent surface is negatively charged, when medium pH is higher than pHpzc vice versa. The surface charge will be neutral as the medium pH coincides with the pHpzc of the material.[9,51] The electrostatic attractions between clay-biochar composite surfaces and cationic and anionic forms of contaminants are governed by this surface charge phenomenon.

Many studies have reported adsorption mechanisms to be facilitated by electrostatic attractions, such as the removal of nutrients (NH_4^+, PO_4^{3-}),[15] antibiotics (Norfloxacin)[11] and heavy metals (As^{5+})[52] from aqueous media with the utilization of clay biochar composites. Negatively charged clay-biochar composite surface encourages the adsorption of positively charged ions (e.g. metal ions, nutrients) through electrostatic interactions as well as ion-exchange mechanisms.[53] Clay mineral enhances the negative charge of biochar surface and increases the ion-exchange capacity, stimulating the contaminant adsorption (e.g. organic dyes).[15]

The size of the metal ion and the available surface functional groups directly influence the adsorption via ion-exchange.[54] Additionally, Aristilde et al.[55] have studied about the intercalation interactions occurred through ion-exchange between contaminants and ions available (e.g. Ca^{2+} and Mg^{2+}) in the interlayer spaces of clays in composites using tetracycline.[55]

Inclusion of clays for biochar-based composites has its own merits and demerits for environmental applications. Clay materials consist of aluminosilicate layers with interlayer spacings, balance out the charge with cations between the layers. This leaves the surface of clays predominantly negatively charged and thereby, providing an excellent active site for adsorbing cationic and hydrophilic contaminants. In addition, adsorption is facilitated due to the higher cation exchange capacity and higher surface area.[26]

The strong electrostatic bonding (cation and anion attractions), and covalent bonding, demonstrate monolayer chemisorption, while weak forces, such as electrostatic bonding (dipolar attraction), Van der Waals forces, hydrophobic effect (hydrophobic interaction), π–π interactions (non-covalent) and hydrogen bonding exhibit multilayer physisorption during adsorptive removal of contaminants using clay-biochar composites.[56–58] In chemisorption, a monolayer coverage with strong interactions have been observed, whereas in physisorption weak forces and multilayers on the adsorbent material surface are apparent (Fig. 1).[59]

In most of the studies, more than one type of removal mechanism has been observed to be involved elaborating the adsorption behavior. Predominantly, both chemisorption and physisorption processes have simultaneously governed the mechanism throughout the adsorption process.[15,60] Ismadji et al.[27] have reported the removal of NH_4^+ using BT-hydrochar composite, where at low concentrations, the Freundlich model was followed, which is a physisorption process. Nevertheless, at high concentrations, the adsorption followed the Langmuir model, which is a chemisorption process.[27] Removal of NH_4^+ and PO_4^{3-} have studied by using an MMT-bamboo biochar composite and the results of the study confirmed that at low concentrations, monolayer adsorption plays a major role while at high concentrations multilayer adsorption is dominant.[15] Hence, both ion-exchange and van der Waals bonding have involved in the adsorption process that follows both chemisorption as well as physisorption.

Furthermore, hematite-magnetic pinewood biochar composite was studied for the sorption of As^{5+}. For example, Langmuir model was found to be the best fit for the isotherm, confirming that monolayer adsorption governed the process.[52] Nevertheless, both chemisorption and physisorption have governed the removal of Cd^{2+} with a BT-plant biochar composite.[60] Arabyarmohammadi et al.[7] investigated the removal of Cu^{2+},

Pb^{2+}, and Zn^{2+} from the soil, using chitosan/cloisite-bark chips biochar nano-biocomposite. The removal of all three metal ions was dominated by multilayer adsorption.[7] However, MMT-corn straw biochar composite demonstrated both monolayer and multilayer adsorption processes in the removal of Zn^{2+}. Differences between surface functional groups of clays and biochars or chitosan might play a role behind the variance of the adsorption mechanisms in the removal of Zn^{2+}.[12]

Goethite-biochar composite has been used to analyze the removal of tylosin, a macrolide group of antibiotic. Many different types of interactions such as hydrophobic, electrostatic, hydrogen bond and π–π interactions have been involved in the removal of tylosin with respect to the pH of the media.[61] Norfloxacin, a fluoroquinolone antibiotic, was removed from aqueous environments using attapulgite clay-potato stem biochar composite. Langmuir isotherm model was identified as the best-fitted model confirmed the chemisorption process.[11] However, MMT-wheat straw biochar composites demonstrated both chemisorption and physisorption mechanisms in the removal of norfloxacin.[24] Similarly, during the removal of tetracycline group antibiotics, using magnetically modified and unmodified clay-biochar composites, both chemisorption and physisorption adsorption mechanisms were predominant.[14,19,21] Yao et al.[13] reported that efficient removal of methylene blue by MMT-bagasse biochar composite was governed principally by electrostatic interactions. However, the findings of the above studies confirm, that other mechanisms such as intraparticle and film diffusion mechanisms may also have been involved in the process.[13]

Hydrogen bonding and complexation between contaminants and clay-biochar composite surfaces, is promoted with the presence of oxygen-containing functional groups.[62] Simultaneously, these functional groups can be bound with cationic metals and anionic metals/metalloids via electrostatic attractions as well.[53] Nevertheless, increased density of functional groups and layered structure clay mineral groups on composite surface sterically hinder cations and anions, while blocking the trajectories to pores in the composite.[63]

Van der Waals and dipole-dipole interactions, which are weak physical bonds/intermolecular forces, also dominate in clay-biochar adsorption mechanisms.[53] The adsorption capacity has been increased with a large amount of surface functional groups from clays, where contaminants undergo pore filling mechanism (Fig. 2).[39]

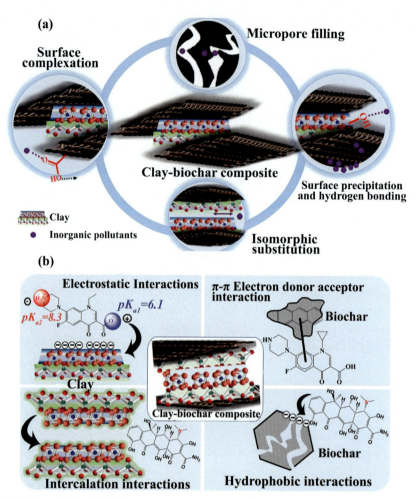

Fig. 2 Overall scheme of adsorption mechanisms of (A) inorganic (B) organic pollutants by clay-biochar composite.

7. Future perspectives

This chapter provides comprehensive details about clay-biochar composites, which can be utilized as a potential novel-material for the remediation of organic and inorganic contaminants in the environment. As discussed above, biochar derived from different feedstock materials have been used to prepare novel composites with different types of clay and were discussed for their efficiency and effectiveness for pollutant removal and retention. Despite the improved sorption abilities, clay-biochar composites possess few limitations in their performance, which need to be addressed in the future.[21,42]

- Clay-biochar composites have exhibited greater adsorption efficiency compared to their respective pristine materials,[22,42] although the fabrication of clay-biochar composites for environmental applications is limited. Therefore, further investigations are required to fabricate novel composites using different types of clay and biochar.
- Environmental stability of clay-biochar composites on both human and animals still remains unknown and studies need to be initiated upon that.
- Future studies are required to assess novel clay-biochar composites for the removal of emerging contaminants such as trace elements (thallium, molybdenum, selenium, etc.), anionic pollutant (halides, nitrate, sulfate, etc.), pharmaceutical compounds (β-blockers, cytostatic drugs, hormones, contrast media, etc.), personal care products (microbial agent, preservatives, etc.), insecticides, and radionuclide from the aquatic environment.
- Adsorptive removal of pollutants from the single-pollutant system by clay-biochar composites have been well understood[11,20] However, the competitive adsorptive removal of pollutants from a multi-pollutant system using clay-biochar composites have not been investigated. This creates a gap in the literature to simulate the effectiveness of the composite in the actual environment.
- The adsorption efficiency of clay-biochar composites depends on the composition ratio of its pristine materials, which has a wide variation used in the literature studied. Ashiq et al.[8] have reported that 1:1 ratio of municipal solid waste biochar and MMT was used to fabricate a composite. Similarly, 1:5 ratio of ATP and potato stem was used to prepared a composite by Li et al.[11] However, further research is still needed to determine the optimum clay: biochar ratio in order to achieve a maximum adsorption capacity of the composite.
- The possibility of using a mixture of clays in a clay-biochar composite instead of a single type of clay can be further studied in the future. The potential of non-smectite clays needs to be addressed to enhance the adsorption affinity of the clay-biochar composites.
- Regeneration capability of composites has been least explored. Experiments on regeneration studies need to be conducted to test the capability for subsequent treatment cycles.

- The slurry method and the biomass–clay pyrolysis method could impose distinctive properties on the biochar surface, but yet remains uncertain. Hence, further research on surface titration studies, ultimate analysis and morphological studies are required to elucidate the surface properties and adsorption effectiveness of the composites prepared of these two different methods.

References

1. Ahmad M, Rajapaksha AU, Lim JE, Zhang M, Bolan N, Mohan D, et al. Biochar as a sorbent for contaminant management in soil and water: a review. *Chemosphere* 2014; **99**:19–33.
2. Liu L, Chen X, Wang Z, Lin S. Removal of aqueous fluoroquinolones with multi-functional activated carbon (MFAC) derived from recycled long-root Eichhornia crassipes: batch and column studies. *Environ Sci Pollut Res* 2019;**26**(33): 34345–56.
3. Chen Z, Xiao X, Xing B, Chen B. pH-dependent sorption of sulfonamide antibiotics onto biochars: Sorption mechanisms and modeling. *Environ Pollut* 2019;**248**:48–56.
4. Wei D, Li B, Huang H, Luo L, Zhang J, Yang Y, et al. Biochar-based functional materials in the purification of agricultural wastewater: fabrication, application and future research needs. *Chemosphere* 2018;**197**. Elsevier.
5. Zhou Y, Gao B, Zimmerman AR, Chen H, Zhang M, Cao X. Biochar-supported zerovalent iron for removal of various contaminants from aqueous solutions. *Bioresour Technol* 2014;**152**:538–42.
6. Han H, Rafiq MK, Zhou T, Xu R, Mašek O, Li X. A critical review of clay-based composites with enhanced adsorption performance for metal and organic pollutants. *J Hazard Mater* 2019;**369**:780–96.
7. Arabyarmohammadi H, Darban AK, Abdollahy M, Yong R, Ayati B, Zirakjou A, et al. Utilization of a novel chitosan/clay/biochar nanobiocomposite for immobilization of heavy metals in acid soil environment. *J Polym Environ* 2018;**26**(5):2107–19.
8. Ashiq A, Sarkar B, Adassooriya N, Walpita J, Rajapaksha AU, Ok YS, et al. Sorption process of municipal solid waste biochar-montmorillonite composite for ciprofloxacin removal in aqueous media. *Chemosphere* 2019;**236**:124384.
9. Premarathna KSD, Rajapaksha AU, Sarkar B, Kwon EE, Bhatnagar A, Ok YS, et al. Biochar-based engineered composites for sorptive decontamination of water: a review. *Chem Eng J* 2019;**372**:536–50.
10. Jiang WT, Wang CJ, Li Z. Intercalation of ciprofloxacin accompanied by dehydration in rectorite. *Appl Clay Sci* 2013;**74**:74–80.
11. Li Y, Wang Z, Xie X, Zhu J, Li R, Qin T. Removal of Norfloxacin from aqueous solution by clay-biochar composite prepared from potato stem and natural attapulgite. *Colloids Surf A Physicochem Eng Asp* 2017;**514**:126–36.
12. Song J, Zhang S, Li G, Du Q, Yang F. Preparation of montmorillonite modified biochar with various temperatures and their mechanism for Zn ion removal. *J Hazard Mater* 2020;**391**:121692.
13. Yao Y, Gao B, Fang J, Zhang M, Chen H, Zhou Y, et al. Characterization and environmental applications of clay-biochar composites. *Chem Eng J* 2014;**242**:136–43.
14. KSD P, Rajapaksha AU, Adassoriya N, Sarkar B, NMS S, Cooray A, et al. Clay-biochar composites for sorptive removal of tetracycline antibiotic in aqueous media. *J Environ Manage* 2019;**238**:315–22.
15. Chen L, Chen XL, Zhou CH, Yang HM, Ji SF, Tong DS, et al. Environmental-friendly montmorillonite-biochar composites: facile production and tunable adsorption-release of ammonium and phosphate. *J Clean Prod* 2017;**156**:648–59.

16. Li L, Zou D, Xiao Z, Zeng X, Zhang L, Jiang L, et al. Biochar as a sorbent for emerging contaminants enables improvements in waste management and sustainable resource use. *J Clean Prod* 2019;**210**. Elsevier.
17. Madejová J, Gates WP, Petit S. IR spectra of clay minerals. In: *Developments in clay science*. vol. 8. Elsevier; 2017. p. 107–49.
18. Kim D, Park KY, Yoshikawa K. Conversion of municipal solid wastes into biochar through hydrothermal carbonization. In: Huang W-J, editor. *Eng Appl Biochar*; 2017.
19. Wang Z, Yang X, Qin T, Liang G, Li Y, Xie X. Efficient removal of oxytetracycline from aqueous solution by a novel magnetic clay–biochar composite using natural attapulgite and cauliflower leaves. *Environ Sci Pollut Res* 2019;**26**(8):7463–75.
20. Sewu DD, Lee DS, Tran HN, Woo SH. Effect of bentonite-mineral co-pyrolysis with macroalgae on physicochemical property and dye uptake capacity of bentonite/biochar composite. *J Taiwan Inst Chem Eng* 2019;**104**:106–13.
21. Liang G, Wang Z, Yang X, Qin T, Xie X, Zhao J, et al. Efficient removal of oxytetracycline from aqueous solution using magnetic montmorillonite-biochar composite prepared by one step pyrolysis. *Sci Total Environ* 2019;**695**:133800.
22. Ashiq A, Adassooriya NM, Sarkar B, Rajapaksha AU, Ok YS, Vithanage M. Municipal solid waste biochar-bentonite composite for the removal of antibiotic ciprofloxacin from aqueous media. *J Environ Manage* 2019;**236**:428–35.
23. Viglašová E, Galamboš M, Danková Z, Krivosudský L, Lengauer CL, Hood-Nowotny R, et al. Production, characterization and adsorption studies of bamboo-based biochar/montmorillonite composite for nitrate removal. *Waste Manag* 2018;**79**:385–94.
24. Zhang J, Lu M, Wan J, Sun Y, Lan H, Deng X. Effects of pH, dissolved humic acid and Cu^{2+} on the adsorption of norfloxacin on montmorillonite-biochar composite derived from wheat straw. *Biochem Eng J* 2018;**130**:104–12.
25. Dou G, Goldfarb JL. In situ upgrading of pyrolysis biofuels by bentonite clay with simultaneous production of heterogeneous adsorbents for water treatment. *Fuel* 2017;**195**:273–83.
26. Wong JTF, Chen Z, Chen X, Ng CWW, Wong MH. Soil-water retention behavior of compacted biochar-amended clay: a novel landfill final cover material. *J Soil Sediment* 2017;**17**(3):590–8.
27. Ismadji S, Tong DS, Soetaredjo FE, Ayucitra A, Yu WH, Zhou CH. Bentonite hydrochar composite for removal of ammonium from koi fish tank. *Appl Clay Sci* 2016;**119**:146–54.
28. El-Naggar A, Lee SS, Rinklebe J, Farooq M, Song H, Sarmah AK, et al. Biochar application to low fertility soils: a review of current status, and future prospects. *Geoderma* 2019;**337**:536–54.
29. Rawal A, Joseph SD, Hook JM, Chia CH, Munroe PR, Donne S, et al. Mineral-biochar composites: molecular structure and porosity. *Environ Sci Technol* 2016;**50**(14):7706–14.
30. Choudalakis G, Gotsis AD. Permeability of polymer/clay nanocomposites: a review. *Eur Polym J* 2009;**45**(4):967–84.
31. Chen S, Zhou M, Wang HF, Wang T, Wang XS, Hou HB, et al. Adsorption of reactive brilliant red X-3B in aqueous solutions on clay-biochar composites from bagasse and natural attapulgite. *Water (Switzerland)* 2018;**10**(6):703.
32. Yin Z, Liu Y, Tan X, Jiang L, Zeng G, Liu S, et al. Adsorption of 17B-estradiol by a novel attapulgite/biochar nanocomposite: Characteristics and influencing factors. *Process Saf Environ Prot* 2019;**121**:155–64.
33. Rafiq MK, Joseph SD, Li F, Bai Y, Shang Z, Rawal A, et al. Pyrolysis of attapulgite clay blended with yak dung enhances pasture growth and soil health: characterization and initial field trials. *Sci Total Environ* 2017;**607–608**:184–94.
34. Wu LM, Zhou CH, Tong DS, Yu WH, Wang H. Novel hydrothermal carbonization of cellulose catalyzed by montmorillonite to produce kerogen-like hydrochar. *Cellul* 2014;**21**(4):2845–57.

35. Fei TX, Guo LY, Ling GY, Xu Y, Ming ZG, Jiang HX, et al. Biochar-based nano-composites for the decontamination of wastewater: a review. *Bioresour Technol* 2016;**212**:318–33.
36. Mao H, Zhou D, Hashisho Z, Wang S, Chen H, Wang H. Preparation of pinewood- and wheat straw-based activated carbon via a microwave-assisted potassium hydroxide treatment and an analysis of the effects of the microwave activation conditions. *BioResources* 2015;**10**(1):809–21.
37. Bilgiç C. Investigation of the factors affecting organic cation adsorption on some silicate minerals. *J Colloid Interface Sci* 2005;**281**(1):33–8.
38. Li H, Dong X, da Silva EB, de Oliveira LM, Chen Y, Ma LQ. Mechanisms of metal sorption by biochars: biochar characteristics and modifications. *Chemosphere* 2017;**178**:466–78.
39. Ahmad M, Lee SS, Rajapaksha AU, Vithanage M, Zhang M, Cho JS, et al. Trichloroethylene adsorption by pine needle biochars produced at various pyrolysis temperatures. *Bioresour Technol* 2013;**143**:615–22.
40. Tareq R, Akter N, Azam MS. *Biochars and biochar composites. Biochar from biomass and waste.* Elsevier; 2019. p. 169–209.
41. Sarkar B, Rusmin R, Ugochukwu UC, Mukhopadhyay R, Manjaiah KM. Modified clay minerals for environmental applications. In: *Modified clay and zeolite nanocomposite materials: environmental and pharmaceutical applications.* Elsevier; 2018. p. 113–27.
42. Zhao Z, Nie T, Zhou W. Enhanced biochar stabilities and adsorption properties for tetracycline by synthesizing silica-composited biochar. *Environ Pollut* 2019;**254**:113015.
43. Belhouchat N, Zaghouane-Boudiaf H, Viseras C. Removal of anionic and cationic dyes from aqueous solution with activated organo-bentonite/sodium alginate encapsulated beads. *Appl Clay Sci* 2017;**135**:9–15.
44. Tareq R, Akter N, Azam MS. Biochars and biochar composites: low-cost adsorbents for environmental remediation. In: *Biochar from biomass and waste: fundamentals and applications.* Elsevier; 2018. p. 169–209.
45. Bannister FA. Clay minerals. In: *Nature.* vol. 170. Elsevier; 1952. p. 337–8.
46. Yu Q, Xia D, Li H, Ke L, Wang Y, Wang H, et al. Effectiveness and mechanisms of ammonium adsorption on biochars derived from biogas residues. *RSC Adv* 2016;**6**(91):88373–81.
47. Sen Gupta S, Bhattacharyya KG. Adsorption of heavy metals on kaolinite and montmorillonite: a review. *Phys Chem Chem Phys* 2012;**14**(19):6698–723.
48. Vithanage M, Sandaruwan L, Samarasinghe G, Jayawardhana Y. Clay-biochar composite for arsenic removal from aqueous media. In: *Environmental arsenic in a changing world—7th international congress and exhibition arsenic in the environment.* vol. 2018; 2018. p. 437–8.
49. Wang H, Tan L, Hu B, Qiu M, Liang L, Bao L, et al. Removal of Cr(VI) from acid mine drainage with clay-biochar composite. *Desalin Water Treat* 2019;**165**:212–21.
50. Vithanage M, Weerasundara L, Ghosh AK. Acid induced arsenic removal from soil amended with clay-biochar composite. In: *Environmental arsenic in a changingworld—7th international congress and exhibition arsenic in the environment.* vol. 2018; 2018. p. 561–2.
51. Bazrafshan AA, Hajati S, Ghaedi M. Synthesis of regenerable Zn(OH)2 nanoparticle-loaded activated carbon for the ultrasound-assisted removal of malachite green: optimization, isotherm and kinetics. *RSC Adv* 2015;**5**(96):79119–28.
52. Wang S, Gao B, Zimmerman AR, Li Y, Ma L, Harris WG, et al. Removal of arsenic by magnetic biochar prepared from pinewood and natural hematite. *Bioresour Technol* 2015;**175**:391–5.
53. Rajapaksha AU, Chen SS, Tsang DCW, Zhang M, Vithanage M, Mandal S, et al. Engineered/designer biochar for contaminant removal/immobilization from soil and water: potential and implication of biochar modification. *Chemosphere* 2016;**148**:276–91.

54. Inyang M, Dickenson E. The potential role of biochar in the removal of organic and microbial contaminants from potable and reuse water: a review. *Chemosphere* 2015; **134**:232–40.
55. Aristilde L, Lanson B, Miéhé-Brendlé J, Marichal C, Charlet L. Enhanced interlayer trapping of a tetracycline antibiotic within montmorillonite layers in the presence of Ca and Mg. *J Colloid Interface Sci* 2016;**464**:153–9.
56. Verhaverbeke S. Cleaning of trace metallic impurities from solid substrates using liquid media. In: Johansson I, Somasundaran PBT-H for C of S, editors. *Handbook for cleaning/ decontamination of surfaces*. Amsterdam: Elsevier Science B.V.; 2007. p. 485–538.
57. Ahmed MB, Zhou JL, Ngo HH, Guo W. Adsorptive removal of antibiotics from water and wastewater: progress and challenges. *Sci Total Environ* 2015;**532**:112–26.
58. Dutta D, Marepally SK, Vemula PK. Noncovalent functionalization of cell surface. In: Karp JM, Zhao WBT-MN of the CS, editors. *Micro-and nanoengineering of the cell surface*. Oxford: William Andrew Publishing; 2014. p. 99–120.
59. Flores RM. Coalification, gasification, and gas storage. In: Flores RMBT-C and CG, editor. *Coal and Coalbed Gas*. Boston: Elsevier; 2014. p. 167–233.
60. Jing Y, Cao Y, Yang Q, Wang X. Removal of cd(II) from aqueous solution by clay-biochar composite prepared from Alternanthera philoxeroides and bentonite. *BioResources* 2020;**15**(1):598–615.
61. Guo X, Dong H, Yang C, Zhang Q, Liao C, Zha F, et al. Application of goethite modified biochar for tylosin removal from aqueous solution. *Colloids Surf A Physicochem Eng Asp* 2016;**502**:81–8.
62. Liu W, Zhang J, Zhang C, Ren L. Sorption of norfloxacin by lotus stalk-based activated carbon and iron-doped activated alumina: mechanisms, isotherms and kinetics. *Chem Eng J* 2011;**171**(2):431–8.
63. Nguyen TH, Cho HH, Poster DL, Ball WP. Evidence for a pore-filling mechanism in the adsorption of aromatic hydrocarbons to a natural wood char. *Environ Sci Technol* 2007;**41**(4):1212–7.

CHAPTER TWO

Remediation of heavy metal contaminated soil: Role of biochar

Lina Gogoi[a,b], Rumi Narzari[a], Rahul S. Chutia[c], Bikram Borkotoki[d], Nirmali Gogoi[b], and Rupam Kataki[a,*]

[a]Department of Energy, Tezpur University, Tezpur, India
[b]Department of Environmental Sciences, Tezpur University, Tezpur, India
[c]Department of Energy Engineering, North Eastern Hill University, Shillong, India
[d]Biswanath College of Agriculture, Assam Agricultural University, Sonitpur, India
*Corresponding author: e-mail address: rupamkataki@gmail.com

Contents

1. Introduction	40
2. Inorganic contaminants in soil	41
3. Biochar as a soil amendment	45
3.1 Physical properties of biochar	46
3.2 Chemical properties of biochar	46
4. Mechanisms of heavy metal remediation by biochar	48
4.1 Electrostatic attraction	48
4.2 Ion exchange and adsorption of cationic π function	49
4.3 Physical attraction	49
4.4 Precipitation	50
4.5 Complexation	50
5. Biochar applications in heavy metal remediation	51
6. Conclusion and recommendations for further research	51
References	57

Abstract

Soil contamination with heavy metal has been a matter of serious concern due to its influences on soil quality, microbial diversity, groundwater and agricultural productivity. Managing soil heavy metal contamination is a challenge owing to its persistent nature in soils which leads to its accumulation and transfer into the food chain via agricultural products and thereby adversely affect human health. Biochar is a promising bio-residue material that can be used for remediation of heavy metal-contaminated soil in a sustainable way. In this chapter, an attempt has been made to understand the biochar induced underlying mechanisms for soil heavy metal remediation. Some of the case studies and the effectiveness of biochar on heavy metal contaminated soil are briefly summarized. The review suggests that significant investigations need to be carried out in the future to evaluate the long-term environmental and economic feasibility of biochar application to remediate heavy metal contamination in soil.

Keyword: Heavy metal, Biochar, Microbial diversity

1. Introduction

Heavy metal pollution is one of the major environmental issues across the globe. From the onset of the industrial revolution, various anthropogenic activities such as smelting, mining, application of herbicides, weedicides, fertilizer and pesticide, wastewater utilization in irrigation, and milling are the prime causes that accumulate or release a huge amount of heavy metals (HMs) into the environment.[1] Soil pollution due to heavy metals has become a matter of global concern over the past few decades. Due to their persistent and accumulative nature, heavy metals possess pernicious threats contrary to the organic pollutant.[2] Metal pollution can disrupt soil microbial population thereby affecting soil fertility which will ultimately lead to loss of biodiversity and crop productivity.[3] Apart from the soil health it also adversely impacts human health, thereby raise concern for food security and safety.[4,5] Although some of these metals are essential for the life process of living organisms, however, accumulation of concerned metals beyond threshold limit due to consumption of heavy metal(loid) enriched produce interrupts the metabolic system, therefore, has detrimental health effects.[6] The presence of cancer villages is a clear warning of an adverse effect of soil contamination with heavy metals.[7] Evaluation of carcinogenic risks based on total cancer risks (TCR) for a number of rice varieties in north-east India's lower Brahmaputra valley has reported a plausible health hazard (as indicated by hazard index (HI)) due to presence of a number of heavy metal(loid)s.[8]

To mitigate soil heavy metal contamination, various remediation technologies has been developed in the past. These includes physical treatment (with thermal desorption, guest land methods, vitrification technology), chemical remediation (with immobilization, elution methods, and electrokinetic methods), and bioremediation (with microorganisms and plants).[9] However, all these traditional approaches have certain limitations such as poor feasibility, complicated technique, high cost, low efficiency, short duration, and high secondary risk.[10] Currently, various published studies suggest that in situ stabilization or immobilization of heavy metals utilizing various organic and inorganic amendments is a promising cost-effective, and environmentally viable approach.[11] It reduces the mobility of the pollutants.[12,13] Stabilization is achieved through mixing contaminated soil with the agents which either by chemical reaction or sorption lower the bioavailability of the contaminant.[14] Different amendments such as lime, iron and ferric salts, phosphate fertilizers, apatite composites, bentonite, zeolite, red mud, silicon and modified silicon have been developed to mitigate heavy metal contamination in soil.[15]

In recent times, biochar has emerged as a new stabilizing agent which has gained much attention in the scientific community. Biochar is a carbon concentrated solid residue obtained after pyrolysis of biomass. It is highly porous, fine-grained, and amorphous.[16] The reason behind biochar's increasing popularity compared to its conventional counterparts is that apart from immobilization of HMs in the soil it also improves soil quality by increasing nutrient retention capacity, water level, soil pH, and thus the crop productivity. Also, the production of biochar is a sustainable and environmentally friendly waste management strategy for organic waste matter.[17] A plethora of studies has been conducted including both pot and field experiments to remediate HMs contaminated soil with a positive response.[14,18] However, various studies have shown that the effectiveness of biochar remediation is influenced by various factors including physicochemical properties of biochar along with soil environmental conditions. Biochar characteristics can be influenced by factors such as feedstock type, pyrolysis temperature, rate of temperature rise, residence time, and modification method.[19,20]

In this chapter, we have discussed different types and sources of heavy metal contamination in soil. Attempt has been made to understand the biochar induced underlying mechanism of soil heavy metal remediation. Finally, some of the case studies and the effectiveness of biochar on heavy metal contaminated soil based on their result are summarized.

2. Inorganic contaminants in soil

2.1.1 Heavy metal as soil inorganic contaminant

Inorganic contaminants comprise of toxic metals and different types of nutrients and salts that mostly occur in the form of dissolved anions and cations. Soil is a major reservoir for Potential Toxic Elements (PTEs). The concentrations formed and availability of PTEs can affect both the soil ecosystem and plant growth. These PTEs may also enter in human body by direct (inhalation or ingestion of soil particles) or indirect (plant consumption) routes and can cause serious health damages. Heavy metals are important in various ways. Many of them have industrial use in technologically progressive nations. Some are physiologically important for plants and animals and consequently have a direct bearing on human health and farm productivity. Many are potential pollutants of ecosystem throughout the globe. In recent years, there has been increasing attention of heavy metal contamination in soils mainly due to the visible effect of irresponsible anthropogenic activities of ever-increasing human population and partly because of rising scientific and public awareness of environmental concerns and advancement of analytical methods with sophisticated instrumentation to measure their concentrations accurately. Atomic Absorption Spectrophotometer (AAS), for example allows many samples to be analyzed rapidly. Conversely, geogenic sources of heavy metals like the arsenic present with pyritic bedrocks are also prodigious concern where lifted groundwater is used for farm irrigation as well as for drinking purpose. Thus, metal contamination is a persistent problem in many polluted soils. The most commonly occurring metals are Cadmium (Cd), Copper (Cu), Nickel (Ni), Lead (Pb), Chromium (Cr), Zinc (Zn), Arsenic (As) and Mercury (Hg).[21,22] Among these metals Cd and As are extremely poisonous; Hg, Pb, Ni are moderately poisonous and Cu, Zn, Mn are relatively lower in toxicity.[23,24]

2.1.2 Sources of heavy metal pollution in soil

Heavy metals are generally characterized by the following factors: Specific gravity >5.2, the shift in their valences, lower solubility of their hydroxides, great capacity to form complex compounds and great affinity for sulfides. Soil and water may be contaminated by these heavy metals both from geogenic and anthropogenic activities as discussed in Fig. 1.

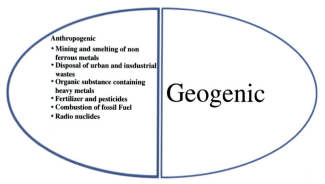

Fig. 1 Sources of heavy metal pollution.

2.1.3 Metallurgical industries can contribute to soil and water pollution in several ways

- By emission of gases and specks of dust containing metals which are transported in the air and in due course deposited into soil and water.
- By effluents which may contaminate soil when waterways were flooded.
- By the establishment of waste dumps from which metals may be leached and thus contaminating the underlying or nearby soils and water bodies.

2.1.4 How heavy metals create problem

Excess concentration of heavy metals in soil and water cause problems in several ways because they decrease the production capacity of the soil, deteriorate soil health and water quality. Heavy metal contaminated soils also cause phytotoxicity to plants. Sources of heavy metals and their cycling in the soil-water-water-air-organism ecosystem is presented in Fig. 2. The concentration of metals in tissues usually builds up from left to right indicating the vulnerability of humans to heavy metal toxicity at the top of the food chain. This is called biomagnification of heavy metal.

2.1.5 Heavy metal polluted soil

Soil properties affect metal accessibility in varied ways. Soil pH is one of the major factors that affect soil metal availability.[25] Significant positive correlations have also been observed between heavy metals and some soil physical properties such as moisture content and water holding capacity.[26] The soil biological properties are affected due to the presence of heavy metals. Toxic level of heavy metals on soil microorganisms depends on a number of factors

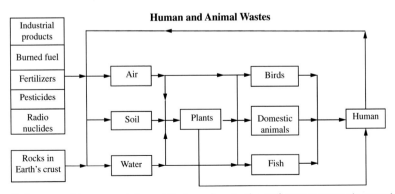

Fig. 2 Sources of heavy metals and their cycling in the soil-water-water-air-organism ecosystem.

such as soil temperature, pH, clay minerals, organic matter, inorganic anions and cations, and chemical forms of the metal.[27,28] Higher concentrations of heavy metals in soils reduce soil microbial biomass carbon, inhibit nitrogen fixation by legumes, decrease specific enzyme activities such as urease, phosphatase, and dehydrogenase etc.[27,29,30] The arbuscular mycorrhizal (AM) fungi symbiosis is susceptible to heavy metals and can be used as a bioindicator in heavy metal polluted soils.[31,32] Plants growing on these soils show a reduction in growth performance, and yield.

Chaney[33] grouped the elements based on the degree of susceptibility on plant uptake and consequent ingestion by humans or animals in the food chain as follows:

Group 1. This group contains the metals, which are insoluble in soil, or root that the plant shoot is not a considerable source of metal transfer/uptake even though the soils are extremely contaminated. Cr is probably the only metal occurring in wastes at high concentrations that falls into this category.

Group 2. This group includes those metals, which can be absorbed into roots, however, are not translocated to shoots at high rates to encompass risk, for example- Hg, Pb and As.

Group 3. This group includes the metals which are easily absorbed by roots and translocated to shoots, but the plant yield is severely reduced as a sign of symptom before humans, livestock, or wildlife are at any danger from prolonged consumption of the crop. Thus, phytotoxicity defends food chain contamination. For example, Cu, Ni and Zn.

Group 4. This group includes metals for whom plants don't provide food chain protection. Therefore, more dangerous compared to other heavy metals for example Selenium (Se), Molybdenum (Mo) and Cd.

2.1.6 Effect on plants

Some of the so-called heavy metals (e.g., Cu, Zn, Ni) are essential plant nutrients. However, at higher concentrations, heavy metals are phytotoxic and result in poor plant stand or complete death.[34,35] Plants grown in pots spiked with soluble metal salts tend to amass the high amount of metals than they grow in the agricultural field with parallel quantities of metals applied as sewage sludge.[36] Eventually, the decline in yield is a most crucial measure of phytotoxicity for crop plants, since it affects the benefit: cost ratio of crop production and therefore, restrict the utility of cropland.[36,37]

2.1.7 Factors governing uptake of heavy metals by crops

The following factors are responsible for uptake of heavy metal from the soil by crops:

- Soil factor: Soil pH, Cation Exchange Capacity, Soil Organic Matter, Soil Texture
- Quality of sludge and waste and
- Nature of crops

Solubility of solid-phase minerals including metal carbonates, phosphates and sulfates is boosted at low pH values.[38–40] Soil reaction also affects the regulatory mechanism of heavy metal uptake by roots. Soil pH value of 6.4 or more has been observed to regulate Ni and Cd uptake by plants grown in soils with higher Ni/Cd content and inhibit the toxicity of these metal cations.[24,41] Higher soil CEC reduces heavy metal induced phytotoxicity due to sorptive protection. This CEC induced impact on phytotoxicity is reduced when organically bound heavy metals are applied in soil.[42] Organic matter can act as a chelating agent to bind heavy metals present in the soil. Heavy metals are more available in light-textured acid soils compared to those containing a large amount of clay with alkaline reactions. The rate at which root absorbs metals from soil depends on the activity of free metal ions in soil solution and at the root surface. The activity at the root surface in turn depends on equilibrium reactions between the soil solid phase and the sludge that are applied in the soil.[35,43] The magnitude of transfer of heavy metals in the food chain through crops depends on the characteristics of plant species to a large extent. Research on vegetables grown on sludge-amended soils revealed that lettuce and other leafy vegetables are good accumulators of heavy metals.[44–46]

3. Biochar as a soil amendment

Biochar is a carbon-rich compound which is produced by thermal decomposition of organic material under a limited supply of oxygen (O_2), and at low temperatures (300–700 °C).[47] Both the physical and chemical properties of biochar depend on feedstock material and the process of carbonization or pyrolysis. The feedstock biomass has a significant influence on the newly produced biochar material. During the conversion of biomass to biochar the volatile organics present in the feedstock material is lost and therefore making a disproportional shrinkage of the original biomass.[48,49]

The chemical composition of the feedstock has a direct influence on the physical nature of the produced biochar. A brief description on both physical and chemical properties of biochar has been presented below.

3.1 Physical properties of biochar

The physical property of biochar mainly deals with the structural configuration of the biochar and properties due to this. The structure of biochar is mostly amorphous in nature though it contains crystalline structure of highly aromatic compounds. These micro-crystallites form the conducting phase of biochar whereas the non-conducting components comprise of the present minerals. Various operating parameters like pretreatment, heating rate, residence time, highest treatment temperature (HTT), pressure etc. affect the biochar physical characteristics.[47] Physical properties of biochar like surface area, pore structure, distribution and charge density, influence the critical functions like nutrient retention, microbial activity and water holding capacity (WHC). The surface area of biochar is generally higher or equal to clay particles which can shift a net change in surface area when added to soil. Higher pyrolysis temperature results in improved surface area and pore development due to the release of volatile particles present in it.[50] With increasing HTT, the surface area of biochar also increases up to a threshold temperature beyond which deformation occurs. The heating rate affects the formation and development of pores in biochar (micro pore and mega pore). Biochar produced at low heating rate were mostly comprised of macro pores whereas high heating rate formed micro pores.[51] Micro pore provides a larger surface area ($750–1360\,m^2\,g^{-1}$) but a higher volume of pore size was reported in case of macro pore ($0.6–1.0\,cm^3\,g^{-1}$). The broader volume of macro pores results the higher functionality and productivity in soil due to proper aeration and hydrology.[52] The particle sizes of biochar produced from the pyrolysis of organic matter are predominantly influenced by the feedstock material. Biochar has a lower particle size compared to its organic feedstock due to both shrinkage and attrition during the process of conversion. The slow pyrolysis can produce much larger particles compared to fast pyrolysis. Bulk density is also an important physical property of biochar which tend to decrease due to development of porosity through process of pyrolysis.

3.2 Chemical properties of biochar

Organic materials present in biomass change during thermal decomposition at a temperature above $120\,°C$. Dehydration and depolymerization are the initial changes that occur during conversion of biomass to biochar.

Between 250 °C and 350 °C, complete cellulose depolymerization occurs resulting the amorphous carbon matrix. During the process of pyrolysis, the aliphatic carbon present in biomass changes into aromatic carbon.[53] Gradually, these aromatic carbons are degraded and form pores. The major constituents of biomass feedstock also volatilize during this phase. With increasing highest temperature treatment (HTT), two carbon phases present in biochar are amorphous, and grapheme packets. Above 600 °C, the graphene packets grow laterally at the expense of the amorphous phase.[54–56] The presence of minerals in biochar is primarily influenced by the feedstock material and the process condition. The surface chemistry of biochar exhibit hydrophilic, hydrophobic, acidic and basic properties which depends on the feedstock material and the process condition. Presence of various functional groups and heteroatoms are found on the surface of the graphene sheets.[57] The oxygen present in atmosphere reacts with the biochar and forms the O-containing functional groups.[58,59] Pyrrolic and pyridinic amines are the N containing functional groups that are present in low temperature biochar whereas equal amount of pyridinic and quaternary groups are present in biochar that are produced at high temperature.[60] In case of S-containing functional groups, sulfonates and sulfates are present in low temperature biochar whereas thiophene and sulfide groups are present in high temperature biochar.[60] H/C and O/C ratio in biochar is important as it defines both the degree of aromaticity and maturation. Stability increased with maximum treatment temperature and decreased H/C and O/C ratios.[61] The pH of biochar is generally alkaline in nature (pH > 7) but it can also vary from 4 to 12. Biochar is widely acknowledged as soil amendment due to its potential to increase and retain the soil nutrients. A range of both macro and micro nutrients are found in biochar. The presence of nutrients in biochar is directly dependent on the feedstock composition and the process condition. The available K content in biochar is high compared to N and available P content.[62,63] At a pyrolysis temperature of 500 °C, almost half of the total S was lost. However, a decrease in inorganic S content was reported with increased temperature. The presence of heavy metals like Cu, Zn, Cr and Ni also depends on the biochar feedstock. The C/N ratio of biochar is very high with an average of about 61. Despite the high C/N ratio, the degree of N immobilization is negligible due to the recalcitrant nature of C present in biochar. The high cation exchange capacity (CEC) of biochars can be attributed due to its high surface area during the process of conversion. Moreover, there is an increase in charge density on the surface of biochar. The CEC of biochar increases with the pyrolysis temperature depending on its surface area.[64]

4. Mechanisms of heavy metal remediation by biochar

Biochar is considered to be a promising material for efficient removal of the toxic HM species from contaminated sites. The mobility and bioavailability of inorganic contaminants are controlled by using biochar through a myriad of reactions, including redox, sorption, and complexation. There are several reported studies that have highlighted its potential effectiveness as remediating agent, which will be discussed in the later section. Biochar enhances the quality of contaminated soil through adsorption of heavy metals. According to the current research the interaction between heavy metals and biochar proceed through various adsorption mechanisms including electrostatic attraction (chemisorption), ion exchange, surface mineral adsorption, cation–π interactions and precipitation of surface functional group.[65] Fig. 3 depicts the various mechanisms through which heavy metal remediation by biochar takes place.

4.1 Electrostatic attraction

The larger surface area and higher surface energy of biochar contribute strongly to adsorb the heavy metal pollutants and remove them from the soil.

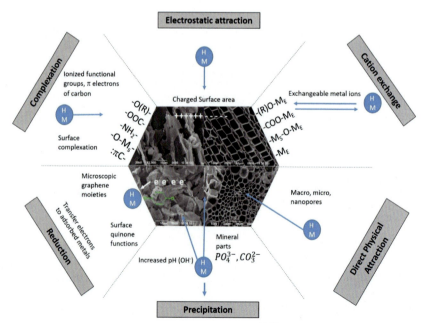

Fig. 3 Representation of biochar sorption for metal.[66,67]

When biochar is applied to soil the electrostatic attraction between heavy metals with positive charge and soil will be enhanced, which increase the adsorption capacity of heavy metals and it could mainly be attributed to electronegativity of biochar surface.[68,69] The intensity of the attraction depends on the surface charge generated by the negatively charged surface groups, which on increasing pH become more negative. It is evident from literature that biochar prepared at lower temperature shows higher adsorption capacity, which could be attributed to the stronger electrostatic attraction of biochar at lower pyrolysis temperature.[70]

4.2 Ion exchange and adsorption of cationic π function

Biochar surface contains various functional groups ($-OH$, $-COOH$, $C\equiv N$, $C=O$, $C-H$, $-CH_3$, etc.) which show acid–base characteristics and develop a variable surface charge in response to the soil solution pH, charge potential of free radicals at interface of the soil. The development of variable surface charges could be due to several chemical reactions including protonation, deprotonation, and ligand exchange by amphoteric functional groups.[71] With time, the increase in functional groups (including mainly carboxyl groups, but also phenolic, hydroxyl, carbonyl or quinone C forms) in biochar leads to an evolution of surface negative charge by replacing the surface positive charge of the particles.[72,73] These negatively charged particles increase the charge density on the surfaces of biochar particles and are responsible for the high cation exchange capacity (CEC) in the pH range of soil.[74]

Due to the presence of redox-active moieties (quinone–hydroquinone moieties) and condensed aromatic structures with conjugated π-electron systems, Biochars' surface functional groups can undergo redox transformations to facilitate electron transport between oxidant and reductant species (redox couples). Cation–π is a complex combination of electrostatic adsorption and π–π conjugation. Generally, the effect of cation–π is dependent mainly on the aromaticity of the biochar surface, which is an indicator of the number of π–π conjugations. Hence the more π-conjugated aromatic structure, the stronger the electron-donating ability.[75]

4.3 Physical attraction

Heavy metal remediation by biochar can proceed through direct physical attraction also. Due to the high surface area and pore volumes, biochar has a greater affinity for metals because metallic ions can be physically sorbed onto the char surfaces and retained within the pores.[76] This process of physical attraction which is basically physical or surface sorption mechanism, could be described by diffusional movement of metal ions into biochar pores without formation of any chemical bonds. It is evident from literature that increase in carbonization temperature leads to increase the specific surface area and pore volumes of biochar, which favors the physical attraction of metallic ions by biochar.

4.4 Precipitation

Precipitation is another proposed mechanism through which biochar can remove heavy metals from contaminated soil. Basically, the mineral matter contained in biochar may precipitate with metals, forming some insoluble precipitates.[77] There are mainly two aspects through which biochar can effectively decrease the activities of heavy metals by precipitation. The increasing pH due to the biochar amendment of soil may lead to decreased heavy metal mobilization, thus forming heavy metal hydroxide precipitation.[78] On the other hand, various phosphate and carbonate precipitations would be formed under different conditions. For instance, a new precipitate was observed solely on Pb-loaded sludge-derived biochar as $5PbO \cdot P_2O_5 \cdot SiO_2$ (lead phosphate silicate) at initial pH 5.0.[79]

4.5 Complexation

The functional group present on the surface of the biochar having low mineral content can immobilize heavy metals via surface complexation mechanism. The functional groups provide site for binding heavy metals to form complexes, increasing the specific adsorption of the metals. The formation of complexes could be observed by shifting of band intensity from $2349\,cm^{-1}$ to $2360\,cm^{-1}$ and $2341\,cm^{-1}$ while adding biochar to Cd (II) contaminated soils.[79] Studies showed that the ability of insoluble and stable complex formation increases with increasing Fe (II), Mn (II) and CO_3^{-2} concentrations. Also, the inorganic ions (e.g., Si, S, and Cl) contained in biochar can form complexes with heavy metals like Cd (II), decreasing their mobility in soil.[80]

In the above section different mechanisms have been briefly discussed that govern heavy metal remediation by biochar from contaminated soil. However, the role of each of mechanism for remediation of each metal varies considerably depending on target metals.

5. Biochar applications in heavy metal remediation

The major driving factors for the development of efficient soil HM remediation techniques are degradation of soil quality and crop productivity to reduce ecological risk.[81,82] The soil remediation technologies can be broadly classified into three categories viz. Physical, chemical, and biological. Thermal treatment, soil washing, phytoremediation, and contaminated soil excavation are some of the traditionally used soil clean up processes.[83,84] However, these techniques have certain limitations such as alternation in soil pH, reduced organic matter, and greenhouse gas (GHG) impact of 365 kg CO_2 eq. per ton of soil is associated with thermal treatment.[85,86] While on the contrary soil washing raises soil's silt and clay content along with organic matter. Therefore, there is an urgent need for extensive research to develop sustainable and effective remediation technology. As of now, the application of soil amendments is regarded as the most suitable method such as coal fly ash, phosphate compounds, organic composts, liming materials, and metal oxides, and biochar.[87] These amendments immobilize heavy metals via complexation, precipitation, adsorption, and reduction.[3] The long-term stability of biochar is due to the abundant organic structure present on its surface which forms organic-inorganic complexes and reducing the risk of decomposition.[88] Below are some of the case studies conducted to study the removal/immobilization efficiency of heavy metal via biochar application (Table 1).

6. Conclusion and recommendations for further research

Soil toxicity owing to heavy metal contamination stands as one of the grave issues for agricultural application. As a result, it has become significant to treat soil contaminated by heavy metals in order to deliver land resources for agricultural application which in turn enhance global food security. For this, a comprehensive understanding of the sources, chemistry, and related potential risks of toxic heavy metals in contaminated soils is necessary for the selection of appropriate remedial/treatment options. Numerous in-situ and ex-situ technologies are available for the treatment (reduction, removal and degradation) of toxic heavy metal in contaminated soils. On the contrary,

Table 1 Case studies conducted to study the removal/immobilization efficiency of heavy metal via biochar.

Metal of concern	Feedstock	Soil type	Location	Production temperature (°C)	Dose	Findings	References
Cd	Cotton stalks	—	—	450	—	Cd bioavailability reduced in soil through adsorption or co-precipitation.	89
As	Hardwood	—	—	400	—	Significant reduction of As in the foliage of *Miscanthus*.	90
As, Cd, Cu, Pb and Zn	Eucalyptus	Sandy	Cobbitty, New South Wales, Australia	550	0, 5, and 15 g/kg and 0, 10, and 50 mg/kg	Decrease in As, Cd, Cu, and Pb in maize shoots.	91
Cd, Cr, Cu, Ni, Pb and Zn	Orchard prune residue	Clay	Raibl mine, Julian Alps, Italy	500	0%, 1%, 5% and 10%	Cd, Pb, Ti and Zn bioavailability was reduced. Biochar increased soil pH, CEC, and water-holding capacity.	92
Cd, Cu and Pb	Chicken manure and green waste	Hills soil and mine soil	Adelaide Hills, South Australia and South Korea	550	1%, 5% and 5%	Reduction of Cd, Cu, and Pb bioavailability.	93
Cd	Wheat straw	Ferric-accumulic stagnic anthrosols	Jiangsu, China	550	0, 10, 20 and 40 t/ha	Soil: Decreased by 5–53% Grain: Decreased by 17–62%.	94
Cu, Cr, Ni and Zn	Wheat straw	—	China Southern Jiangsu Province	—	0, 1, 3, 5, 7 and 10 g/kg	Biochar modified the forms of heavy metals in the soil and significantly immobilized them.	95

Metal	Biomass source	Soil type	Location	Temp	Application rate	Remarks	Ref
Cd, Zn, Pb and As	Rice straw, husk and bran	–	Zhuzhou city, Hunan province, China	500	0% and 5%	Decreased concentrations of Cd, Pb and Zn by up to 98%, 72% and 83% in rice shoot; decreased concentrations of Cd and Zn in the pore water.	96
Cd	Oil mallet and wheat chaff	Sandy loam	Gingin, Western, Australia	550	0.5% and 5 % (w/w)	Significantly reduced EDTA-extractable Cd in soil.	97
Pb	Soybean stover	Loamy sand	Busan Metropolitan city, Korea	700	0%, 1%, 2%, 3%, 4%, 5%, 10% and 20%	Leachability decreased with increasing biochar content. A reduction of over 90% leachability was achieved.	98
Cd	Wheat straw	–	Jiangsu, Hunan, Sichuan and Fujian, China	350 and 550	0, 20 and 40 t/ha	Reduction in extractable Cd pool decreased Cd concentration by 20–90% in rice grain.	99
Cd and Pb	Wheat straw	Hydroagric Stagnic Anthrosol	Jingtang Village Jiangsu, China	350 and 550	0, 10, 20 and 40 t/ha	Immobilization primarily due to the precipitation and surface adsorption.	100
As, Cd, Co, Cu, Mn, Pb and Zn	Sewage sludge	–	Fujian, China	–	5% and 10%	As, Cd, Cu and Pb concentrations decreased, no difference in Zn.	101
Cd and Ni	Unfertilized dates	Sandy Loam	Hada El-Sham, Jeddah, Saudi Arabia	500	0.5, 1 and 2% (w/w)	Soil Cd and Ni were lower due to increased soil pH. NH_4NO_3-extractable soil Ni was reduced by 53%.	102

Continued

Table 1 Case studies conducted to study the removal/immobilization efficiency of heavy metal via biochar.—cont'd

Metal of concern	Feedstock	Soil type	Location	Production temperature (°C)	Dose	Findings	References
Cd and Zn	Cassava stem	Silty clay loam	Mae Tao subcatchment, Thailand	350	0%, 5%, 10%, and 15%	Cd and Zn bioavailability in soil was reduced by cassava stem-derived BC.	103
Pb(II)	Peanut, soybean, canola, rice	Oxisol and Ultisol	Chengmai and Kunlun, Hainan, China	400	5% (w/w)	All BC types increased adsorption capacity of Pb (II) due to increased soil cation exchange capacity (CEC) and pH.	104
Cd, Cu, Pb, and Zn	Rice straw	Red loam	Guangdong, China	–	1:10 (biochar: soil)	DTPA-extractable heavy metals bioavailability decreased significantly, thereby reducing its accumulation in the crop.	105
Cd, Cu, Pb and Zn	Rice hull	Sandy loam	Mining area, Keongsangbuk-do, South Korea	500	0%, 0.5%, 1%, 2%, 5% and 10% (v/v)	Significant decrease in NH_4NO_3-extractable heavy metal in soil.	106
Cd	Bamboo, coconut shell, pine wood shavings, sugarcane bagasse	Ultisol	Changsha City, Hunan, China	450	–	Bamboo-derived BC showed the highest effect on Cd(II) immobilization in soil.	107
Zn, Cd, Cu and Pb	Miscanthus	Sandy loam	Großbeeren, Berlin	860	0%, 1%, 2.5% and 5%	Reduces plant availability and concentrations of Zn and Cd. However, increases concentrations of Cu and Pb in the soil solution.	108

Cd and Pb	Wheat straw	Anthrosol	Yifeng, Jiangsu, China	350–550	0, 10, 20, and 40 t/ha	Cd: decreased by 40% at biochar rates of 20 and 40 t/ha. Pb: Significantly decreased	109
Pb and As	Soybean stover and pine needles	Sandy loam	Cheorwongun, Gangwondo, Korea	300 and 700	0% and 10%	Pb was immobilized while As was mobilized in biochar treated soils.	110
Cd and Pb	Wheat straw	Ferric-accumulic Stagnic Anthrosols	Yixing Jiangsu, China	450	0, 10, 20 and 40 t/ha	Decreased bioavailable Cd and Pb in a contaminated paddy field during 5 years.	111
Cu	Chicken manure	Alfisol	Puchuncaví Valley of Central Chile	500	0%, 5% and 10%	Through increase in organic matter and residual faction and diminishing exchangeable fraction; Cu mobility is reduced.	112
Cd	Rice straw	Sandy loam	Wuhan, China	500	0, 10 and 20 t/ha	Exchangeable Cd fraction: decreased from 14.5% to 1.4%.	113
Cd, Cu and Pb	Sugarcane bagasse	Sandy loam	Guangdong, China	450	0, 1.5, 2.25 and 3.0 t/ha	Cd decreased by 62–76%, Pb decreased by 17.3–49.1% and Cu decreased by 15–38% in shoot.	114
Cu	Coconut shell, Orange bagasse and Sewage sludge	Ultisol	Northeast Brazil	500	30 and 60 t/ha	None of the biochars was shown to be suitable as soil amendment to 38 reduce the uptake of Cu.	115
Cd, Cr, Cu, Ni, Pb and Zn	Phyllostachys pubescens	Silt	Niuwei River, southern Hebei Province, China	600	0%, 0.5%, 1%, 3%, 5%, 8%, 10% and 15%	Significant reduction in bioavailable fraction of metals (except for Cr).	116

Continued

Table 1 Case studies conducted to study the removal/immobilization efficiency of heavy metal via biochar.—cont'd

Metal of concern	Feedstock	Soil type	Location	Production temperature (°C)	Dose	Findings	References
Cu	Oat hull (OH) and chicken manure (CM)	Sedimentary alfisol	Puchuncaví Valley of Central Chile	300	1% and 5%	Reduces exchangeable Cu by 5 and 10 times (for OHB and CMB, respectively).	117
Cd, Pb, and Zn	Sewage sludge (SS) Eucalyptus Wood (EW)	–	Vazante, State of Minas Gerais (MG), Brazil	SS: 500 EW: 350	EW: 30 and 60 g/kg; SS: 10 and 20 g/kg	Increased leachate and soil pH; reduced the concentration of bioavailable Cd, Pb and Zn concentration.	118
Multiple heavy metals	Rabbit manure waste	Loam	The Cu-mining area of Riotinto, Spain	450 and 600	0% and 10%	Reduced the accumulation of heavy metals by *Brassica napus*. Decreased the amount of As, Cu, Co, Cr, Se and Pb in the soil.	119
Pb	Poultry litter and Biosolids	Sandy loam	Urban Victoria, Australia.	300, 400 and 500	0%, 1% and 3%	Biochars were able to outperform phosphate amendments for Pb immobilization.	20
Cu	Orange bagasse	Ultisol	Brazil	500	30 and 60 t/ha	Reduced copper availability.	120
Cd	Rice straw	Sandy Loam	Xiangyin, Hunan, China	450	0, 10 and 20 t/ha	Soil: Cd immobilization induced and Rice plant: Cd accumulation decrease.	121
Cu and Cd	Wood, bamboo, cornstalk and rice husk	–	Huangshi, Hubei Province	Modified with K_3PO_4 Temp: 550	0%, 3%, 5% and 10%	Transforms Cu(II) and Cd(II) from acid soluble to more stable forms. Extractability of Cu(II) and Cd(II) reduced by 2–3 times.	122

Temp: Temperature, RT: Residence time

the chances of harming the water table are paramount, as chelating agents or surfactants used in these technologies can be leached to the groundwater. In addition, the amendments used for restraining the contaminants may be specific for a particular metal which can lead to release of a toxic metal into the soil. Hence, further pilot scale studies are required to develop new approaches/methods that can effectively remove toxic heavy metals from contaminated soil. Biochar is a promising bio-residue material that can be used for remediation of heavy metal-contaminated soil. Type of biomass resource and the pyrolysis temperature are two significant components that affect the physico-chemical properties of biochar. Both these components also determine the remediation effect of heavy metal contaminated soil. Further, the properties of biochar can be modified through activation, magnetization, oxidation and digestion in order to significantly improve its adsorption of heavy metals, thereby reducing heavy metals mobility and bioavailability. Practically, on site soil conditions and associated environmental factors are ideal for selecting appropriate remedial technologies. Significant investigations are required to assess the long-term environmental and economic feasibility of biochar application to remediate heavy metal contaminated soils.

References

1. Houben D, Evrard L, Sonnet P. Beneficial effects of biochar application to contaminated soils on the bioavailability of Cd, Pb and Zn and the biomass production of rapeseed (*Brassica napus* L.). *Biomass Bioenergy* 2013;**57**:196–204.
2. Yang D, Zeng DH, Li LJ, Mao R. Chemical and microbial properties in contaminated soils around a magnesite mine in Northeast China. *Land Degrad Dev* 2012;**23**:256–62.
3. Bolan N, Kunhikrishnan A, Thangarajan R, Kumpiene J, Park J, Makino T, et al. Remediation of heavy metal(loid)s contaminated soils–to mobilize or to immobilize? *J Hazard Mater* 2014;**266**:141–66.
4. Roy M, McDonald LM. Metal uptake in plants and health risk assessments in metal-contaminated smelter soils. *Land Degrad Dev* 2015;**26**(8):785–92.
5. Vacca A, Bianco MR, Murolo M, Violante P. Heavy metals in contaminated soils of the Rio Sitzerri floodplain (Sardinia, Italy): characterization and impact on pedodiversity. *Land Degrad Dev* 2012;**23**:250–364.
6. Feng X, Li P, Qiu G, Wang S, Li G, Shang L, et al. Human exposure to methylmercury through rice intake in mercury mining areas, Guizhou Province, China. *Environ Sci Technol* 2008;**42**(1):326–32.
7. Palansooriya KN, Shaheen SM, Chen SS, Tsang DCW, Hashimoto Y, Hou D, et al. Soil amendments for immobilization of potentially toxic elements in contaminated soils: a critical review. *Environ Int* 2020;**134**:105046.
8. Gohain Baruah S, Ahmed I, Das B, Ingtipi B, Boruah H, Gupta SK, et al. Heavy metal(loid)s contamination and health risk assessment of soil-rice system in rural and peri-urban areas of lower brahmaputra valley, Northeast India. *Chemosphere* 2021;**266**:129150.

9. Dhaliwal SS, Singh J, Taneja PK, Mandal A. Remediation techniques for removal of heavy metals from the soil contaminated through different sources: a review. *Environ Sci Pollut Res* 2019;**27**:1–15.
10. Lahori AH, Guo ZY, Zhang ZQ, Li RH, Mahar A, Awasthi MK, et al. Use of biochar as an amendment for remediation of heavy metal-contaminated soils: prospects and challenges. *Pedosphere* 2017;**27**:991–1014.
11. Ghosh M, Singh SP. A review on phytoremediation of heavy metals and utilization of it's by products. *Asian J Energy Environ* 2005;**6**(4):18.
12. Beiyuan J, Tsang DC, Ok YS, Zhang W, Yang X, Baek K, et al. Integrating EDDS-enhanced washing with low-cost stabilization of metal-contaminated soil from an e-waste recycling site. *Chemosphere* 2016;**159**:426–32.
13. Liu L, Li W, Song W, Guo M. Remediation techniques for heavy metal-contaminated soils: principles and applicability. *Sci Total Environ* 2018;**633**:206–19.
14. O'Connor D, Peng T, Zhang J, Tsang DCW, Alessi DS, Shen Z, et al. Biochar application for the remediation of heavy metal polluted land: a review of in situ field trials. *Sci Total Environ* 2018;**619**:815–26.
15. Wang Y, Liu Y, Zhan W, Zheng K, Wang J, Zhang C, et al. Stabilization of heavy metal-contaminated soils by biochar: challenges and recommendations. *Sci Total Environ* 2020;**729**:39060.
16. Inyang MI, Gao B, Yao Y, Xue Y, Zimmerman A, Mosa A, et al. A review of biochar as a low-cost adsorbent for aqueous heavy metal removal. *Criti Rev Environ Sci Technol* 2016;**46**:406–33.
17. Ahmad M, Lee SS, Lim JE, Lee SE, Cho JS, Moon DH, et al. Speciation and phytoavailability of lead and antimony in a small arms range soil amended with mussel shell, cow bone and biochar: EXAFS spectroscopy and chemical extractions. *Chemosphere* 2014;**95**:433–41.
18. Netherway P, Reichman SM, Laidlaw MA, Scheckel KG, Pingitore NE, Gasco G. Phosphorus-rich biochars can transform lead in an urban contaminated soil. *J Environ Qual* 2019;**48**(4):1091–9.
19. Uchimiya M, Llma IM, Thomas Klasson K, Chang S, Wartelle LH, Rodgers JE. Immobilization of heavy metal ions (CuII, CdII, NiII, PbII) by broiler litter-derived biochars in water and soil. *J Agric Food Chem* 2010;**58**:5538–44.
20. Qian L, Zhang W, Yan J, Han L, Gao W, Liu R, et al. Effective removal of heavy metal by biochar colloids under different pyrolysis temperatures. *Bioresour Technol* 2016;**206**:217–24.
21. Wuana RA, Okieimen FE. Heavy metals in contaminated soils: a review of sources, chemistry, risks and best available strategies for remediation. *ISRN Ecol* 2011;**2011**.
22. Tiller KG. Heavy metals in soils and their environmental significance. In: *Advances in Soil Science*. New York, NY: Springer; 1989.
23. Kabata-Pendias A, Pendias H. *Trace Elements in Soils and Plants*. Boca Raton: CRC Press; 1992.
24. Tangahu BV, Sheikh Abdullah SR, Basri H, Idris M, Anuar N, Mukhlisin M. A review on heavy metals (As, Pb, and Hg) uptake by plants through phytoremediation. *Int J Chem Eng* 2011.
25. Harter RD. Effect of soil pH on adsorption of lead, copper, zinc, and nickel. *Soil Sci Soc Am J* 1983;**47**(1):47–51.
26. Sharma MR, Raju NS. Correlation of heavy metal contamination with soil properties of industrial areas of Mysore, Karnataka, India by cluster analysis. *Int J Environ Sci* 2013;**2**(10):22–7.
27. Baath E. Effects of heavy metals in soil on microbial processes and populations. *Water Air Soil Pollut* 1989;**47**:335–79.

28. Giller KE, Witter E, Mcgrath SP. Toxicity of heavy metals to microorganisms and microbial processes in agricultural soils. *Soil Biol Biochem* 1998;**30**:1389–414.
29. Tyler G. Heavy metals in soil biology and biochemistry. In: *E.A. Paul and J.N. Ladd, (Eds.), Soil Biochemistry.* New York, USA: Marcel Dekker; 1981.
30. McGrath S. Effects of heavy metals from sewage sludge on soil microbes agricultural ecosystems. In: Ross SM, editor. *Toxic Metals in Soil-Plant Systems.* Chichester: John Wiley and Sons; 1994.
31. Gucwa-Przepióra E, Nadgórska-Socha A, Fojcik B, Chmura D. Enzymatic activities and arbuscular mycorrhizal colonization of *Plantago lanceolata* and *Plantago major* in a soil root zone under heavy metal stress. *Environ Sci Pollut Res* 2016;**23**(5):4742–55.
32. Weissenhorn I, Leyval C. Root colonization of maize by a Cd-sensitive and a Cd-tolerant Glomus mosseae and cadmium uptake in sand culture. *Plant and Soil* 1995;**175**:233–8.
33. Chaney RL. Toxic element accumulation in soils and crops: protecting soil fertility and agricultural food-chains. In: Bar-Yosef B, Barrow NJ, Goldshmid J, editors. *Inorganic Contaminants in the Vadose Zone. Ecological Studies (Analysis and Synthesis)*, Berlin, Heidelberg: Springer; 1989.
34. Purves D. *Trace Element Contamination of the Environment.* Amsterdam, The Netherlands: Elsevier Science Publisher; 1985.
35. Chibuike GU, Obiora SC. Heavy metal polluted soils: effect on plants and bioremediation methods. *Appl Environ Soil Sci* 2014.
36. Chang AC, Granato TC, Page AL. A methodology for establishing phytotoxicity criteria for chromium, copper, nickel and zinc in agricultural land application of municipal sewage sludges. *J Environ Qual* 1992;**21**:521–36.
37. Kebrom TH, Woldesenbet S, Bayabil HK, Garcia M, Gao M, Ampim P, et al. Evaluation of phytotoxicity of three organic amendments to collard greens using the seed germination bioassay. *Environ Sci Pollut Res* 2019;**26**(6):5454–62.
38. Förstner U. Land contamination by metals—Global scope and magnitude of problem. In: Allen HE, Huang CP, Bailey GW, Bowers AR, editors. *Metal Speciation and Contamination of Soil.* USA: CRC Press; 1995.
39. Lindsay WL. *Chemical Equilibria in Soils.* New York: John Wiley and Sons; 1979.
40. Neina D. The role of soil pH in plant nutrition and soil remediation. *Appl Environ Soil Sci* 2019.
41. Kukier U, Peters CA, Chaney RL, Angle JS, Roseberg RJ. The effect of pH on metal accumulation in two Alyssum species. *J Environ Qual* 2004;**33**(6):2090–102.
42. Tica D, Udovic M, Lestan D. Immobilization of potentially toxic metals using different soil amendments. *Chemosphere* 2011;**85**:577–83.
43. Alloway BJ. *Heavy metals in soils.* New York: Wiley; 1990.
44. Khan S, Cao Q, Zheng YM, Huang YZ, Zhu YG. Health risks of heavy metals in contaminated soils and food crops irrigated with wastewater in Beijing, China. *Environ Pollut* 2008;**152**(3):686–92.
45. Liu WH, Zhao JZ, Ouyang ZY, Söderlund L, Liu GH. Impacts of sewage irrigation on heavy metal distribution and contamination in Beijing, China. *Environ Int* 2005;**31**(6):805–12.
46. Singh A, Sharma RK, Agrawal M, Marshall FM. Risk assessment of heavy metal toxicity through contaminated vegetables from wastewater irrigated area of Varanasi, India. *Trop Ecol* 2010;**51**:375–87.
47. Lehmann J, Joseph S. *Biochar for Environmental Management: Science and Technology.* London: Earthscan; 2009.
48. Laine J, Simoni S, Calles R. Preparation of activated carbon from coconut shell in a small scale concurrent flow rotary kiln. *Chem Eng Commun* 1991;**99**:15–23.

49. Wildman J, Derbyshire F. Origins and functions of macroporosity in activated carbons from coal and wood precursors. *Fuel* 1991;**70**:655–61.
50. Lua AC, Yang T, Guo J. Effects of pyrolysis conditions on the properties of activated carbons prepared from pistachio-nut shells. *J Anal Appl Pyrolysis* 2004;**72**:279–87.
51. Cetin E, Moghtaderi B, Gupta R, Wall TF. Influence of pyrolysis conditions on the structure and gasification reactivity of biomass chars. *Fuel* 2004;**83**:2139–50.
52. Troeh FR, Thompson LM. *Soils and Soil Fertility*. Iowa, USA: Blackwell Publishing; 2005.
53. Rutherford DW, Wershaw RL, Cox LG. *Changes in Composition and Porosity Occurring During the Thermal Degradation of Wood and Wood Components*. Denver, CO, USA: US Department of the Interior. US Geological Survey; 2004. p. 2004.
54. Cohen-Ofri I, Popovitz-Biro R, Weiner S. Structural characterization of modern and fossilized charcoal produced in natural fires as determined by using electron energy loss spectroscopy. *Chem A Eur J* 2007;**13**:2306–10.
55. Cohen-Ofri I, Weiner L, Boaretto E, Mintz G, Weiner S. Modern and fossil charcoal: aspects of structure and diagenesis. *J Archaeol Sci* 2006;**33**:428–39.
56. Kercher AK, Nagle DC. Microstructural evolution during charcoal carbonization by X-ray diffraction analysis. *Carbon* 2003;**41**:15–27.
57. Brennan JK, Bandosz TJ, Thomson KT, Gubbins KE. Water in porous carbons. *Colloids Surf A Physicochem Eng Asp* 2001;**187–188**:539–68.
58. Shafizadeh F. Introduction to pyrolysis of biomass. *J Anal Appl Pyrolysis* 1982;**3**(4):283–305.
59. Bourke J, Manley-Harris M, Fushimi C, Dowaki K, Nunoura T, Antal MJ. Do all carbonized charcoals have the same chemical structure? A model of the chemical structure of carbonized charcoal. *Ind Eng Chem Res* 2007;**46**:5954–67.
60. Koutcheiko S, Monreal CM, Kodama H, McCracken T, Kotlyar L. Preparation and characterization of activated carbon derived from the thermo-chemical conversion of chicken manure. *Bioresour Technol* 2007;**98**:2459–64.
61. Hammes K, Smernik RJ, Skjemstad JO, Herzog A, Vogt UF, Schmidt MWI. Synthesis and characterisation of laboratory- charred grass straw (*Oryza sativa*) and chestnut wood (*Castanea sativa*) as reference materials for black carbon quantification. *Org Geochem* 2006;**37**:1629–33.
62. Lehmann J, da Silva JP, Steiner C, Nehls T, Zech W, Glaser B. Nutrient availability and leaching in an archaeological Anthrosol and a Ferralsol of the Central Amazon basin: fertiliser, manure and charcoal amendments. *Plant and Soil* 2003;**249**:343–57.
63. Chan KY, Van Zwieten L, Meszaros I, Downie A, Joseph S. Agronomic values of green waste biochar as a soil amendment. *Aust J Soil Res* 2007;**45**:629–34.
64. Lehmann J. Bio-energy in the black. *Front Ecol Environ* 2007;**5**:381–7.
65. Inyang M, Gao B, Yao Y, Xue Y, Zimmerman AR, Pullammanappallil P, et al. Removal of heavy metals from aqueous solution by biochars derived from anaerobically digested biomass. *Bioresour Technol* 2012;**110**:50–6.
66. Li H, Dong X, da Silva EB, de Oliveira LM, Chen Y, Ma LQ. Mechanisms of metal sorption by biochars: biochar characteristics and modifications. *Chemosphere* 2017;**178**:466–78.
67. Ok YS, Uchimiya SM, Chang SX, Bolan N. *Biochar: Production, Characterization, and Applications*. CRC Press; 2015.
68. Uchimiya M, Klasson KT, Wartelle LH, Lima IM. Influence of soil properties on heavy metal sequestration by biochar amendment: 1. Copper sorption isotherms and the release of cations. *Chemosphere* 2011;**82**(10):431–1437.
69. Yang Y, Lin X, Wei B, Zhao Y, Wang J. Evaluation of adsorption potential of bamboo biochar for metal-complex dye: equilibrium, kinetics and artificial neural network modeling. *Int J Environ Sci Technol* 2014;**11**(4):1093–100.

70. Zengli A, Yanwei H, Chao CAI. Lead (II) adsorption characteristics on different biochars derived from rice straw. *Environ Chem* 2011;**30**(11):1851–7.
71. Chintala R, Subramanian S, Fortuna AM, Schumacher TE. Examining biochar impacts on soil abiotic and biotic processes and exploring the potential for pyrosequencing analysis. In: Ralebitso-Senior TK, Orr CH, editors. *Biochar Application Essential Soil Microbial Ecology*. Elsevier; 2016.
72. Cheng CH, Lehmann J, Thies JE, Burton SD, Engelhard MH. Oxidation of black carbon by biotic and abiotic processes. *Org Geochem* 2006;**37**(11):1477–88.
73. Cheng CH, Lehmann J, Engelhard MH. Natural oxidation of black carbon in soils: changes in molecular form and surface charge along a climosequence. *Geochim Cosmochim Acta* 2008;**72**(6):1598–610.
74. Liang B, Lehmann J, Solomon D, Kinyangi J, Grossman J, O'neill B, et al. Black carbon increases cation exchange capacity in soils. *Soil Sci Soc Am J* 2006;**70**(5):1719–30.
75. Ma JC, Dougherty DA. The Cationminus signpi interaction. *Chem Rev* 1997;**97** (5):1303–24.
76. Kumar S, Loganathan VA, Gupta RB, Barnett MO. An assessment of U (VI) removal from groundwater using biochar produced from hydrothermal carbonization. *J Environ Manage* 2011;**92**(10):2504–12.
77. Cao X, Harris W. Properties of dairy-manure-derived biochar pertinent to its potential use in remediation. *Bioresour Technol* 2010;**101**(14):5222–8.
78. Shi H, Zhou Q. Research progresses in the effect of biochar on soil-environmental behaviors of pollutants. *Chin J Plant Ecol* 2014;**33**(2):486–94.
79. Lu H, Zhang W, Yang Y, Huang X, Wang S, Qiu R. Relative distribution of Pb^{2+} sorption mechanisms by sludge-derived biochar. *Water Res* 2012;**46**(3):854–62.
80. Tan Z, Wang Y, Zhang L, Huang Q. Study of the mechanism of remediation of Cd-contaminated soil by novel biochars. *Environ Sci Pollut Res* 2017;**24**(32):24844–55.
81. Ma Y, Du X, Shi Y, Hou D, Dong B, Xu Z, et al. Engineering practice of mechanical soil aeration for the remediation of volatile organic compound-contaminated sites in China: advantages and challenges. *Front Environ Sci Eng* 2016;**10**(6):6.
82. Wang N, Xue XM, Juhasz AL, Chang ZZ, Li HB. Biochar increases arsenic release from an anaerobic paddy soil due to enhanced microbial reduction of iron and arsenic. *Environ Pollut* 2017;**220**:514–22.
83. Chou J, Wey M, Chang SH. Study on Pb and PAHs emission levels of heavy metals- and PAHs-contaminated soil during thermal treatment process. *J Environ Eng* 2010;**136**:112–8.
84. Zhang W, Tong L, Yuan Y, Liu Z, Huang H, Tan F, et al. Influence of soil washing with a chelator on subsequent chemical immobilization of heavy metals in a contaminated soil. *J Hazard Mater* 2010;**178**:578–87.
85. Ma L, Zhong H, Wu YG. Effects of metal soil contact time on the extraction of mercury from soils. *Bull Environ Contam Toxicol* 2015;**94**:399–406.
86. Hou D, Gu Q, Ma F, O'Connell S. Life cycle assessment comparisons of thermal desorption and stabilization/solidification of mercury contaminated soil on agricultural land. *J Clean Prod* 2016;**139**:949–56.
87. Karna RR, Luxton T, Bronstein KE, Hoponick Redmon J, Scheckel KG. State of the science review: potential for beneficial use of waste by-products for in situ remediation of metal-contaminated soil and sediment. *Crit Rev Environ Sci Technol* 2017;**47**:65–129.
88. Pan GX, Zhou P, Li LQ, Zhang HX. Core issues and research progressen of soil science of soil science of sequestration. *Acta Pedol Sin* 2007;**44**:327–37.
89. Zhou JB, Deng CJ, Chen JL, Zhang QS. Remediation effects of cotton stalk carbon on cadmium(Cd) contaminated soil. *Ecol Environ* 2008;**17**:1857–60.
90. Hartley W, Dickinson NM, Riby P, Lepp NW. Arsenic mobility in brown field soils amended with green waste compost or biochar and planted with Miscanthus. *Environ Pollut* 2009;**157**:2654–62.

91. Namgay T, Singh B, Singh BP. In fluence of biochar application to soil on the availability of As, Cd, Cu, Pb, and Zn to maize (*Zea mays* L.). *Soil Res* 2010;**48**:638–47.
92. Fellet G, Marchiol L, Vedove GD, Peressotti A. Application of biochar on mine tailings: effects and perspectives for land reclamation. *Chemosphere* 2011;**83**(9):1262–7.
93. Park JH, Choppala GK, Bolan NS, Chung JW, Chuasavathi T. Biochar reduces the bioavailability and phytotoxicity of heavy metals. *Plant and Soil* 2011;**348**:439–51.
94. Cui LQ, Li LQ, Zhang AF, Pan GX, Bao DD, Chang AC. Biochar amendment greatly reduces rice cd uptake in a contaminated paddy soil: a two-year field experiment. *Bioresources* 2011;**6**(3):2605–18.
95. Gan WJ, He Y, Zhang XF, Zhang ST, Liu YS. Effects and mechanisms of straw biochar on remediation contaminated soil in electroplating factory. *J Ecol Rural Environ* 2012;**28**(3):305–9.
96. Zheng RL, Cai C, Liang JH, Huang Q, Chen Z, Huang YZ, et al. The effects of biochars from rice residue on the formation of iron plaque and the accumulation of Cd, Zn, Pb, As in rice (*Oryza sativa* L.) seedlings. *Chemosphere* 2012;**89**:856–63.
97. Zhang Z, Solaiman ZM, Meney K, Murphy DV, Rengel Z. Biochars immobilize soil cadmium, but do not improve growth of emergent wetland species Juncus subsecundus in cadmium-contaminated soil. *J Soils Sediment* 2013;**3**(1):140–51.
98. Moon DH, Park JW, Chang YY, Ok YS, Lee SS, Ahmad M, et al. Immobilization of lead in contaminated firing range soil using biochar. *Environ Sci Pollut Res* 2013;**20**(12):8464–71.
99. Bian R, Chen D, Liu X, Cui L, Li L, Pan G, et al. Biochar soil amendment as a solution to prevent cd-tainted rice from China: results from a cross-site field experiment. *Ecol Eng* 2013;**58**:378–83.
100. Bian R, Joseph S, Cui L, Pan G, Li L, Liu X, et al. A three-year experiment confirms continuous immobilization of cadmium and lead in contaminated paddy field with biochar amendment. *J Hazard Mater* 2014;**272**:121–8.
101. Khan S, Reid BJ, Li G, Zhu Y. Application of biochar to soil reduces cancer risk via rice consumption: a case study in Miaoqian village, Longyan, China. *Environ Int* 2014;**68**:154–61.
102. Ehsan M, Barakat MA, Husein DZ, Ismail SM. Immobilization of Ni and Cd in soil by biochar derived from unfertilized dates. *Water Air Soil Pollut* 2014;**225**:2123.
103. Prapagdee S, Piyatiratitivorakul S, Petsom A, Tawinteung N. Application of biochar for enhancing cadmium and zinc phytostabilization in *Vignaradiata* L cultivation. *Water Air Soil Pollut* 2014;**225**:2233.
104. Jiang TY, Xu RK, Gu TX, Jiang J. Effect of crop-straw derived biochars on Pb(II) adsorption in two variable charge soils. *J Integr Agric* 2014;**13**:507–16.
105. Niu LQ, Jia P, Li SP, Kuang JL, He XX, Zhou WH, et al. Slash-and-char: an ancient agricultural technique holds new promise for management of soils contaminated by Cd, Pb and Zn. *Environ Pollut* 2015;**205**:333–9.
106. Kim HS, Kim KR, Kim HJ, Yoon JH, Yang JE, Ok YS, et al. Effect of biochar on heavy metal immobilization and uptake by lettuce (*Lactuca sativa* L.) in agricultural soil. *Environ Earth Sci* 2015;**74**(2):1249–59.
107. Tan XF, Liu YG, Gu YL, Zeng GM, Wang X, Hu XJ, et al. Immobilization of Cd(II) in acid soil amended with different biochars with a long term of incubation. *Environ Sci Pollut Res* 2015;**22**:12597–604.
108. Wagner A, Kaupenjohann M. Biochar addition enhanced growth of D actylis glomerata L. and immobilized Zn and Cd but mobilized Cu and Pb on a former sewage field soil. *Eur J Soil Sci* 2015;**66**:505–15.
109. Zhang A, Bian R, Li L, Wang X, Zhao Y, Hussain Q, et al. Enhanced rice production but greatly reduced carbon emission following biochar amendment in a metal-polluted rice paddy. *Environ Sci Pollut Res* 2015;**22**:18977–86.

110. Ahmad M, Ok YS, Kim BY, Ahn JH, Lee YH, Zhang M, et al. Impact of soybean stover- and pine needle-derived biochars on Pb and as mobility, microbial community, and carbon stability in a contaminated agricultural soil. *J Environ Manage* 2016;**166**:131–9.

111. Cui LQ, Pan GX, Li LQ, Bian RJ, Liu XY, Yan JL, et al. Continuous immobilization of cadmium and lead in biochar amended contaminated paddy soil: a five-year field experiment. *Ecol Eng* 2016;**93**:1–8.

112. Meier S, Curaqueo G, Khan N, Bolan N, Cea M, Eugenia GM, et al. Chicken-manure-derived biochar reduced bioavailability of copper in a contaminated soil. *J Soils Sediment* 2017;**17**:741–50.

113. Zhang RH, Li ZG, Liu XD, Wang B, Zhou GL, Huang XX, et al. Immobilization and bioavailability of heavy metals in greenhouse soils amended with rice straw-derived biochar. *Ecol Eng* 2017;**98**:183–8.

114. Nie CR, Yang X, Niazi NK, Xu XY, Wen YH, Rinklebe J, et al. Impact of sugarcane bagasse-derived biochar on heavy metal availability and microbial activity: a field study. *Chemosphere* 2018;**200**:274–81.

115. Gonzaga MIS, Mackowiak C, de Almeida AQ, Wisniewski Jr A, de Souza DF, da Silva Lima I, et al. Assessing biochar applications and repeated *Brassica juncea* L production cycles to remediate Cu contaminated soil. *Chemosphere* 2018;**201**:278–85.

116. Zhang C, Shan BQ, Zhu YY, Tang WZ. Remediation effectiveness of Phyllostachys pubescens biochar in reducing the bioavailability and bioaccumulation of metals in sediments. *Environ Pollut* 2018;**242**:1768–76.

117. Moore F, González ME, Khan N, Curaqueo G, Sanchez-Monedero M, Rilling J, et al. Copper immobilization by biochar and microbial community abundance in metal-contaminated soils. *Sci Total Environ* 2018;**616**:960–9.

118. Penido ES, Martins GC, Mendes TB, Melo LC, Guimaraes ID, Guilherme LR. Combining biochar and sewage sludge for immobilization of heavy metals in mining soils. *Ecotoxicol Environ Saf* 2019;**172**:326–33.

119. Gascó G, Álvarez ML, Paz-Ferreiro J, Méndez A. Combining phytoextraction by *Brassica napus* and biochar amendment for the remediation of a mining soil in Riotinto (Spain). *Chemosphere* 2019;**231**:562–70.

120. Gonzaga MIS, MIDAS M, Andrade KR, AND J, GDC C, RSD A, et al. Aged biochar changed copper availability and distribution among soil fractions and influenced corn seed germination in a copper-contaminated soil. *Chemosphere* 2020;**240**:124828.

121. Sui F, Wang J, Zuo J, Joseph S, Munroe P, Drosos M, et al. Effect of amendment of biochar supplemented with Si on cd mobility and rice uptake over three rice growing seasons in an acidic Cd-tainted paddy from central South China. *Sci Total Environ* 2020;**709**:136101.

122. Zhang H, Shao JG, Zhang SH, Zhang X, Chen HP. Effect of phosphorus modified biochars on immobilization of Cu (II), Cd (II), and As (V) in paddy soil. *J Hazard Mater* 2020;**390**:121349.

CHAPTER THREE

Application of biochar for emerging contaminant mitigation

Elsa Antunes[a,*,†], Arun K. Vuppaladadiyam[a,b,†], Ajit K. Sarmah[c], S.S.V. Varsha[d], Kamal Kishore Pant[b,*], Bhagyashree Tiwari[e], and Ashish Pandey[b]

[a]College of Science and Engineering, James Cook University, Townsville, QLD, Australia
[b]Catalytic Reaction Engineering Laboratory, Department of Chemical Engineering, IIT Delhi, New Delhi, India
[c]Department of Civil and Environmental Engineering, The Faculty of Engineering, The University of Auckland, Auckland, New Zealand
[d]Department of Civil Engineering, AVN institute of Engineering & Technology, Hyderabad, Telengana, India
[e]INRS-Eau, Terre et Environment, Quebec City, QC, Canada
*Corresponding authors: e-mail address: elsa.antunes1@jcu.edu.au; Kamal.Kishore.Pant@chemical.iitd.ac.in

Contents

1. Introduction	66
2. Emerging contaminants classes	68
3. Adsorption technology for the removal of ECs	74
3.1 Mechanism of adsorption	74
3.2 Artificial neural networks for adsorption studies	77
4. Adsorption of ECs	79
5. Technoeconomic feasibility of Biochar's use as adsorbent	83
6. Conclusions and future prospects	84
References	86

Abstract

The detection of trace concentrations of emerging contaminants in the environment has been a serious concern in the past decades. Emerging contaminants (ECs) are a group of pollutants that can have lethal impact on human and wildlife endocrine systems, even if available at trace quantities. Conventional treatments and natural attenuation are not able to remove these pollutants from the environment and therefore, they eventually end up in human food chain. An in-depth review has been done on the engineering perspective of adsorption process for the removal of ECs from aqueous solutions. The results of the review indicate that adsorption is a potential technology for

---

[†] The symbol indicates both the authors share the first authorship for contributing equal work.

Advances in Chemical Pollution, Environmental Management and Protection, Volume 7
ISSN 2468-9289
https://doi.org/10.1016/bs.apmp.2021.08.003

Copyright © 2021 Elsevier Inc.
All rights reserved.

65

the removal of a wide range of contaminants of emerging concern. This book chapter provides a holistic assessment on the sources and categories of ECs, adsorption as a potential technology for their removal, along with the mechanisms of adsorption. Furthermore, we made an attempt to critically discuss various adsorption mechanisms involved during the removal of the contaminants from water streams as well as provided a perspective of the use and importance of artificial neural networks in adsorption technology. Finally, an attempt has been made to critically discuss the potential of widely acknowledged adsorbents in removing ECs and to perform an assessment of the techno-economic feasibility of using biochar as an effective sorbent.

Keywords: Adsorbate, Adsorption, Artificial neural networks, Biochar, Emerging contaminants, Pharmaceuticals

1. Introduction

Emerging contaminants (ECs) are a diverse group of compounds that have become a global concern due to their potential to significantly threaten aquatic ecosystems and human health. Despite ubiquitous distribution in water sources and sediments globally, standard regulations do not exist for most ECs and they are not effectively removed by current conventional wastewater treatment processes. Typical ECs include pharmaceuticals and personal care products (PPCPs), per- and polyfluoroalkyl substances (PFAS), flame retardants, microplastics and surfactants. Emerging contaminants such as PPCPs and PFAS have low biodegradability and long half-lives, which allow them to circulate through drinking water and wastewater systems causing mass contamination. The simplified classification of ECs is as shown in Fig. 1. Aquatic and drinking water contamination have led to the bioaccumulation of certain ECs in humans and animals across the world, which has been linked to multiple diseases, cancers and long-term developmental and behavioral effects due to endocrine disruption. According to the United States Geological Survey, "any chemical of natural or synthetic origin or any microorganism that is not generally monitored but can pose potential risk to human health and/or ecology" can be considered as emerging contaminants.[1]

Microbial contaminants in the environment, more precisely, antibiotic resistant genes/bacteria produced as a result of mutation due to antibacterial drugs are also considered as ECs and specifically called as emerging microbial contaminants (EMCs).[2] In addition, horizontal gene transfer phenomena allow the transference of genetic material between microorganisms, which imply that antibiotic resistant genes can be further transferred in between

Application of biochar for emerging contaminant mitigation

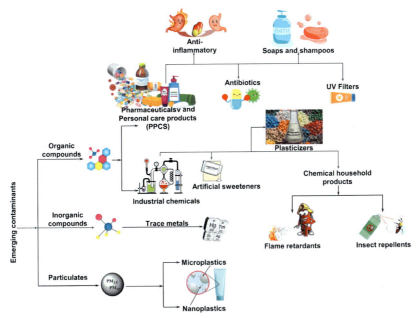

Fig. 1 Simplified classification of ECs considered for this chapter.

microbial population.[3] This demands mitigation strategies to prevent dissemination of antibiotic resistant genes. Few examples of EMCs are sapoviruses,[2] *Waddlia chondrophila*,[2] and *Streptococcus parauberis*.[4] The existence of ECs is not new and dated back to 2000 years when lead mines were over exploited by the Romans and Greeks.[5] Though the first documented report on ECs and associated environmental hazards was released in 1962, most of the research on the detection and occurrence in the environment was carried out in the last two decades. Gas chromatography with mass spectrometry is widely used in ECs analysis; however, the majority of ECs are insensitive to these techniques. The liquid chromatography with mass spectrometry (LC/MS) has been adopted due to the ubiquitous nature of these compounds.

Developing technologies for EC removal in wastewater treatment plants (WWTPs) is a complicated and challenging task due to the large chemical variation between families of compounds as well as the cost limitations associated with implementation. Currently, adsorption is one of the most promising technologies for the removal of organic and inorganic micropollutants. The advantages of adsorption include simplicity in operation, low cost, no by-products and the possibility of adsorbent regeneration and reuse.[6]

Adsorbents that are both economically and environmentally sustainable as well as highly effective for the removal of ECs must be identified to develop a standard infrastructure for the remediation of ECs in environmental systems. At present, there are a number of studies that report the efficiency of adsorbents in removing emerging contaminates from wastewater. However, the physicochemical properties vary with type of adsorbent; therefore, the efficiency of a particular adsorbent varies with respect to adsorbate. The research on few ECs is still in its initial stages and so, the adsorption mechanism is still unclear. Therefore, it is crucial to systematically review the recent progress in using adsorption technique for the removal of ECs. In this chapter, an attempt has been made to systematically summarize the recent progress on adsorption of ECs using different adsorbents. Important aspects, with respect to ECs as well as adsorbents are critically discussed. Special attention is given to the mechanism of adsorption and different modeling techniques to understand the adsorption process efficiency. Further, recent studies in which adsorption of ECs using various adsorbents have been presented and discussed.

2. Emerging contaminants classes

The recognition of so-called emerging contaminants in the biosphere in the recent years has turned out to be a grave environmental concern. ECs mainly originate from the day-to-day activities such as the use of personal care products, pesticides, pharmaceuticals, plasticizers, surfactants, and so, eliminating them from the system is not possible in the near future. Synthetic organic compounds have transformed the contemporary life and their use is undoubtedly a necessary component to maintain a modern society. In the recent years, ECs have been detected at trace levels in surface water and groundwater, effluents of sewage treatment plants and sporadically in drinking water. In addition, these compounds are also found in suspended solids[3] and river sediments.[7] Irrespective of quantity, either at trace or ultra-trace concentrations (below ng/L), ECs are a potential threat to the ecosystem. With new drugs and replacements in the market, ECs and their unknown transformation products detected in the environment are virtually limitless. Most of the researchers believe that legislative intervention by the government can be one possible way to control the contamination. Currently, there are no set regulations in many countries for the release of ECs, but few countries such as North America and Canada have made attempts to prioritize a list of ECs and their release from the wastewater treatment plants (WWTPs).[8]

The effluents from WWTPs are one of the primary sources of ECs. The effluents are normally released into surface waters; eventually, end up in soil, groundwater and seas. As the availability of ECs in wastewater is at very low concentration (normally micro- or nanogram per liter), the WWTPs are often ineffective in treating such compounds at low concentrations.[7] Moreover, the complexity of the matrices and their availability at low concentrations have hindered the development of standard and efficient methods for the determination of ECs, which subsequently led to the unavailability of data on the occurrence, pathways and ecotoxicity of ECs.[9] It is a common misconception that ECs pose no harm due of their availability at low concentrations being aware that their toxicity is chronic and often travel across generations. Furthermore, among all the ECs, antibiotics need special attention as they are expected to play a key role in the development of antibiotic resistant bacteria. At a very low concentration, the antibiotics cannot kill bacteria and at the same time allow bacteria to enter into mutagenic reactions forcing them to produce genes that protect them against antibiotics.[10] Also, these genes can be transferred across other strains of bacteria.[11] Similarly, other ECs such as pesticides and endocrine disruptors need to be studied as their toxicity and chronic effects are not clear yet. Therefore, the impacts of ECs in the environment have to be looked as a matter of urgent priority to understand and mitigate their risks. Table 1 shows that WWTPs are a common source for the release of most of ECs that can also pose huge risk to humans as well as plants and aquatic life. The impacts vary from simple, such as allergies, to severe and adverse effects, such as chronic diseases and cancer. These adverse effects can be attributed to the persistent and bioaccumulative nature of the EC compounds in the environment. ECs can enter the environment through various routes, such as domestic wastewater, effluents from hospitals, industries, agricultural activities and livestock farming.

Pharmaceuticals and Personal Care Products (PPCPs) are a diverse group of compounds including antibiotics, anti-inflammatory agents, steroidal hormones and the active ingredients in soaps, detergents and perfumes.[7] The use of PPCPs has been increasing overtime and due to their low biodegradability and high persistence, especially in WWTPs, they have become a global problem requiring careful consideration and the development of management solutions to tackle this problem is critical.[7] Kinetic studies performed on pharmaceuticals have shown that a considerable portion of an administered drug is not adsorbed within the body; hence, pharmaceuticals inevitably enter the environment through biological waste and WWTP effluent discharge.[12] Water samples collected from 114

Table 1 General information on prominent emerging contaminants.[3,7–24]

Contaminant class	Function	Important subgroup	Nature in the environment	Sources	Adverse effects	Examples
Pharmaceuticals and personal care products (PPCPs) — Pharmaceuticals	Substances that stop infection by killing or inhibiting growth of bacteria	Broad-spectrum antibiotics (A1), Hormones (A2), Nonsteroidal anti-inflammatory drugs (A3), β-blockers (A4), Blood lipid regulators (A5)	Slightly-very persistent	Pharmaceutical industry effluent, hospital effluent, domestic wastewater, effluent from livestock farms and aqua culture	Antibiotic resistance in microbial strains. Alter microbial community structure, and cause harm to other species. Increased risk of gastrointestinal ulcers, kidney diseases, gill alterations of rainbow trout	A1-Levofloxacin, Penicillin; A2–17-β-Estradiol (E2), Estriol (E3), Estrone (E1); A3-Diclofenac, Ibuprofen, Naproxen; A4- Metoprolol, Propranolol; A5-Clofibric acid, Gemfibrozil
Personal care products (PCPs)	Used as a fragrance ingredient in a wide range of consumer products including perfumes, cosmetics, shampoos	Preservatives (B1), Bactericides/disinfectants (B2), Insect repellents (B3), Fragrances (B4), Sunscreen UV filters (B5)	Slightly-very persistent	Surface water, WWTP effluent, and landfill leachate	Toxic to aquatic organisms. Cause oxidation stress to gold fish. Carcinogenic and may damage the human nervous system. Responsible for weak estrogenic activity	B1-Parabens; B2-Methyltriclosan, Triclocarban (TCC); B3-N,N-diethyl-m-toluamide (DEET); B4-Galaxolide fragrance (HHCB), B5-2-Ethyl-hexyl-4-trimethoxycinnamate (EHMC)
Industrial chemicals — Fire retardants	Used as flame retardant chemicals in paints, plastics, televisions, building materials, to prevent a fire	Organohalogen compounds (C1), Organophosphorus compounds (C2), Compounds containing both halogens and phosphorus (C3) and Organic compounds (C4)	Persistent, bioaccumulative	Industrial effluent and domestic wastewater	Affect brain and nervous system, hormone activity, reproduction and fertility	C1-Organochlorines, Organobromines, Decabromodiphenyl ethane, Polymeric brominated compounds; C2-Organophosphates, Phosphonates; (C3)-Tris (1-chloro-2-propyl) phosphate (TCPP); C4-Carboxylic acid, Dicarboxylic acid
Plasticizers		For polymers (C5) and for nonpolymers (C6)	Low to medium persistent, atmospheric deposition			C5-Dicarboxylic/tricarboxylic esters, Trimellitates, Adipates, Sebacates, Maleates; C6- Polycarboxylate ether

Category	Use	Type	Persistence	Source	Health/Environmental effects	Examples
Endocrine disrupting chemicals (EDCs)	Group of chemicals used as plastics, plasticizers, industrial solvents/ lubricants	Xenohormone (D1), Bisphenol (D2) and Phthalates (D3)	Moderately persistent	Surface water, drinking water, secondary sludge, soil, and sediments	Interfere with endocrine system Estrogenic effects in men Birth defects and developmental delays	D1–Xenoestrogen; D2–Bisphenol A (BPA), and D3–Dioctyl phthalate (DOP)
Biocides		Fungicide (E1), Herbicide (E2) and Molluscicide (E3)	Slightly-very persistent, bioaccumulative	Agricultural runoff, aquaculture effluent, and surface water	Carcinogenic effect	E1–Epoxiconazole; E2–Butachlor and E3–Metaldehyde
Regulated compounds (RCs)	Used in manufacture of pesticides, plastics, dyes	Poly aromatic hydrocarbons (F1) and Pesticides (F2)	Extremely persistent, bioaccumulative	Sewage treatment plants, agricultural runoffs, surface water, soils, and sediments	Carcinogenic effect, cardio vascular diseases Poor fetal development	F1–Naphthalene, Biphenyl; F2–Phenanthrene, Chlorpyrifos
Surfactants	Used in a variety of manufacturing processes and products	Ionic surfactants (G1) and nonionic surfactants (G2)	Medium persistent, bioaccumulative	Domestic wastewater and industrial effluents	Environmental toxicity	G1–Sodium lauryl sulfate (SLS); G2–Tweens (Polysorbates)
Artificial sweeteners	Used in food industry	–	Extremely persistent	Domestic wastewater, sewage treatment plant effluent, wastewater, and landfill leachate	Transformation and/or degradation can result in the formation of toxic compounds	Aspartame, Saccharin and, Sucralose
Perfluorinated alkylated substances (PFAs)	Used in emulsion polymerization, polishes, paints, and coatings	Perfluorosulfonic acids (H1) and Perfluorocarboxylic acids (H2)	Nondegradable and persistent compounds, bioacculumative and possess toxic characteristics	Surface water, ground water, wastewater, and sediments	Thyroid disease, liver damage, kidney cancer Reduced response to vaccines Developmental effects on unborn child	H1–Perfluorooctanesulfonate (PFOS); and H2–Perfluorooctanoic acid (PFOA)
Dyes	Used in textile industry as coloring agents	Disperse dyes (I1), Direct dyes (I2) and Reactive dyes (I3)	Ecotoxic, bioaccumulative and ubiquity in surface water	Domestic wastewater, sewage treatment plant effluent and industrial wastewater	Harmful to plants and aquatic life Can cause cancer, skin diseases and allergies in humans They are mutagenic and carcinogenic in nature	I1–Artisil; I2–Blue 2B; I3–Copper phythalocyanine dye Reactive Blue 21

Australian sites for analysis of 28 types of antibiotics revealed that one or more PPCPs were detected in over 90% of samples, which is of particular concern as discharged antibiotics can prompt the development of antibiotic resistant microorganisms.[13] Other PPCPs such as diclofenac, ibuprofen and triclosan are frequently detected in surface waters, and have been shown to cause toxicological problems.[7] Caffeine and carbamazepine have been proposed as anthropogenic markers in water streams since they are two of the most frequently detected PPCPs.[14] A study of 100 drinking water samples analysis collected from 61 locations in Brazil showed that caffeine was detected in 93% of samples at concentrations ranging from 1.8 ng/L to over 2000 ng/L.[15] Several PPCPs act as endocrine disrupting chemicals (EDCs) and might impact on the reproductive health and hormonal functions in fish and humans, highlighting the need for adequate remediation in WWTPs.[16]

Surfactants, or surface-active agents, are organic chemicals used to minimize surface tension in water or other liquids. Surfactants have a wide range of applications, including manufacturing of soaps and shampoos, dish cleaners and laundry detergents, lubricants, mining flocculates, textile industries, wastewater treatment and petroleum recovery.[25] Depending on the head groups of the surfactant, they are classified into four main categories: anionic, cationic, nonionic and amphoteric.[26] The most commonly used surfactants include alkyl sulfates, alkylethoxylates, alkylethoxy sulfates, alkylphenol ethoxylates, linear alkylbenzene sulfonates and quaternary ammonium based compounds.[26] Out of the 15 million tons of surfactants produced annually, anionic surfactants account for ca. 60%.[27] The nonionic surfactants contribute to ca. 30% and, both, cationic and amphoteric surfactants together account for the remaining 10%.[28]

Microplastics are solid synthetic materials or polymeric matrix with irregular size, generally in the range of 1–5 μm, and insoluble in water. In addition, nanoplastics can display colloidal behaviors with a size varying from 1 to 1000 nm.[29] However, no current definition exists for the term "nanoplastics" and the size mentioned above is based on the reports mentioned in the literature.[30] Lately, both micro- and nanoplastics have gained immense attention and numerous studies have been reported on the environmental behavior of these materials in the terrestrial and water ecosystems.[29] Nanoplastics may result from the fragmentation of aged-plastic materials, degradation of plastic materials, during the manufacturing process or during their application.[29] Microplastics are primarily categorized into two categories depending on the source of generation; primary microplastics, which are originally manufactured in the microsize, and secondary microplastics, which are generated from the decomposition of large sized plastics.[29]

Flame retardant (FR) materials are extremely important to reduce the number of victims during a fire accident. According to World Health Organization (WHO), ca. 180,000 deaths per year occur due to fire-related accidents.[31] FRs are increasingly required in many fields such as construction, transportation, electrical and electronic industries and are designed to extend the time-of-escape in fires. However, FRs are toxic and have lethal effects on humans and ecosystems. FRs are mostly lipophilic and do not easily dissolve in water.[32] Therefore, in aquatic ecosystems, sediments act as major sink for FRs that are transported via atmosphere and enter the aquatic ecosystem. This process is considered as major pathway for FRs in many aquatic species and sediments.[32] As the most hydrophobic compounds bind to food and the less hydrophobic compounds remain dissolved in water, plants and animals are generally exposed to FRs via uptake of contaminated water and food. To date, there are ca. 175 compounds or group of compounds with the properties of a flame retardant.[32] FRs are all grouped into four main categories: (i) halogenated FRs, (ii) phosphorus containing FRs, (iii) nitrogen containing FRs and (iv) inorganic FRs.[33] The most commonly used FRs are brominated flame retardants (BFRs) and organophosphate flame retardants (OPFRs). The sources of FRs into the environment include industries, wastewater treatment plants (WWTPs), incinerators, plastic and e-waste recycling facilities.[32]

Polyfluoroalkyls (PFAS) are a group of synthetic compounds which have been extensively used for over six decades in a variety of applications such as oil-repelling containers, water-proof fabrics, nonstick cookware, paints, aqueous film-forming foams used in firefighting, industrial emulsifiers and surfactants.[34] Polyfluoroalkyls are composed of a hydrophobic carbon chain of varying length where at least one hydrogen atom has been replaced by fluorine, whereas perfluoroalkyls require all hydrogen atoms to have been replaced by fluorine.[35] Both per- and polyfluoroalkyl substances also include a hydrophilic functional group attached at one end, typically carboxylic or sulfonic acid. Two PFAS in particular, perfluorooctanoic acid (PFOA) and perfluorooctanesulfonic acid (PFOS), have been studied extensively due to their bioaccumulative, toxic and transport potential in the environment.[34]

Endocrine disruptors (EDs) are a segment of compounds that can have severe negative impacts on human and/or animal health either by blocking or imitating natural hormones that take care of specific functions in human body.[36] Endocrine disrupting chemicals (EDCs) are highly heterogeneous and are classified into two categories: (i) natural chemicals and (ii) synthetic chemicals. The natural chemicals are found in food of humans and animals, for example, phytoestrogen: genistein and coumestrol. The synthetic chemicals are those that are synthesized and can be further grouped into industrial solvents, lubricants, plastics, pesticides and some pharmaceuticals.[37] The typical human exposure to EDCs occurs via food chain and contaminated dust from households.[38] Other exposure paths include exposure to a myriad of chemical compounds available with household products. Bisphenol A (BPA) has been found in materials such as polycarbonate plastics, epoxy resins, ink used for thermal paper receipts, flame retardants, medical drugs. Thus, human exposure to EDCs is inevitable.

3. Adsorption technology for the removal of ECs

Adsorption is the adhesion of a substance (the adsorbate) from a gas or liquid phase to the surface of another substance (the adsorbent), which is typically a solid material. This process is widely used for the extraction of compounds, especially for pollutant removal in industrial water treatment.[16] Adsorbents that are suitable for industrial use are typically highly porous with a large surface area and must be easily regenerated to be economically sustainable. Environmental factors that can affect the adsorption capacity of an adsorbent include pH, temperature, pressure and the presence of other molecules in the sample matrix competing for adsorption sites.[39] The magnitude of adsorption is also affected by adsorbate and adsorbent initial concentration as well as contact time.[39] Adsorption can be operated as a continuous process, which is performed within systems such as fixed-bed columns and is commonly utilized in industry due to the high throughput potential. Alternatively, adsorption can be operated as a batch process, which is often used on a laboratory scale for examining adsorption capacity and mechanisms.[39]

3.1 Mechanism of adsorption

The study of adsorption mechanism can provide insights into the process performance of adsorbents to adsorbates. Adsorption is a phase transfer process that involves the accumulation of material at the boundary of two phases, such as gas–liquid, liquid–liquid, liquid–solid or gas–solid interface. Adsorption involves intermolecular transfer of micropollutants onto the surface of solid sorbents and the process mostly depends upon the chemical nature and composition of adsorbent and adsorbate.[16] If the interaction between adsorbed molecules and solid surface has a physical nature, the process can be referred as *physisorption*. On the other side, if the force of attraction between adsorbent and adsorbate are due to chemical bonding, then the process can be termed as *chemisorption*. In the former case, the process results are reversible as the forces of attraction are weak Van der Waals forces. In the latter case, due to the stronger forces at the stake the chemisorbed substances can hardly be removed.[40]

Chemisorption involves forces between the adsorbent and adsorbate that are similar in strength to chemical bonds, such as covalent bonds or very strong electrostatic interactions (ionic bonds). This process is not easily reversible and is limited to a monomolecular layer of adsorbate due to the short reach of valence forces. Chemisorption commonly occurs between metallic ions and adsorbents with several functional groups.[16] Also, chemisorption is influenced by the charge of the functional group via alteration of electron contributing behavior, stability of the colloids, the nature of the electrostatic interactions and wettability.[41] Nevertheless, chemisorption potential depends on other system parameters, for instance, the pH of the solution largely influences the adsorption mechanism that involves hydrogen bonding via dipole–dipole interaction.[42] In physisorption, there are no chemical bonds between the adsorbent and adsorbate, only very weak attractive forces such as Van der Waals forces and hydrogen bonding. Hydrophobic interactions, which involve the partitioning of hydrophobic compounds out of aqueous solution due to polarity differences, also fall under this category.[43] Another unique interaction is π-π stacking, in which the π bonds within aromatic rings attract each other and the resulting complex is stabilized by Van der Waals forces. Physical adsorption or physisorption can be reversed by heat or pressure changes and is directly proportional to the available adsorbent surface area, but is not restricted to a monomolecular layer of adsorbate. A schematic presentation of different adsorption mechanisms available in the literature is shown in Fig. 2.

Primarily driven by entropy, hydrophobic interactions are nonspecific interactions that occur due to the inclination of nonpolar compounds to aggregate in water to avoid their contact with water molecules.[44] Also, the adsorption of nonpolar chemicals on porous materials is proportional to the K_{ow} values implying hydrophobic interactions are the dominant mechanism.[45,46] However, when biomass derived biochar is used for adsorption of phenanthrene this correlation was not observed, implying a

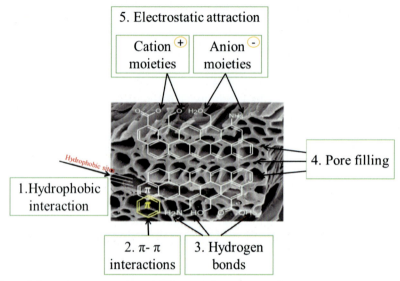

Fig. 2 Schematic presentation of different adsorption mechanisms.

mechanism other than hydrophobic interaction may have occurred.[47] The π–π interaction is weaker than hydrogen bonds and is one type of dipole interaction that allows the interaction of electron-rich π-system with neutral organic molecules, metal ions or with another π-system. Villaescusa et al. investigated the adsorption of paracetamol onto biochar produced from vegetable waste.[48] The authors explained the occurrence of π–π interaction between aromatic rings and syringyl groups of paracetamol and lignin, respectively. In a solid–liquid sorption system, external or intraparticle diffusion process may decide the sorption regulating step.[16]

To explain the adsorption of organics onto metal-organic framework, researchers have proposed acid base interactions, electrostatic interactions, H-bonding, and p-complexation. Recently, to explain the adsorption of PPCPs, H-bonding has been considered as an important mechanism. For instance, Seo et al. (2016) and Song et al. (2016) considered the H-bonding to explain the mechanism of PPCS, such as naproxen, ibuprofen, oxybenzone[49] and triclosan.[50] Hydrogen bonds with energies in the range of 5–18 kJ mol^{-1} are formed between electronegative atoms (like F, N, O) and the electropositive H nucleus of functional groups, such as –NH and –OH.[51] The probability of the formation of H-bond may be reduced because of the availability of several functional groups on the adsorbent, for example, in materials such as biochar.[45] During the sorption of organic compounds using carbonaceous materials, H-bonds are expected to play an important role. When compared to ordinary H-bond, seen among water and different functional groups such as alcohols, esters and ethers, the charge assisted H-bonds are often much stronger, as explained in a study conducted by Pignatello et al.[52]

Electrostatic interactions (EIs) include the attraction and repulsion forces between charged adsorbent and adsorbate.[6] In case of pharmaceutical pollutants, ionizable functional groups may interact with chemical functional groups on the adsorbents via electrostatic attractions or repulsions.[53] Anion exchange is the affinity of an anion to a site that is positively charged on the surface of the sorbent and involves the exchange of anion at the binding site. This mechanism is expected to play an important role in the adsorption of negatively charged adsorbate. On the other side, the cation exchange can occur between positively charged adsorbate and negatively charged adsorbent.[45]

3.2 Artificial neural networks for adsorption studies

Artificial neural networks (ANN), inspired by biological nervous processing, can be used to solve and model numerous complex environmental systems owing to its advantages such as simplicity, reliability nonlinearity and robustness. Recently, ANN models have been successfully employed in modeling adsorption process.[54] The ANN was first introduced by McCulloch and Pitts,[55] and resembles a simple concept of human brain. Multilayer perception is most widely recognized type of ANNs and consists of one input and output layers along with one or more hidden layers. The ANN models work by creating a nonlinear relationship between dependent and independent variables depending on a set of experimental data. A simple ANN model for adsorption process can be visualized in Fig. 3.

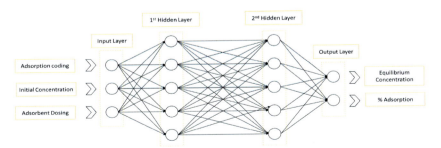

Fig. 3 Simple schematic of ANN model with two hidden layers for modeling adsorption process.

A detailed procedure, including selection of parameters and training process of ANN model, with reference to adsorption of contaminants from aqueous solutions can be found elsewhere.[56] For instance, ANNs are used for the prediction of adsorption of emerging contaminants such as aromatic hydrocarbons and methyl tert-butyl ether (MTBE).[57] With the help of multilayer perception structure, the information regarding performance error can also be obtained by adopting back-propagating technique. However, ANNs can be complex as they generally require a huge number of networks. An ANN with 3-10-1 configuration was used to model the removal efficiency of black 5 dye on TiO_2 surface by Dutta and coworkers.[58] Similarly, Mahmoodi and his team[59] evaluated the removal efficiency of CuO–NiO nanocomposite in adsorbing basic red 18 and basic red 41 dyes from wastewater.

The investment in AI in the field of wastewater treatment is expected to reach \$6.3 million by 2030.[60] The ANN model is the most used in water treatment to predict the adsorption of emerging contaminants, which can be divided in four types: Recurrent Neural Network (RNN), Fuzzy Neural Network (FNN), Convoluted Neural Network (CNN) and Deep Neural Network (DNN). According to literature, ANN models can predict well the adsorption capacity of emergent contaminants by the adsorbent, but prediction accuracy can be further improved. Fig. 3 is a simple representation of the ANN model, but the number of input variables can be much higher including, initial concentration of the emerging contaminants, initial concentrations of other contaminants, pH, operating temperature, contaminant and adsorbent ratio, adsorbent dosage, adsorbent properties, contact time, flowrate and bed depth. The removal efficiency and the equilibrium concentration are the main output variables, but they are related.

ANN models may contribute to a better understanding of the adsorption process of contaminants of emerging concern. ANN models can be used to predict the adsorption capacity of a specific system (adsorbent-sorbate) and replacing adsorption kinetic studies since they have an excellent prediction performance.[56] The ANN models can be even more important for real scenarios, especially in situations where some of the emerging contaminants breakdown into metabolites during the wastewater treatment processes. In these cases that are characterized by a nonlinear behavior, the ANN model can be used to optimize the operating conditions in order to maximize the adsorption of these contaminants and their metabolites. Further, ANN models can be used to better understand the main adsorption processes as well as the breakdown mechanisms.

4. Adsorption of ECs

Various adsorbents, ranging from simple biochar and activated carbon to new generation nanoadsorbents, have been proposed to remove a wide range of pollutants from the water ecosystem. A list of studies using various adsorbents for the removal of ECs from aqueous solutions is given in Table 2. Biochar that is a carbon rich material has gained huge popularity in the removal of contaminants from wastewater. Pyrolysis of biomass is the general route to produce the biochar from a variety of feedstock.[71,72] Depending on the operating conditions and type of feedstock, the biochar may contain many noncarbonized fractions with functionalities that could impact on the adsorption capacity.[6] For instance, slow pyrolysis favors a high yield of biochar, increase in pyrolysis temperature and residence time could increase the surface area and pore volume of biochar.[42,71–73] Biochar obtained with temperatures below 600 °C can have properties from their parent feedstock.[71] In addition, few compounds such as dioxins and furans are bioavailable and are prone to leach into the solution. In such conditions, preconditioning of biochar by washing with methanol is recommended to remove leachable components and increase the adsorption capacity.[74] However, from an economic point of view, using biochar with a cost 350–1200 USD per ton when compared to 1100–1700 USD per ton for activated carbon,[75] can significantly minimize the wastewater treatment costs. Biochar has been effectively used in the adsorption of various organic and inorganic pollutants. For instance, biochar obtained from pine needle and peanut shell were used for adsorption of volatile organic compounds (VOCs) such as nitrobenzene[76] and trichloroethylene,[77] respectively. In addition, biochar has been successfully used in the sorption of dyes using bamboo,[78] pharmaceuticals such as doxycycline hydrochloride and tetracycline using peanut shell[79] and swine manure[80] derived biochar, respectively; pesticides such as dimethoate and thiacloprid using biochar derived from sugarcane bagasse[81] and maize straw,[82] respectively; solvents such as benzene and nitrobenzene using MSW[83] and spent black tea waste[84] derived biochar, respectively; heavy metals and personal care products using swine manure[85] and camphor leaves[86] derived biochar, respectively.

Activated carbon (AC) has been widely used for adsorption because of its high specific area and high porosity. The activation of the carbon can be done either physically or chemically, to produce activated carbon.[6] Chemical activation involves pyrolysis of organic precursors impregnated

Table 2 Summary of recent adsorption studies for various ECs using different adsorbents.

Sorbent	Type of ECs	Compound	Residence time (h)	Initial concentration of the adsorbate (mgL^{-1})	Adsorbent dosage (mgL^{-1})	Sorption capacity (mg/g)	Isotherms	Kinetic model	Remarks	References
Corn straw biochar	Biphenyls	Propranolol	48	0.8	2	7.6±9.2	Freundlich model	PSO	Multiple mechanisms were noticed	61
Crop residues biochar	Pharmaceuticals	Levofloxacin	24	40–75	4	1.49–7.72	Langmuir model	PSO	Multiple mechanisms were noticed	62
Pine chips biochar	Pharmaceuticals	Diclofenac	168	20	0.1–0.8	372	Langmuir models	PSO	π–π mechanism was dominant	63
Tea wastes biochar	Pharmaceuticals	Sulfamethazine	72	0–50	1000	33.81	–	PSO	π–π electron donor–acceptor interaction, cation–π interaction and cation exchange at low pH were the primary mechanisms	64
Modified rice husk biochar	Antibiotics	Tetracycline	1	50	30	98.33	Freundlich model	PSO	Adsorption occurs mainly by hydrogen bonding and pore-filling effect	65
Bamboo biochar- Modified using Magnetic CuZnFe$_2$O$_4$	Endocrine disrupting chemicals	Bisphenol A	24	250	50	263.2	Freundlich model	PSO	The immobilization was due to hydrophobic and π–π type interactions	66

Granular activated carbon (GAC)	Pharmaceuticals	Diclofenac	NA	40	NA	233.9	Freundlich model	PSO	No interactions between adsorbed compounds on diffusion	67
MWCNTs	Pharmaceuticals	Ciprofloxacin			2000–3000	77.8	Langmuir isotherm	PSO	π–π interactions, hydrophobic interactions were dominant	68
Silica aerogel	Pharmaceuticals	Nimesulide	0.5–48		250	79	Freundlich model	PSO	π–π interactions, hydrophobic interactions were dominant	69
Natural clays	Fungicides	Metalaxyl and tricyclazole	4	5	1000	83.7–98.2	Freundlich model	PSO	The adsorption of these fungicides was onto the external surfaces of clay minerals	70

with various inorganics, such as KOH, $ZnCl_2$, $FeCl_3$, H_3PO_4, and the inorganics are later removed to produce activated carbon with high surface area.[6] Recently, Moussavi et al.[87] used NH_4Cl^- activated carbon for the adsorption of diazinon, an organophosphate insecticide. Similarly, H_3PO_4 activated carbon prepared from peach stones was used to study the sorption of carbamazepine with a maximum sorption capacity of 335 mg/L.[14] In a study from Torrellas research group, the authors compared H_3PO_4 activated carbon prepared from low-cost carbon sources, such as rice husk, with commercially available ACs such as carbon calgon F400 and multiwalled carbon nanotubes (MWCNTs). They found that the chemically activated carbon had a higher sorption capacity than the commercially available ACs.[88] Literature pointed out the significant amount of success in the application of activated carbon for removal of hydrophobic and charged pharmaceuticals in water.[46] Activated carbons produced from agro-industrial wastes, such as coffee wastes, have been used for the removal of nonsteroidal anti-inflammatories, such as, diclofenac and naproxen, and hormones, such as, 17 β-estradiol (E2).[46] For instance, PPCPs such as naproxen and diclofenac were removed using plant derived carbon and micelle–clay complex[89] and cocoa shell carbon chemically activated with $ZnCl_2 + FeCl_3 + lime$,[90] respectively. Pesticides, such as atrazine were removed using oak charcoal chemically activated with mixture of $ZnCl_2 + FeCl_3 + KOH + lime$.[91] Industrial chemicals, such as 2-nitrophenol and, synthetic industrial effluents containing very high concentrations of 10 phenols, humic acid and 5 inorganic salts were removed using Tucuma shell carbon chemically activated with $ZnCl_2$.[92] However, using AC for adsorption is a costly process as AC are hardly regenerated after adsorption.[6] These reasons hinder the large-scale application of AC. Though, thermal regeneration of AC has been explored for the removal of ECs, but the regenerated AC lost adsorption capacity and was not effective in removing ECs.[93]

Carbon nanotubes (CNTs), given their relatively high surface area, large pore size and small area, are considered promising contestants among the family of adsorbents.[94] CNTs are extensively used in the removal of organic pollutants from wastewater.[16] However, considering the difficulties associated with ECs with regard to their separation from aqueous phase, small particle size and reduced dispersibility, CNTs have been chemically modified to suit the adsorption process.[95] The type of CNTs, either single-walled or multiwalled, can significantly impact on the affinity to ECs. Oxidized multiwalled CNTs, with varying oxygen levels, were used in the adsorption of tetracycline from aqueous solution. The maximum adsorption capacity was noticed to be 269.25 mg/g for CNTs with 3.2% of oxygen.[96] In another study, the removal of antibiotics form wastewater using CNTs was evaluated by Yu and his team.[94] Peng and coworkers studied the removal of antibiotics such as norfloxacin and ofloxacin onto multiwalled CNTs and modified single-walled CNTs with added functional groups such as hydroxyl (-OH) and carbonyl (C=O).[95]

Nanocomposite membranes with single- and multiwalled CNTs were used in the removal of personal care products such as triclosan and pharma products such as paracetamol.[97] Usually single-walled CNTs are more effective than the multiwalled CNTs because of their high surface areas. CNTs have been successfully employed in the adsorption of metals and metalloids such as mercury[98] and lead,[99] pharmaceuticals and personal care products,[100] pesticides such as diazinon,[101] dichlorodiphenyltrichloroethanes (DDTs),[102] endocrine disruptors such as (Cd^{2+})[103] and bisphenol A.[104] However, the cost of the process could be a severe obstacle for the progress of CNTs to be considered for urban wastewater treatments. In addition, unwanted sludge particles entering CNTs after the treatment process should be avoided as they may pose risk to the environmental organisms. Graphene, mesoporous carbons, phenolic resins, clays and minerals based sorbents have been studied extensively in the adsorption of ECs.[46]

5. Technoeconomic feasibility of Biochar's use as adsorbent

Biochar production and use as an adsorbent for removal of contaminants has been successfully demonstrated in literature. However, just a few studies evaluated the technoeconomic feasibility of using biochar for contaminant adsorption, which is fundamental to scaling-up this technology. Biomass waste such as sewage sludge, agriculture residues is usually considered as the precursor for biochar production. The selection of biomass could significantly impact on the biochar yield as well as the economic viability, but in general the production cost of biochar is much lower than the activated carbon. Further, the type of biomass precursor and pyrolysis conditions have also a significant impact on reducing pollution, in particular reducing greenhouse gas (GHG) emissions. Biochar could reduce up to 870 kg CO_2 equivalent per dry ton of biomass precursor.[105] Biochar production is simultaneously a good approach for GHG emissions mitigation as well as environmental alternative to landfilling. The biochar production from biomass waste is also economically attractive due to existing high landfilling costs and could be even more attractive with the implementation of emissions reduction subsidies.

The technoeconomic feasibility of biochar adsorption application depends on several factors: biomass feedstock and preprocessing, pyrolysis conditions, energy cost, volume of water to treat and pollutant concentration, degree of water purification required, type of contaminant and recyclability of biochar. The biomass production cost is relatively low and can be zero if using biomass waste as a precursor for biochar production. However, biomass preprocessing costs depend on the moisture content, type of biomass as well as the requirement for the pyrolysis reactor and conditions. The energy balance depends on the energy of the biomass that depends on the biomass composition, energy required for pyrolysis including the preprocessing stage and the energy of the biofuel obtained from pyrolysis. Biomass with low ash content and high oxygen/carbon ratio is economically preferred.

The market cost for biochar is about USD$750 per ton that is roughly 50% of the price of activated carbon (USD$1400 per ton).[75] However, the adsorption capacity of biochar is similar to activated carbon, which makes biochar a competitive adsorbent. For example, the estimated adsorption costs for pharmaceuticals adsorption on a fixed-bed column was about USD$1.6 per cubic meter of wastewater.[106] The biochar can be recycled and reused between three to five times while keeping a stable and efficient adsorption capacity.[107] The cost of biochar regeneration is also an important variable to take in consideration in the overall economic analysis of biochar adsorption. The regeneration of biochar can be carried out by a thermal, chemical or a combination of both processes, therefore not just the cost of regeneration should be taken in consideration, but its environmental impact should be carefully analyzed. More pilot studies are required to accurately assess the economics of adsorption.

6. Conclusions and future prospects

The continuing circulation and pervasive reach of harmful emerging contaminants accentuates the need for cost-effective and sustainable treatment technologies. Adsorption has been demonstrated to be an efficient, reusable and low-cost approach to the effective remediation of a diverse array of ECs in water systems and the environment. Activated carbons and multiwalled carbon nanotubes have exhibited good adsorption capacities for a broad range of contaminants such as PPCPs and PFAS, however their medium costs and poor regenerative capability restrict their use for industrial processes. As an alternative, several renewable biochar adsorbents have the capacity to outperform commercial activated carbons, displaying great promise as economically and environmentally sustainable options for ECs treatment. Biochar pyrolysis conditions and feedstock type require more investigation in order to establish a consistent production of high capacity biochars with standardized properties. Quaternized cotton and bamboo-derived biochar are some of the most promising renewable adsorbents for removal of PFOS and PFOA, whereas pine chip-derived biochar has a high capacity for the adsorption of PPCPs such as diclofenac and ibuprofen. However, it is unknown if these adsorbents are suitable for the adsorption of other, less common compounds from the PFAS and PPCPs families. The most prominent adsorption mechanisms of PFAS are hydrophobic and electrostatic interactions, hence they are well-suited to adsorbents that facilitate anion-exchange. PPCPs are commonly adsorbed through electrostatic, hydrophobic and π–π interactions as well as hydrogen bonding, so they are effectively removed by carbonaceous adsorbents with diverse surface functional groups. Performance of adsorbents is difficult to compare between emerging contaminant studies, due to a lack of consistent experimental conditions and poor reporting of influential factors such as temperature, pH and initial concentration. Future studies should also focus on adsorption performance under realistic industrial conditions, such as using continuous fixed-bed columns and sample matrices that reflect common ECs reservoirs. Future studies can be directed to more research into adsorption of ECs using low-cost adsorbents. In addition, more focus should be shed on developing adsorbents with minimal environmental footprint. In the recent times, researchers have focused on exploring hybrid systems for dealing with ECs. These hybrid systems include adsorption in combination with oxidation or electrocoagulation or biological systems such as membrane bioreactors and activated sludge process. More research is likely to be carried out in this area of hybrid system of adsorption. Limited number of studies are available on artificial neural networks approach combined with optimization algorithms. The optimization of network configuration with an evolutionary computation method such as invasive weed optimization, teaching-learning-based optimization, harmony search, ant colony algorithm, tabu search, simulated annealing, artificial bee colony, firefly algorithm and shuffled frog-leaping algorithm is highly desirable.

References

1. USGS. *Contaminants of emerging concern in the environment. Environmental health—toxic substances hydrology program. U.S. Geological Survey*; 2017. Available at https://toxics.usgs.gov/investigations/cec/index.php.
2. Hartmann J, van Driezum I, Ohana D, Lynch G, Berendsen B, Wuijts S, et al. The effective design of sampling campaigns for emerging chemical and microbial contaminants in drinking water and its resources based on literature mining. *Sci Total Environ* 2020;**742**:140546.
3. Rout PR, Zhang TC, Bhunia P, Surampalli RY. Treatment technologies for emerging contaminants in wastewater treatment plants: a review. *Sci Total Environ* 2021;**753**:141990.
4. Hartmann J, Wuijts S, van der Hoek JP, de Roda Husman AM. Use of literature mining for early identification of emerging contaminants in freshwater resources. *Environ Evid* 2019;**8**(1):33.
5. Sauvé S, Desrosiers M. A review of what is an emerging contaminant. *Chem Cent J* 2014;**8**(1):15.
6. Islam MA, Jacob MV, Antunes E. A critical review on silver nanoparticles: from synthesis and applications to its mitigation through low-cost adsorption by biochar. *J Environ Manag* 2021;**281**:111918.
7. Chaturvedi P, Shukla P, Giri BS, Chowdhary P, Chandra R, Gupta P, et al. Prevalence and hazardous impact of pharmaceutical and personal care products and antibiotics in environment: a review on emerging contaminants. *Environ Res* 2021;**194**:110664.
8. Morales-Caselles C, Gao W, Ross P, Fanning L. *Emerging contaminants of concern in Canadian harbours: a case study of Halifax harbour*. Halifax, NS, Canada: Marine Affairs Program Technical Report Dalhousie University; 2016.
9. Dosis I, Ricci M, Majoros L, Lava R, Emteborg H, Held A, et al. Addressing analytical challenges of the environmental monitoring for the water framework directive: ERM-CE100, a new biota certified reference material. *Anal Chem* 2017;**89**(4):2514–21.
10. Nazaret S, Aminov R. Role and prevalence of antibiosis and the related resistance genes in the environment. *Front Microbiol* 2014;**5**:520.
11. Rizzo L, Manaia C, Merlin C, Schwartz T, Dagot C, Ploy MC, et al. Urban wastewater treatment plants as hotspots for antibiotic resistant bacteria and genes spread into the environment: a review. *Sci Total Environ* 2013;**447**:345–60.
12. Appa R, Mhaisalkar VA, Naoghare PK, Lataye DH. Adsorption of an emerging contaminant (primidone) onto activated carbon: kinetic, equilibrium, thermodynamic, and optimization studies. *Environ Monit Assess* 2019;**191**(4):1–16.
13. Watkinson A, Murby E, Kolpin DW, Costanzo S. The occurrence of antibiotics in an urban watershed: from wastewater to drinking water. *Sci Total Environ* 2009;**407**(8):2711–23.
14. Torrellas SÁ, Lovera RG, Escalona N, Sepúlveda C, Sotelo JL, García J. Chemical-activated carbons from peach stones for the adsorption of emerging contaminants in aqueous solutions. *Chem Eng J* 2015;**279**:788–98.
15. Machado KC, Grassi MT, Vidal C, Pescara IC, Jardim WF, Fernandes AN, et al. A preliminary nationwide survey of the presence of emerging contaminants in drinking and source waters in Brazil. *Sci Total Environ* 2016;**572**:138–46.
16. Sophia AC, Lima EC. Removal of emerging contaminants from the environment by adsorption. *Ecotoxicol Environ Saf* 2018;**150**:1–17.
17. Yang Y, Ok YS, Kim K-H, Kwon EE, Tsang YF. Occurrences and removal of pharmaceuticals and personal care products (PPCPs) in drinking water and water/sewage treatment plants: a review. *Sci Total Environ* 2017;**596–597**:303–20.

18. Ouda M, Kadadou D, Swaidan B, Al-Othman A, Al-Asheh S, Banat F, et al. Emerging contaminants in the water bodies of the Middle East and North Africa (MENA): a critical review. *Sci Total Environ* 2021;**754**:142177.
19. Stefanakis AI, Becker JA. A review of emerging contaminants in water: classification, sources, and potential risks. In: *Impact of water pollution on human health and environmental sustainability.* IGI Global; 2016. p. 55–80.
20. Wei G-L, Li D-Q, Zhuo M-N, Liao Y-S, Xie Z-Y, Guo T-L, et al. Organophosphorus flame retardants and plasticizers: sources, occurrence, toxicity and human exposure. *Environ Pollut* 2015;**196**:29–46.
21. Kokotou MG, Asimakopoulos AG, Thomaidis NS. Artificial sweeteners as emerging pollutants in the environment: analytical methodologies and environmental impact. *Anal Methods* 2012;**4**(10):3057–70.
22. Lewis DM. 9—The chemistry of reactive dyes and their application processes. In: Clark M, editor. *Handbook of textile and industrial dyeing.* vol. 1. Woodhead Publishing; 2011. p. 303–64.
23. Gulrajani ML. 10—Disperse dyes. In: Clark M, editor. *Handbook of textile and industrial dyeing.* vol. 1. Woodhead Publishing; 2011. p. 365–94.
24. Sekar N. 12—Direct dyes. In: Clark M, editor. *Handbook of textile and industrial dyeing.* vol. 1. Woodhead Publishing; 2011. p. 425–45.
25. Bautista-Toledo MI, Rivera-Utrilla J, Méndez-Díaz JD, Sánchez-Polo M, Carrasco-Marín F. Removal of the surfactant sodium dodecylbenzenesulfonate from water by processes based on adsorption/bioadsorption and biodegradation. *J Colloid Interface Sci* 2014;**418**:113–9.
26. Siyal AA, Shamsuddin MR, Low A, Rabat NE. A review on recent developments in the adsorption of surfactants from wastewater. *J Environ Manag* 2020;**254**:109797.
27. Ramprasad C, Philip L. Surfactants and personal care products removal in pilot scale horizontal and vertical flow constructed wetlands while treating greywater. *Chem Eng J* 2016;**284**:458–68.
28. Aloui F, Kchaou S, Sayadi S. Physicochemical treatments of anionic surfactants wastewater: effect on aerobic biodegradability. *J Hazard Mater* 2009;**164**(1):353–9.
29. Wang C, Zhao J, Xing B. Environmental source, fate, and toxicity of microplastics. *J Hazard Mater* 2020;**407**:124357.
30. Gigault J, Halle AT, Baudrimont M, Pascal P-Y, Gauffre F, Phi T-L, et al. Current opinion: what is a nanoplastic? *Environ Pollut* 2018;**235**:1030–4.
31. WHO. *Global burn registry*; 2020 https://www.who.int/violence_injury_prevention/burns/en/. Accessed on 31st December, 2020.
32. Iqbal M, Syed JH, Katsoyiannis A, Malik RN, Farooqi A, Butt A, et al. Legacy and emerging flame retardants (FRs) in the freshwater ecosystem: a review. *Environ Res* 2017;**152**:26–42.
33. Delva L, Hubo S, Cardon L, Ragaert K. On the role of flame retardants in mechanical recycling of solid plastic waste. *Waste Manag* 2018;**82**:198–206.
34. Emery I, Kempisty D, Fain B, Mbonimpa E. Evaluation of treatment options for well water contaminated with perfluorinated alkyl substances using life cycle assessment. *Int J Life Cycle Assess* 2019;**24**(1):117–28.
35. Gagliano E, Sgroi M, Falciglia PP, Vagliasindi FG, Roccaro P. Removal of poly-and perfluoroalkyl substances (PFAS) from water by adsorption: role of PFAS chain length, effect of organic matter and challenges in adsorbent regeneration. *Water Res* 2020;**171**:115381.
36. Vieira WT, de Farias MB, Spaolonzi MP, da Silva MGC, Vieira MGA. Removal of endocrine disruptors in waters by adsorption, membrane filtration and biodegradation. A review. *Environ Chem Lett* 2020;**18**(4):1113–43.

37. Kabir ER, Rahman MS, Rahman I. A review on endocrine disruptors and their possible impacts on human health. *Environ Toxicol Pharmacol* 2015;**40**(1):241–58.
38. Darbre PD. The history of endocrine-disrupting chemicals. *Curr Opinion Endocr Metab Res* 2019;**7**:26–33.
39. de Andrade JLR, Oliveira MF, da Silva MG, Vieira MG. Adsorption of pharmaceuticals from water and wastewater using nonconventional low-cost materials: a review. *Ind Eng Chem Res* 2018;**57**(9):3103–27.
40. De Gisi S, Lofrano G, Grassi M, Notarnicola M. Characteristics and adsorption capacities of low-cost sorbents for wastewater treatment: a review. *Sustain Mater Technol* 2016;**9**:10–40.
41. Cai N, Larese-Casanova P. Sorption of carbamazepine by commercial graphene oxides: a comparative study with granular activated carbon and multiwalled carbon nanotubes. *J Colloid Interface Sci* 2014;**426**:152–61.
42. Antunes E, Jacob MV, Brodie G, Schneider PA. Isotherms, kinetics and mechanism analysis of phosphorus recovery from aqueous solution by calcium-rich biochar produced from biosolids via microwave pyrolysis. *J Environ Chem Eng* 2018;**6**(1):395–403.
43. Zhao L, Deng J, Sun P, Liu J, Ji Y, Nakada N, et al. Nanomaterials for treating emerging contaminants in water by adsorption and photocatalysis: systematic review and bibliometric analysis. *Sci Total Environ* 2018;**627**:1253–63.
44. Ouyang J, Zhou L, Liu Z, Heng JYY, Chen W. Biomass-derived activated carbons for the removal of pharmaceutical mircopollutants from wastewater: a review. *Sep Purif Technol* 2020;**253**:117536.
45. Kah M, Sigmund G, Xiao F, Hofmann T. Sorption of ionizable and ionic organic compounds to biochar, activated carbon and other carbonaceous materials. *Water Res* 2017;**124**:673–92.
46. Rathi BS, Kumar PS, Show P-L. A review on effective removal of emerging contaminants from aquatic systems: current trends and scope for further research. *J Hazard Mater* 2021;**409**:124413.
47. Pan B, Xing B. Adsorption mechanisms of organic chemicals on carbon nanotubes. *Environ Sci Technol* 2008;**42**(24):9005–13.
48. Villaescusa I, Fiol N, Poch J, Bianchi A, Bazzicalupi C. Mechanism of paracetamol removal by vegetable wastes: the contribution of π–π interactions, hydrogen bonding and hydrophobic effect. *Desalination* 2011;**270**(1):135–42.
49. Seo PW, Bhadra BN, Ahmed I, Khan NA, Jhung SH. Adsorptive removal of pharmaceuticals and personal care products from water with functionalized metal-organic frameworks: remarkable adsorbents with hydrogen-bonding abilities. *Sci Rep* 2016; **6**(1):34462.
50. Song JY, Ahmed I, Seo PW, Jhung SH. UiO-66-type metal–organic framework with free carboxylic acid: versatile adsorbents via H-bond for both aqueous and nonaqueous phases. *ACS Appl Mater Interfaces* 2016;**8**(40):27394–402.
51. Suh SB, Kim JC, Choi YC, Yun S, Kim KS. Nature of one-dimensional short hydrogen bonding: bond distances, bond energies, and solvent effects. *J Am Chem Soc* 2004;**126**(7):2186–93.
52. Li X, Pignatello JJ, Wang Y, Xing B. New insight into adsorption mechanism of ionizable compounds on carbon nanotubes. *Environ Sci Technol* 2013;**47**(15):8334–41.
53. Huerta-Fontela M, Galceran MT, Ventura F. Occurrence and removal of pharmaceuticals and hormones through drinking water treatment. *Water Res* 2011;**45**(3):1432–42.
54. Pauletto PS, Dotto GL, Salau NPG. Optimal artificial neural network design for simultaneous modeling of multicomponent adsorption. *J Mol Liq* 2020;**320**:114418.
55. McCulloch WS, Pitts W. A logical calculus of the ideas immanent in nervous activity. *Bull Math Biophys* 1943;**5**(4):115–33.

56. Ghaedi AM, Vafaei A. Applications of artificial neural networks for adsorption removal of dyes from aqueous solution: a review. *Adv Colloid Interf Sci* 2017;**245**:20–39.

57. Giwa A, Yusuf A, Balogun HA, Sambudi NS, Bilad MR, Adeyemi I, et al. Recent advances in advanced oxidation processes for removal of contaminants from water: a comprehensive review. *Process Saf Environ Prot* 2021;**146**:220–56.

58. Dutta S, Parsons SA, Bhattacharjee C, Bandhyopadhyay S, Datta S. Development of an artificial neural network model for adsorption and photocatalysis of reactive dye on TiO2 surface. *Expert Syst Appl* 2010;**37**(12):8634–8.

59. Mahmoodi NM, Hosseinabadi-Farahani Z, Bagherpour F, Khoshrou MR, Chamani H, Forouzeshfar F. Synthesis of CuO–NiO nanocomposite and dye adsorption modeling using artificial neural network. *Desalin Water Treat* 2016;**57**(37):17220–9.

60. Alam G, Ihsanullah I, Naushad M, Sillanpää M. Applications of artificial intelligence in water treatment for the optimization and automation of the adsorption process: recent advances and prospects. *Chem Eng J* 2021;**427**:130011.

61. Wang F, Ren X, Sun H, Ma L, Zhu H, Xu J. Sorption of polychlorinated biphenyls onto biochars derived from corn straw and the effect of propranolol. *Bioresour Technol* 2016;**219**:458–65.

62. Yi S, Gao B, Sun Y, Wu J, Shi X, Wu B, et al. Removal of levofloxacin from aqueous solution using rice-husk and wood-chip biochars. *Chemosphere* 2016;**150**:694–701.

63. Jung C, Boateng LK, Flora JRV, Oh J, Braswell MC, Son A, et al. Competitive adsorption of selected non-steroidal anti-inflammatory drugs on activated biochars: experimental and molecular modeling study. *Chem Eng J* 2015;**264**:1–9.

64. Rajapaksha AU, Vithanage M, Zhang M, Ahmad M, Mohan D, Chang SX, et al. Pyrolysis condition affected sulfamethazine sorption by tea waste biochars. *Bioresour Technol* 2014;**166**:303–8.

65. Dai J, Meng X, Zhang Y, Huang Y. Effects of modification and magnetization of rice straw derived biochar on adsorption of tetracycline from water. *Bioresour Technol* 2020;**311**:123455.

66. Heo J, Yoon Y, Lee G, Kim Y, Han J, Park CM. Enhanced adsorption of bisphenol A and sulfamethoxazole by a novel magnetic CuZnFe2O4–biochar composite. *Bioresour Technol* 2019;**281**:179–87.

67. Sotelo JL, Ovejero G, Rodríguez A, Álvarez S, Galán J, García J. Competitive adsorption studies of caffeine and diclofenac aqueous solutions by activated carbon. *Chem Eng J* 2014;**240**:443–53.

68. Álvarez-Torrellas S, Peres JA, Gil-Álvarez V, Ovejero G, García J. Effective adsorption of non-biodegradable pharmaceuticals from hospital wastewater with different carbon materials. *Chem Eng J* 2017;**320**:319–29.

69. Caputo G, Scognamiglio M, De Marco I. Nimesulide adsorbed on silica aerogel using supercritical carbon dioxide. *Chem Eng Res Des* 2012;**90**(8):1082–9.

70. Azarkan S, Peña A, Draoui K, Sainz-Díaz CI. Adsorption of two fungicides on natural clays of Morocco. *Appl Clay Sci* 2016;**123**:37–46.

71. Antunes E, Schumann J, Brodie G, Jacob MV, Schneider PA. Biochar produced from biosolids using a single-mode microwave: characterisation and its potential for phosphorus removal. *J Environ Manag* 2017;**196**:119–26.

72. Antunes E, Jacob MV, Brodie G, Schneider PA. Microwave pyrolysis of sewage biosolids: dielectric properties, microwave susceptor role and its impact on biochar properties. *J Anal Appl Pyrolysis* 2018;**129**:93–100.

73. Antunes E, Jacob MV, Brodie G, Schneider PA. Silver removal from aqueous solution by biochar produced from biosolids via microwave pyrolysis. *J Environ Manag* 2017;**203**:264–72.

74. Jing X-R, Wang Y-Y, Liu W-J, Wang Y-K, Jiang H. Enhanced adsorption performance of tetracycline in aqueous solutions by methanol-modified biochar. *Chem Eng J* 2014;**248**:168–74.
75. Feizi F, Reguyal F, Antoniou N, Zabaniotou A, Sarmah AK. Environmental remediation in circular economy: end of life tyre magnetic pyrochars for adsorptive removal of pharmaceuticals from aqueous solution. *Sci Total Environ* 2020;**739**:139855.
76. Ahmad M, Lee SS, Dou X, Mohan D, Sung J-K, Yang JE, et al. Effects of pyrolysis temperature on soybean stover- and peanut shell-derived biochar properties and TCE adsorption in water. *Bioresour Technol* 2012;**118**:536–44.
77. Chen B, Chen Z, Lv S. A novel magnetic biochar efficiently sorbs organic pollutants and phosphate. *Bioresour Technol* 2011;**102**(2):716–23.
78. Yang Y, Lin X, Wei B, Zhao Y, Wang J. Evaluation of adsorption potential of bamboo biochar for metal-complex dye: equilibrium, kinetics and artificial neural network modeling. *Int J Environ Sci Technol* 2014;**11**(4):1093–100.
79. Li H, Dong X, da Silva EB, de Oliveira LM, Chen Y, Ma LQ. Mechanisms of metal sorption by biochars: biochar characteristics and modifications. *Chemosphere* 2017;**178**:466–78.
80. Chen T, Luo L, Deng S, Shi G, Zhang S, Zhang Y, et al. Sorption of tetracycline on H3PO4 modified biochar derived from rice straw and swine manure. *Bioresour Technol* 2018;**267**:431–7.
81. Sun Y, Qi S, Zheng F, Huang L, Pan J, Jiang Y, et al. Organics removal, nitrogen removal and N2O emission in subsurface wastewater infiltration systems amended with/without biochar and sludge. *Bioresour Technol* 2018;**249**:57–61.
82. Zhang P, Sun H, Min L, Ren C. Biochars change the sorption and degradation of thiacloprid in soil: insights into chemical and biological mechanisms. *Environ Pollut* 2018;**236**:158–67.
83. Jayawardhana Y, Mayakaduwa S, Kumarathilaka P, Gamage S, Vithanage M. Municipal solid waste-derived biochar for the removal of benzene from landfill leachate. *Environ Geochem Health* 2019;**41**(4):1739–53.
84. Tariq M, Durrani AI, Farooq U, Tariq M. Efficacy of spent black tea for the removal of nitrobenzene from aqueous media. *J Environ Manag* 2018;**223**:771–8.
85. Jiang B, Lin Y, Mbog JC. Biochar derived from swine manure digestate and applied on the removals of heavy metals and antibiotics. *Bioresour Technol* 2018;**270**:603–11.
86. Wang C, Wang H, Cao Y. Pb(II) sorption by biochar derived from Cinnamomum camphora and its improvement with ultrasound-assisted alkali activation. *Colloids Surf A Physicochem Eng Asp* 2018;**556**:177–84.
87. Moussavi G, Hosseini H, Alahabadi A. The investigation of diazinon pesticide removal from contaminated water by adsorption onto NH4Cl-induced activated carbon. *Chem Eng J* 2013;**214**:172–9.
88. Álvarez-Torrellas S, Rodríguez A, Ovejero G, García J. Comparative adsorption performance of ibuprofen and tetracycline from aqueous solution by carbonaceous materials. *Chem Eng J* 2016;**283**:936–47.
89. Qurie M, Khamis M, Malek F, Nir S, Bufo SA, Abbadi J, et al. Stability and removal of naproxen and its metabolite by advanced membrane wastewater treatment plant and micelle–clay complex. *Clean–Soil Air Water* 2014;**42**(5):594–600.
90. Saucier C, Adebayo MA, Lima EC, Cataluña R, Thue PS, Prola LDT, et al. Microwave-assisted activated carbon from cocoa shell as adsorbent for removal of sodium diclofenac and nimesulide from aqueous effluents. *J Hazard Mater* 2015;**289**:18–27.
91. Shirmardi M, Alavi N, Lima EC, Takdastan A, Mahvi AH, Babaei AA. Removal of atrazine as an organic micro-pollutant from aqueous solutions: a comparative study. *Process Saf Environ Prot* 2016;**103**:23–35.

92. Umpierres CS, Thue PS, Lima EC, Reis GSD, de Brum IAS, Alencar WSD, et al. Microwave-activated carbons from tucumã (Astrocaryum aculeatum) seed for efficient removal of 2-nitrophenol from aqueous solutions. *Environ Technol* 2018; **39**(9):1173–87.
93. Marques SCR, Marcuzzo JM, Baldan MR, Mestre AS, Carvalho AP. Pharmaceuticals removal by activated carbons: role of morphology on cyclic thermal regeneration. *Chem Eng J* 2017;**321**:233–44.
94. Yu F, Sun S, Han S, Zheng J, Ma J. Adsorption removal of ciprofloxacin by multi-walled carbon nanotubes with different oxygen contents from aqueous solutions. *Chem Eng J* 2016;**285**:588–95.
95. Peng H, Pan B, Wu M, Liu Y, Zhang D, Xing B. Adsorption of ofloxacin and norfloxacin on carbon nanotubes: hydrophobicity- and structure-controlled process. *J Hazard Mater* 2012;**233–234**:89–96.
96. Yu F, Ma J, Han S. Adsorption of tetracycline from aqueous solutions onto multi-walled carbon nanotubes with different oxygen contents. *Sci Rep* 2014; **4**(1):5326.
97. Wang Y, Zhu J, Huang H, Cho H-H. Carbon nanotube composite membranes for microfiltration of pharmaceuticals and personal care products: capabilities and potential mechanisms. *J Membr Sci* 2015;**479**:165–74.
98. Bandaru NM, Reta N, Dalal H, Ellis AV, Shapter J, Voelcker NH. Enhanced adsorption of mercury ions on thiol derivatized single wall carbon nanotubes. *J Hazard Mater* 2013;**261**:534–41.
99. Nyairo WN, Eker YR, Kowenje C, Akin I, Bingol H, Tor A, et al. Efficient adsorption of lead (II) and copper (II) from aqueous phase using oxidized multiwalled carbon nanotubes/polypyrrole composite. *Sep Sci Technol* 2018;**53**(10):1498–510.
100. Mousavi SA, Janjani H. Antibiotics adsorption from aqueous solutions using carbon nanotubes: a systematic review. *Toxin Rev* 2020;**39**(2):87–98.
101. Dehghani MH, Kamalian S, Shayeghi M, Yousefi M, Heidarinejad Z, Agarwal S, et al. High-performance removal of diazinon pesticide from water using multi-walled carbon nanotubes. *Microchem J* 2019;**145**:486–91.
102. Hua S, Gong J-L, Zeng G-M, Yao F-B, Guo M, Ou X-M. Remediation of organochlorine pesticides contaminated lake sediment using activated carbon and carbon nanotubes. *Chemosphere* 2017;**177**:65–76.
103. Zhang C, Wang W, Duan A, Zeng G, Huang D, Lai C, et al. Adsorption behavior of engineered carbons and carbon nanomaterials for metal endocrine disruptors: experiments and theoretical calculation. *Chemosphere* 2019;**222**:184–94.
104. Yang D, Gao P, Ren X, Niu Y, Wu Z, Gu Z, et al. The role of solvents and oxygen-containing functional groups on the adsorption of bisphenol A on carbon nanotubes. *Environ Technol* 2020;**41**:1–9.
105. Gupta S, Kua HW, Dai Pang S. Biochar-mortar composite: manufacturing, evaluation of physical properties and economic viability. *Constr Build Mater* 2018;**167**:874–89.
106. Feizi F, Sarmah AK, Rangsivek R. Adsorption of pharmaceuticals in a fixed-bed column using Tyre-based activated carbon: experimental investigations and numerical modelling. *J Hazard Mater* 2021;**417**:126010.
107. Cheng N, Wang B, Wu P, Lee X, Xing Y, Chen M, et al. Adsorption of emerging contaminants from water and wastewater by modified biochar: a review. *Environ Pollut* 2021;**273**:116448.

CHAPTER FOUR

Role of biochar as a cover material in landfill waste disposal system: Perspective on unsaturated hydraulic properties

Sanandam Bordoloi[a,*], Janarul Shaikh[b,c], Ján Horák[b], Ankit Garg[d], S. Sreedeep[a], and Ajit K. Sarmah[e]

[a]Department of Civil Engineering, Indian Institute of Technology Guwahati, Assam, India
[b]Department of Biometeorology and Hydrology, Faculty of Horticulture and Landscape Engineering, Slovak University of Agriculture in Nitra, Nitra, Slovakia
[c]Department of Civil Engineering, C.V. Raman Global University, Bhubaneswar, Odisha, India
[d]Department of Civil and Environmental Engineering, Shantou University, Shantou, China
[e]Department of Civil and Environmental Engineering, The Faculty of Engineering, The University of Auckland, Auckland, New Zealand
[*]Corresponding author: e-mail address: sanandam@alumni.iitg.ac.in

Contents

1. Introduction	94
2. Materials and methods	96
2.1 Soil properties	96
2.2 Biochar characterization	96
2.3 Instrumentation and experimental methodology	98
3. Results and discussion	101
3.1 Soil water retention characteristics	101
3.2 Effect of biochar on gas permeability in soil	102
3.3 Effects of biochar on infiltration rate on soil	103
4. Conclusions	104
References	105
Further reading	106

Abstract

Biochar amended soil (BAS) has emerged as a sustainable and alternate material as a final cover system for landfills. This final cover is often subjected to change in matric suction due to seasonal change in evapotranspiration rates. The water retention characteristics, gas permeability and water infiltration vary for unsaturated soil depending on the suction in the soil. Main objective of this study was to investigate these unsaturated soil properties of BAS which are of major concern in landfill cover functioning. The biochar was produced from water hyacinth inside an in-house pyrolysis unit and mixed with a silty sand at 5% application rate. Water retention capacity increased with

---

Advances in Chemical Pollution, Environmental Management and Protection, Volume 7
ISSN 2468-9289
https://doi.org/10.1016/bs.apmp.2021.08.004

Copyright © 2021 Elsevier Inc.
All rights reserved.

93

biochar amendment to silty sand. The air entry value increased by almost 100%, while the desorption rate marked by *n* parameter had no significant change, which was attributed to entrapped clay particles within the biochar and consequent alteration of soil porosity. Beyond suction range of 1 MPa, gas permeability of BAS decreased by almost 50% as compared to bare soil. This may be due to suppression of desiccation cracks in biochar amended soil. Infiltration rate was found to be relatively higher in case of biochar amended soil in comparison to bare soil. However, infiltration rate at different suction range were in same order, indicating marginal change in soil infiltration rate for actual field application.

Keywords: Biochar amended soil, Partial saturation, Landfill cover

1. Introduction

Biochar is a solid carbonaceous porous material produced by heating organic biomass at temperature range of 335–700 °C in the absence of or under limited amount of oxygen[1] through a process known as pyrolysis. Use of biochar in agriculture dates to *Terra Preta* times, wherein ancient Amazonians utilized biochar for increasing yield of crops.[2] Through biochar production and amendment in soil, carbon is locked within soil mass and can remain stable for hundreds and thousands of years.[1] Biochar is a carbon-sink material, which has garnered popularity for being used to ameliorate soil degradation.[3,4] Biochar increases mesoporosity and water availability of the soil mass.[5] Biochar has also been found to favor forest restoration by increasing soil nutrients (e.g., P, K and Ca) in the shorter term, and nutrient retention in the long run.[6] Biochar can promote the abundance and diversity of bacterial and fungal establishment, which are essential to ecological restoration.[7,8] These properties indicate that biochar is a potential candidate to improve soil quality of degraded lands including newly restored landfills.

Upon completion of a landfill, a cover layer is constructed (Fig. 1A) atop the waste layer to protect it from water ingress due to rainfall events,[9] while the surrounding areas are also protected from gas emitted from the waste below.[10] Biochar amended soil (BAS) has recently gained momentum in the field of geo-environmental applications as a sustainable cover material for degraded landfill sites.[11] BAS has potential for enhancing methane oxidation by increasing methane retention time.[12] Biochar possesses inherent physical and chemical properties, which have been found to enhance micro-porosity, adsorption capacity and to alter the hydraulic characteristics of amended soil.[13] Few studies[14,15] have been conducted so far to focus on mechanical properties of BAS using direct shear test.

Role of biochar as a cover material in landfill waste disposal system

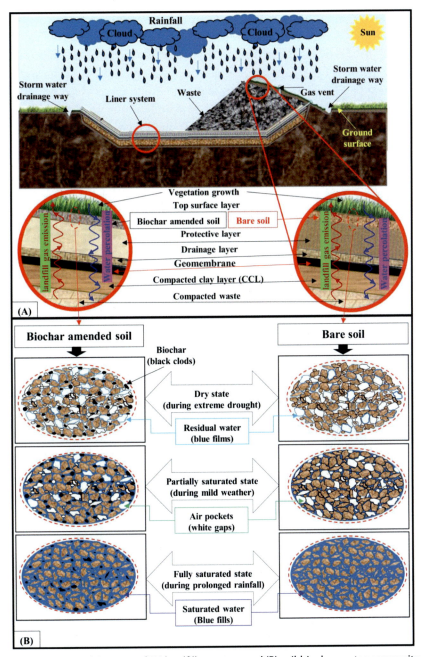

Fig. 1 Conceptual diagram of (A) landfill structure and (B) soil-biochar-water composite.

For instance, an increase in compressive shear strength parameters for BAS was reported in literature.[14] Incorporation of biochar in sand resulted an increase in shear resistance under monotonic and cyclic conditions.[15] However, there have been limited number of studies that have measured unsaturated hydraulic properties simultaneously such as soil water retention characteristics (SWRC), gas permeability and infiltration rate. Knowledge of SWRC is essential for predicting hydraulic conductivity function and shear strength component in soil.[16] Unsaturated gas permeability rate can foretell the amount of gas that might be emitted from landfill system.[17] Furthermore, knowledge of infiltration rate in surface soil is essential for estimating water balance of a cover during rainfall events.[18,19]

The overarching objective of this study is to investigate the unsaturated soil properties (SWRC, gas permeability, and infiltration rate) which play important roles in landfill cover functioning. The biochar was produced from water hyacinth inside an in-house pyrolysis unit and mixed with a silty sand at 5% application rate. All parameters were subsequently measured at different suctions indicating to the state of unsaturation (Fig. 1B).

2. Materials and methods

2.1 Soil properties

Soil used in current study was collected from the campus of Indian Institute of Technology (IIT) Guwahati, Assam, India. The soil was tested for various basic geotechnical properties as per ASTM-D422–63-07[20] and ASTM-D4318-10.[21] The soil constitutes of 70% silt, 16% of sand and nominal clay content. Its liquid limit, plastic limit and shrinkage limit were 42%, 24% and 19%, respectively. According to Unified soil classification system, the soil can be described as low plastic inorganic silt. Based on standard Proctor's test, its maximum dry density (MDD) is 16.5kN/m^3, corresponding to an optimum moisture content of 16% (by mass). Basic properties of the soil are summarized in Table 1.

2.2 Biochar characterization

Eichhornia crassipes also known as water hyacinth (WH) was selected as feedstock for producing biochar. The WH plants were selected from the same water body to minimize the effect of any genetic variation and cut into small slender pieces of 4 cm length and fed into the pyrolizer. Pyrolysis was carried out a temperature range of 350–400 °C. Biochar thus produced was crushed, and resultant particle size distribution was determined (Table 2).

Table 1 Basic characterizations of soil and biochar.

Soil properties	Standard	Soil	WHBC
Particle size distribution	ASTM D 422		
Coarse sand (2–4.75 mm)		23.11	0.00
Medium sand (0.425–2 mm)		37.28	000
Fine sand (0.075–0.425 mm)		20.84	30.48
Silt (0.002–0.075 mm)		17	68.52
Clay (<0.002 mm)		1.77	1
Atterberg limits	ASTM D 4318		
Liquid limit (LL)		42.4	ND
Plastic limit (PL)		26.0	ND
Plastic index (PI)		16.4	ND
Max. dry density (kN/m^3)	ASTM D 698	15.6	ND
OMC (% by mass)	ASTM D 698	16.5	ND
Specific gravity	ASTM D 854	2.46	0.8
pH	ASTM D 4972	6.7	9.25

WHBC, water hyacinth biochar; *ND*, not determined; *OMC*, optimum moisture content.

Table 2 Production conditions, elemental composition, and other chemical properties.

Feedstock	Water hyacinth stem
Pyrolysis process	Slow pyrolysis
Pyrolysis temperature (°C)	350–400
Elemental analysis	
C (%)	53.39
O (%)	42.80
H (%)	1.99
N (%)	1.82
Ash content (%)	39
CEC (cmol kg^{-1})	21.95

C, carbon; *O*, oxygen; *H*, hydrogen; *N*, nitrogen; *CEC*, cation exchange capacity.

Surface morphology of the biochar is shown in Fig. 2A and can be visualized that its surface is composed of intrapore which is resulted from thermal degradation of cellulose and hemicellulose. The intrapore when mixed with soil can facilitate in storing or retaining water. Surface functional groups of the biochar were measured by Fourier Transform Infra-Red (FTIR) spectroscopy.[22] The FTIR analysis exhibited a high peak spectral band at 3618 cm^{-1} which can be assigned as –OH (hydrophilic) group. Spectral peak at 1708 cm^{-1} was attributed to stretching vibration of C=O bonds of carboxyl groups (–COOH, –COOCH$_3$). These can be linked to carboxylic acids groups.[22,23] Moreover, mild peak at 1395 cm^{-1} was credited to aromatic (C=C–C) vibrations.[24]

2.3 Instrumentation and experimental methodology

Fig. 3 shows instrumented column setup used for measuring gas permeability and SWRC. The cylindrical column having diameter of 20 cm and height of 25 cm consisted of two chambers. The lower chamber having a height of 5 cm measured the gas discharge and corresponding pressure. On the other hand, upper chamber was filled with compacted soil specimen. The compaction state for soil sample was chosen as 0.8 times of

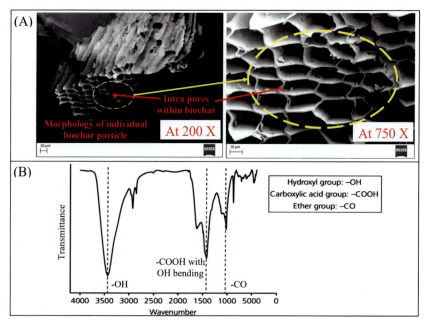

Fig. 2 Microanalysis of biochar based on (A) FE-SEM image and (B) FTIR.

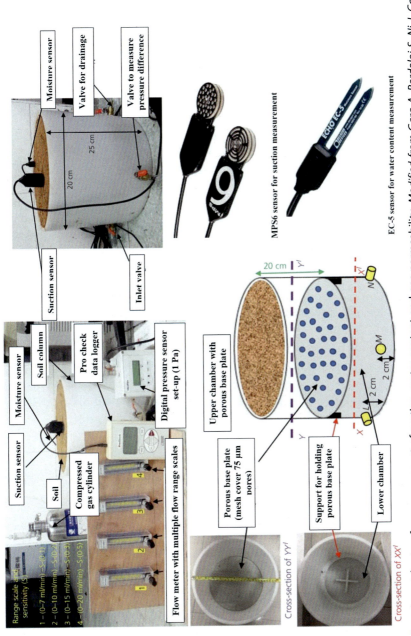

Fig. 3 Experimental setup for measurement of suction, water content and gas permeability. Modified from Garg A, Bordoloi S, Ni J, Cai W, Maddibiona PG, Mei G, Poulsen TG, Lin P. Influence of biochar addition on gas permeability in unsaturated soil. Geotech Lett 2019;**9**(1):66–71.

the maximum dry density (MDD) and corresponding water content along the wet side of compaction curve considered for each individual soil used (Fig. 4). The upper chamber was provided with a porous base (mesh size less than 75 μm). The setup was connected with flow meter to measure gas flow rate and digital pressure sensor to measure the gas pressure. The sensors (MPS-6 for suction measurement; EC-5 for measuring volumetric moisture content) from Meter/Decagon group (DDI, 2016) were installed at the soil surface.

According to Darcy's law, flow velocity is expressed as

$$v = \frac{q}{a} = k_g i = k_g \frac{\Delta h}{L} \tag{1}$$

where q: flow rate (m³/s); a: soil cross sectional area (m²); k_g: gas permeability (m²); i: hydraulic gradient; Δh: head-difference; and L: length of soil (m). Therefore,

$$\Delta q = k_g \frac{\Delta p}{\gamma L} A \Rightarrow k_g = \frac{\Delta q \gamma}{\Delta p A} L \tag{2}$$

where Δp: pressure difference (Pa); γ: specific weight of flow medium (15×10^{-6} N s/m²).

For the gas permeability experiments, applied pressure was slightly higher than general atmospheric pressure. Thus, gas dissolution could be reasonably ignored when gas passes through the sample. The setup was initially calibrated to account for expected resistance against gas flow due to bottom porous base.

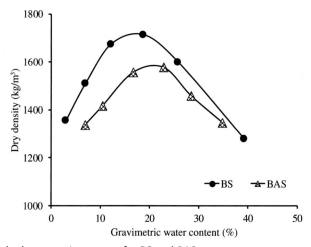

Fig. 4 Standard compaction curves for BS and BAS.

Compressed CO_2 gas was used in for the current study. The soil sample was initially ponded to ensure near saturation, after which it was exposed to evaporation in a temperature-controlled room (at 24 °C and 50% RH). Gas permeability (k_g) was measured periodically at regular intervals so that requisite measurements were obtained between 10 and 1500 kPa. After gas permeability measurements, mini disk infiltrometer was used to measure the infiltration rate at random points within the soil surface such that the water boundary did not converge for the individual measurements.

3. Results and discussion
3.1 Soil water retention characteristics

SWRCs of biochar amended soil (BAS) and bare soil (BS) have been shown in Fig. 5, and measured data points are fitted using van Genuchten equation.[26] The SWRC for both BS and BAS showed three sections: saturated zone, transition zone, and residual zone. The saturated zone maintains a constant saturated volumetric water content corresponding to any increase in soil suction. This saturated volumetric water content (θ_S; refer Table 3) is theoretically related to porosity of the material. θ_S was higher for BAS as compared to bare soil and essentially attributed to the fine-grained biochar particles and additional pore space provided by the intrapore of biochar.[27] The transition zone or desaturation zone starts after suction increases beyond the air entry value (AEV) where air intrudes the soil matrix. The AEV of BAS was much higher than that of BS by almost 100%. This can be attributed to finer pore throats and ability of biochar to retain additional water in their intrapore.

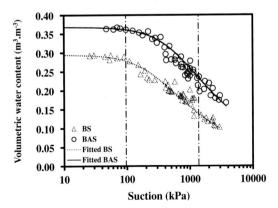

Fig. 5 Soil water retention characteristics of BS and BAS.

Table 3 Summary of various parameters for fitting SWRC of bare soil and biochar amended soil.[a]

Parameter	θ_s	θ_r	α	n	m	AEV
Unit	[%]	[%]	[kPa⁻¹]	[-]	[-]	[kPa]
Bare soil (BS)	0.29	1	0.0036	1.48	0.32	279
Biochar amended soil (BAS)	0.37	1	0.0016	1.47	0.32	600

[a]θ_s is volumetric water content at saturation, θ_r is residual volumetric water content, α is fitting parameter primarily dependent on the air entry value (AEV), n = fitting parameter depending on the rate of desaturation of water from the soil, m = 1-(1/n), BAS is biochar amended soil.

Thereafter, upon continued drying beyond AEV there is an expulsion of water and suction increases correspondingly. From n parameter reported in Table 3, rate of desaturation had very marginal change in both bare soil and BAS. However, there have been similar studies which reported lower desaturation rate for BAS than that of bare soil.[27] Nevertheless, for a particular moisture content the suction in BAS was always higher in this desaturation stage. This could be attributed to the lower pore throats and finer soil particle getting lodged in the biochar intrapore.[5] Both the mechanisms increase the air-water menisci formation in BAS, thus increasing the matric suction component. This additional suction component has implication in slope stability and transient seepage of the slopes during a rainfall event.[28] The third stage named as residual water content stage was not observed in this study due to sensor limitation (less than 2000 kPa).

3.2 Effect of biochar on gas permeability in soil

Fig. 6 presents gas permeability variations in BS and BAS as a function of suction and a general trend is that the gas permeability increases linearly with soil suction in semi-log scale. At higher magnitude of suction, a greater number of pores are filled with air cavity. Therefore, more available flow channels would increase gas permeability. At low suction range (i.e., less than 30 kPa), there was no difference in gas permeability and thus at this suction range, even though there was 40% difference in volumetric water content (Fig. 5), biochar had minor effects on gas permeability. This is probably because of negligible difference in air pore connectivity of BAS. However, gas permeability for BS increased at a greater rate than that of BAS. When suction become higher than this range, water in large pores would dissipate first without any significant changes in smaller pores or intra-pores (Fig. 2A). Additional water in these intra-pores of biochar, would then cause reduction

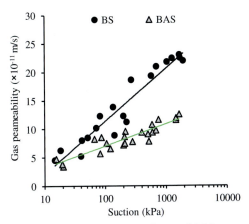

Fig. 6 Gas permeability variation with suction of BS and BAS.

in air conductive channels for gas migration resulting in further decrease in gas permeability as compared to BS. Moreover, biochar forms aggregate with clay particles over time and may further hinder the pore conductivity.[29] At around 2 MPa, gas permeability of BAS was lower by almost 50% than that of BS. This is also attributed to suppression of desiccation cracks in BAS which otherwise acted as preferential flow channels for gas migration in bare soil.[30] Similar trend for gas permeability decreasing at high suction was observed for biochar-amended clay soil at 90% degree of compaction.[31] The lower gas permeability in the unsaturated state has direct implication of other greenhouse gases such as methane and carbon-monoxide.

3.3 Effects of biochar on infiltration rate on soil

Variation of infiltration rate with suction for BS and BAS is shown in Fig. 7. It can be observed that change in hydraulic conductivity at different suction range varied within the same order of 10^{-3} m/s for both soils. This indicates marginal change in surface infiltration rate for field application. Nevertheless, the infiltration rate was relatively higher in case of BAS than that of BS. This can be attributed to higher head loss caused by increase in suction head in case of BAS (Fig. 5). Moreover, biochar is hydrophilic in nature (Fig. 2) and has a higher affinity to absorb water molecules. Relatively higher decrease in infiltration rate from lower to high suction in case of BAS could be attributed to path tortuosity.[32] Tortuosity is caused by intrapore of the biochar at high suction range. In engineering practice, the infiltration rate for bare soil and BAS do not have significant effect on the seepage of a slope or landfill cover, as they

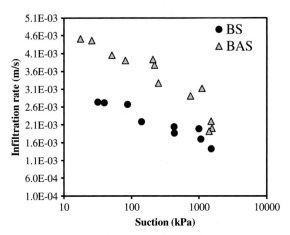

Fig. 7 Infiltration rate variation with suction of BS and BAS.

fall within the same order. However, literature extensively shows that the hydraulic conductivity may vary in more than one order for BAS in case of clays and silty clay.[27] It is also well established that the biochar production conditions also play a vital role in the flow of water in compacted BAS.[33] This is due to change in distribution of intrapore and aromaticity of biochar, which ultimately influences the unsaturated hydraulic properties.

4. Conclusions

We investigated the unsaturated soil properties including soil water retention characteristics, gas permeability and surface infiltration rate by conducting a series of laboratory experiments. Major conclusions derived from the study are:

(1) Water retention capacity increased with biochar incorporation to silty sand. Its air entry value increased by almost 100%, while n value had no significant change. This might be attributed to entrapped clay particles within the biochar and consequent alteration of soil porosity.

(2) Beyond suction range of 1 MPa, gas permeability of biochar amended soil decreased by almost 50% as compared to bare soil, which is likely due to suppression of desiccation cracks in biochar amended soil.

(3) Infiltration rate was found to be relatively higher in case of biochar amended soil in comparison to bare soil. However, hydraulic conductivity changed at different suction range of same order indicating very marginal change in soil infiltration rate for any field application.

References

1. Joseph S, Lehmann J. *Biochar for environmental management: science and technology*. London, GB: Earthscan; 2009.
2. Smetanová A, Dotterweich M, Diehl D, Ulrich U, Dotterweich NF. Influence of biochar and terra preta substrates on wettability and erodibility of soils. *Z Geomorphol* 2013;**57**(1):111–34.
3. Atkinson CJ, Fitzgerald JD, Hipps NA. Potential mechanisms for achieving agricultural benefits from biochar application to temperate soils: a review. *Plant and Soil* 2010;**337**(1):1–8.
4. Barrow CJ. Biochar: potential for countering land degradation and for improving agriculture. *Appl Geogr* 2012;**34**:21–8.
5. Wong JT, Chen Z, Chen X, Ng CW, Wong MH. Soil-water retention behavior of compacted biochar-amended clay: a novel landfill final cover material. *J Soil Sediment* 2017;**17**(3):590–8.
6. Thomas SC, Gale N. Biochar and forest restoration: a review and meta-analysis of tree growth responses. *New For* 2015;**46**(5):931–46.
7. Kolton M, Harel YM, Pasternak Z, Graber ER, Elad Y, Cytryn E. Impact of biochar application to soil on the root-associated bacterial community structure of fully developed greenhouse pepper plants. *Appl Environ Microbiol* 2011;**77**(14):4924–30.
8. Bhaduri D, Saha A, Desai D, Meena HN. Restoration of carbon and microbial activity in salt-induced soil by application of peanut shell biochar during short-term incubation study. *Chemosphere* 2016;**148**:86–98.
9. Shaikh J, Bordoloi S, Yamsani SK, Sekharan S, Rakesh RR, Sarmah AK. Long-term hydraulic performance of landfill cover system in extreme humid region: field monitoring and numerical approach. *Sci Total Environ* 2019;**688**:409–23.
10. Wong JT, Chen X, Deng W, Chai Y, Ng CW, Wong MH. Effects of biochar on bacterial communities in a newly established landfill cover topsoil. *J Environ Manage* 2019;**236**:667–73.
11. Xie T, Reddy KR, Wang C, Yargicoglu E, Spokas K. Characteristics and applications of biochar for environmental remediation: a review. *Crit Rev Environ Sci Technol* 2015;**45**(9):939–69.
12. Reddy KR, Yargicoglu EN, Yue D, Yaghoubi P. Enhanced microbial methane oxidation in landfill cover soil amended with biochar. *J Geotech Geoenviron Eng* 2014;**140**(9): 04014047.
13. Sohi SP, Krull E, Lopez-Capel E, Bol R. A review of biochar and its use and function in soil. *Adv Agron* 2010;**105**:47–82.
14. Bora MJ, Bordoloi S, Kumar H, Gogoi N, Zhu HH, Sarmah AK, et al. Influence of biochar from animal and plant origin on the compressive strength characteristics of degraded landfill surface soils. *Int J Damage Mech* 2021;**30**(4):484–501. 1056789520925524.
15. Pardo GS, Sarmah AK, Orense RP. Mechanism of improvement of biochar on shear strength and liquefaction resistance of sand. *Géotechnique* 2019;**69**(6):471–80.
16. Bachmann J, van der Ploeg RR. A review on recent developments in soil water retention theory: interfacial tension and temperature effects. *J Plant Nutr Soil Sci* 2002;**165**(4):468–78.
17. Moldrup P, Poulsen TG, Schjønning P, Olesen T, Yamaguchi T. Gas permeability in undisturbed soils: measurements and predictive models. *Soil Sci* 1998;**163**(3):180–9.
18. Shaikh J, Bordoloi S, Leung AK, Yamsani SK, Sekharan S, Rakesh RR. Seepage characteristics of three-layered landfill cover system constituting fly-ash under extreme ponding condition. *Sci Total Environ* 2021;**758**:143683.
19. Ng CW, Coo JL, Chen ZK, Chen R. Water infiltration into a new three-layer landfill cover system. *J Environ Eng* 2016;**142**(5): 04016007.

20. ASTM D422-63. *Standard test method for particle-size analysis of soils*; 2007.
21. ASTM D4318. *Standard test methods for liquid limit, plastic limit, and plasticity index of soils*; 2010.
22. Abid M, Niazi NK, Bibi I, Farooqi A, Ok YS, Kunhikrishnan A, et al. Arsenic (V) biosorption by charred orange peel in aqueous environments. *Int J Phytoremediation* 2016;**18**(5):442–9.
23. Mohan D, Rajput S, Singh VK, Steele PH, Pittman Jr CU. Modeling and evaluation of chromium remediation from water using low cost bio-char, a green adsorbent. *J Hazard Mater* 2011;**188**(1–3):319–33.
24. El-Banna MF, Mosa A, Gao B, Yin X, Ahmad Z, Wang H. Sorption of lead ions onto oxidized bagasse-biochar mitigates Pb-induced oxidative stress on hydroponically grown chicory: experimental observations and mechanisms. *Chemosphere* 2018;**208**:887–98.
26. Van Genuchten MT. A closed-form equation for predicting the hydraulic conductivity of unsaturated soils. *Soil Sci Soc Am J* 1980;**44**(5):892–8.
27. Hussain R, Ravi K, Garg A. Influence of biochar on the soil water retention characteristics (SWRC): potential application in geotechnical engineering structures. *Soil Tillage Res* 2020;**204**:104713.
28. Ng CW, Shi Q. A numerical investigation of the stability of unsaturated soil slopes subjected to transient seepage. *Comput Geotech* 1998;**22**(1):1–28.
29. Moldrup P, Olesen T, Komatsu T, Schjønning P, Rolston DE. Tortuosity, diffusivity, and permeability in the soil liquid and gaseous phases. *Soil Sci Soc Am J* 2001;**65**(3):613–23.
30. Bordoloi S, Garg A, Sreedeep S, Lin P, Mei G. Investigation of cracking and water availability of soil-biochar composite synthesized from invasive weed water hyacinth. *Bioresour Technol* 2018;**263**:665–77.
31. Wong JT, Chen Z, Ng CW, Wong MH. Gas permeability of biochar-amended clay: potential alternative landfill final cover material. *Environ Sci Pollut Res* 2016;**23**(8):7126–31.
32. Hussain R, Bordoloi S, Garg A, Ravi K, Sreedeep S, Sahoo L. Effect of biochar type on infiltration, water retention and desiccation crack potential of a silty sand. *Biochar* 2020;**2**(4):465–78.
33. Ganesan SP, Bordoloi S, Ni J, Sizmur T, Garg A, Sekharan S. Exploring implication of variation in biochar production on geotechnical properties of soil. *Biomass Convers Biorefin* 2020;**1**:1.

Further reading

25. Garg A, Bordoloi S, Ni J, Cai W, Maddibiona PG, Mei G, et al. Influence of biochar addition on gas permeability in unsaturated soil. *Geotech Lett* 2019;**9**(1):66–71.

CHAPTER FIVE

Effects of modified biochar on As-contaminated water and soil: A recent update

Jingzi Beiyuan[a,†], Yiyin Qin[a,†], Qiqi Huang[a], Hailong Wang[a], Daniel C.W. Tsang[b], and Jörg Rinklebe[c,*]

[a]School of Environmental and Chemical Engineering, Foshan University, Foshan, China
[b]Department of Civil and Environmental Engineering, Hong Kong Polytechnic University, Hong Kong, China
[c]University of Wuppertal, Institute of Foundation Engineering, Waste and Water Management, School of Architecture and Civil Engineering, Soil and Groundwater Management, Wuppertal, Germany
*Corresponding author: e-mail address: rinklebe@uni-wuppertal.de

Contents

1. Introduction	108
2. Biochar for As contamination: Limitation and challenge	112
2.1 Biochar	112
2.2 Biochar application in As-contaminated water and soil	112
2.3 Limitations of biochar in its use of As-contaminated soil	113
2.4 Modification of biochar for As	114
3. Effect of Fe/Mn-based modified biochar	116
3.1 Effect of Fe-modified biochar	116
3.2 Effect of Mn-modified biochar	123
3.3 Effect of Fe-Mn-modified biochar	124
4. Effects of other modified biochar in As-contaminated soil	124
4.1 Rare earth elements-modified biochar	125
4.2 Zn-modified biochar	125
4.3 Si-modified biochar	128
4.4 Acid-base-modified biochar	129
5. The way forward	129
References	130

Abstract

Arsenic is deemed to be a highly toxic and non-metallic element, which is widely distributed in nature. Due to natural and anthropogenic factors, more areas across the globe are suffering As contamination, and this has attracted international attention.

[†] The authors contributed equally to this work.

Biochar, as a low-cost and environmentally friendly adsorbent, has become a popular solution for As remediation. However, the unmodified biochar still has limitations as a sorbent and immobilizer for As in environment. For example, stronger affinity to As, higher removal efficiency of As, and better capacity of adaption to the varied environment on the surface of biochar by varied mechanisms should be improved by different methods. This chapter reviews the latest findings of modified biochar for As remediation and shows that the current trend is mainly for Fe/Mn-containing minerals embedment on biochar. Multiple new modification methods for As removal/immobilization in both water and soil were reviewed, such as Fe-modified biochar, Mn-modified biochar, Zn-modified biochar, Si-modified biochar, rare earth elements-modified biochar, and acid-base-modified biochar. The metal-based modified biochar facilitates the adsorption of As, especially for Fe/Mn-containing minerals, by increasing the active sites of adsorption. Besides, the acid/base-modified biochar increase pore volume and oxygen-containing groups to improve adsorption capacity of As. Nonetheless, further studies are suggested to include engineered biochar for higher stability to overcome alteration in soil environment and promote removals of organic As in both water and soil. A combination of modification methods is proposed for better As removal/immobilization.

Keywords: Potentially toxic elements, Heavy metal, Tailored biochar, Soil remediation, Water purification

1. Introduction

Arsenic (As) is one of the common potentially toxic elements (PTEs) as environmental pollutants, which is one of the teratogenic and carcinogenic toxins in the world.[1] As a good sink for As and other PTEs, the average contents of As in the earth's crust are about $1.5–3\,mg\,kg^{-1}$, and the typical contents are below $10\,mg\,kg^{-1}$ in non–contaminated soil.[2,3] There are main four As species in both water and soil, including arsenite (As^{III}), arsenate (As^V), methylarsonic acid (MMAA), and dimethylarsinic acid (DMAA), while a small part of the methylated As exists in the water.[4] Both inorganic and organic forms of As are toxic, however, the inorganic forms are over 100 times toxic than the organic As compounds in the natural environment. There are two common inorganic species, As^{III} and As^V, which have been recognized as important toxic species of As in the natural environment. Among the inorganic forms, As^{III} has the highest toxicity. The common organic As forms of environmental contaminants with detectable contents in soil are only MMAA and DMAA.[2] The two organic forms are normally found in paddy soil and normally produced under anaerobic conditions.[5] Once As contamination occurs, it can cause negative impact on animals, plants, humans and the environment nearby. For example, skin contact or inhalation of As compounds, or long-term exposure to trace As compounds, can cause abdominal pain, nausea, vomiting, diarrhea, dizziness, headaches, and breathing difficulties, and even risk of cancer to human.[6] The As limitation in the standard for drinking water is set as $10\,\mu g\,L^{-1}$ by the World Health Organization (WHO), which is widely adopted by the Council of European Communities, Environmental Protection Agency USA (EPA USA), and Ministry of Ecology and Environment of China in their national drinking water standards.[7]

The main source of As is mother rock in nature, such as volcanic eruption, rock weathering, soil erosion, and other natural processes which can release As into the environment. Anthropogenic activities, such as agricultural and industrial activities can enhance As contamination. For agricultural contamination, As-containing materials such as pesticides, herbicides, insecticides, fungicides, wood preservatives, and crop desiccants were widely used in agricultural production activities in the last century.[2,8] Besides, irrigation by As-containing sewage and the use of As-containing sludge in agricultural activities are two significant causes for As contamination.[9] Industrial activities can also cause serious As contamination, for instance, mining, dressing, smelting, and processing of As-bearing ore, which can discharge large amounts of As-containing fume, wastewater, exhaust gas, and offscum which can further contaminate the water and soil.[10] Among them, metallurgy and chemical industry are the most important sources of industrial contamination, which have the highest amounts of As discharge.

Arsenic contamination can be found globally, and the most serious As contamination happens in the South and Southeast Asian countries. As contamination in the water has been reported more than in 70 countries, like Bangladesh, India, West Bengal, Pakistan, Vietnam, Nepal, Cambodia and China. For drinking As-contaminated water, 94 million to 220 million people are being suffered from excessive As.[11] The mean levels of As in the soil of various countries are about $6\,mg\,kg^{-1}$, and the average As contents in the soil in Japan, China, and Bangladesh are 10, 11.2, and $22.1\,mg\,kg^{-1}$, respectively.[3] In detail, the range As content in West Bengal, India is $10-196\,mg\,kg^{-1}$ in 2235 samples, and in China is $0.01-626\,mg\,kg^{-1}$ in 4095 samples. A national report of soil survey in China in 2014 found that inorganic contamination is the dominant type of contamination, while As ranks top 5 of the PTEs in the collected samples.[12]

The As content of different water bodies is various, which the content of As in groundwater is higher.[13] The accumulation of As in groundwater is greatly affected by the mineralogical characteristics of the aquifer. As can enter the groundwater through the reduction and dissolution of As–rich iron oxide in the dispersed phase of the aquifer. In geothermal activity, As and a large amount of hydrogen sulfide is present in high–temperature hydrothermal vents. Together with a large amount of hydrogen sulfide, they exist in high–temperature hydrothermal vents. When magma rushes from the mantle to the upper rocks, heating the surrounding rock, warm groundwater can obtain As from As-bearing rock aquifers or impermeable bottom rocks.[14] Additionally, Zakhar, Derco[15] the distribution of As species in the water is mainly dependent on redox potential (Eh) and pH of the water. Under oxidizing conditions, the main form is As^V, on the contrary, As^{III} is assumed to be stable under hypoxic conditions.

Remediation of As in the water can mainly through chemical, physical, and biological technologies. The physical and chemical technologies include oxidation–reduction, adsorption, coagulation precipitation, ion exchange, membrane filtration, etc. Biological technologies mainly include phytoremediation and microbial remediation.[14,16] The As removal efficiency from water can be remarkably affected by the environmental conditions, for example, the initial As concentration, pH, temperature, and other chemical substances.[14]

To remove As in water, chemical technologies are commonly applied. Recently, to enhance the feasibility and removal efficiencies, chemical technologies are frequently combined with physical technologies, whose mechanisms include oxidation–reduction, adsorption, coagulation precipitation, ion exchange, and filtration.[13,16] Due to the higher toxicity of As^{III}, a pre-oxidation process is recommended to be carried out. For example, oxidants, such as Fenton and $KMnO_4$, can be used to oxidize As^{III} to As^V as pretreatment, which can effectively alleviate the toxicity of As.[16] Among the technologies, adsorption is the most cost-effective one. Generally, the larger the surface area, the higher porosity of the adsorbent, the better the adsorption rate.[17] Currently, biochar, as a cost-effective adsorbent, are frequently reported in removing As from water. Both As^V and As^{III} can be successfully removed by biochar produced from wood, rice husks, sewage sludge, etc. Modification upon the biochar surface for loading zero valence iron, metal(loid)s, Fe/Mn/Al minerals, and rare earth elements are proposed with satisfying outcomes.[18,19]

Biological technologies are use plants or microorganisms to control As contamination.[20] Phytoremediation is an environmentally-friendly method can be used to remove As from water bodies and soil, with limited requirements of nutrients and manageable technological factors.

Hyperaccumulator plants are often applied to mobilize PTE from the soil to the root through migration, aggregation, distribution, exclusion, and osmotic adjustment. Some As hyperaccumulators have a high accumulation capacity of a minimum content of $1000\,mg\,kg^{-1}$ As in the aboveground biomass without high accumulation in the root.[21] The removal efficiency of phytoremediation can also be improved by assisting with other technologies, including the use of chelating agents, soil amendments, and nanoparticles.[17]

In soil, the bioavailability and mobility of As and other PTEs are strongly controlled by soil physicochemical properties, especially by their geochemical distributions.[22–24] It should be noted that the long-term flooded environment, especially in the paddy soil, can substantially affect the soil redox conditions, promoting the reduction of As^V to As^{III} and increasing the toxicity of As in crops and soil.[25] Besides, the flooded conditions strongly affect the reduction and oxidation status of Fe and Mn oxides, which consequently release As that bond to the Fe/Mn oxides.[18,26,27]

The principle of As-contaminated soil repair technology in the soil is similar to the As in the water. Chemical, physical and biological technologies can also use to control As contamination in soil.[28,29] There are two typical mechanisms of the chemical techniques for treating the As contamination: one is to change the As to stable forms, which indirectly reduces the mobility and bioavailability of As in soil, for example, soil amendments; the other is to change it to mobilized forms, which can be directly removed from the soil with help of further treatments, for example, soil washing.[30,31] Soil immobilization is applying amendments that can immobilize As via adsorption, ion exchange, complexation, and precipitation, which decreases the potential risks.[29] Metal oxides, iron oxide-based sorbents, zero-valent, carbonaceous materials, clay minerals, iron compounds, aluminum oxides, and manganese oxides were frequently used for immobilizing As in soil in the past 10 years.[32,33] Among them, biochar, as soil amendments, has been proposed for immobilizing As in soil.[18,34] On the other hand, soil mobilization techniques usually use phosphoric acid and phosphate as cleaning reagents to remove As from the soil because they can replace As^V for their similarity of chemical properties.[35] Nowadays, research works found that a single technology can hardly meet its requirements. To achieve better remediation effects, combined technologies of various chemical, physical, and biological technologies are frequently studied.[36]

In recent 10 years, among the technologies for As remediation in both water and soil, organic adsorbents/immobilizers, such as biochar which is produced from organic waste, are frequently studied. The effects of As remediation by the pristine biochar were found not stable enough especially under changing environment and some of the biochar might cause unsatisfied effects. Therefore, many modification methods on the biochar surface were proposed to enhance the remediation effects and stability through various methods. In this chapter, we reviewed and compared some emerging biochar modification methods for As remediation in both water and soil. As a result, the limitations and future trends of biochar modification for As remediation would be summarized.

2. Biochar for As contamination: Limitation and challenge

2.1 Biochar

Biochar, has a rich content of carbon (C), is a cost-effective material produced by pyrolysis under oxygen-limited conditions.[18,37] It converts C from biomass into a stable form, allowing it can be remained in the soil for hundreds or even years, which attracts intensive research interests recently.[38] In recent years, biochar has shown excellent adsorption capacity to immobilize PTEs, due to its advantages such as simple production, low cost, and excellent performance in environmental management.[39] It has a high specific surface area, numerous pores, and a large number of carboxyl groups, phenolic hydroxyl groups, carbonyl groups, anhydride hydroxyl groups, as well as alkyl groups on the surface, which make it to possess a certain adsorption capacity for PTEs, affecting their distribution, migration, and availability in soil.[40–42] The common feedstocks of biochar can be agricultural wastes, municipal wastes, industrial wastes, and food wastes, which makes it a low-cost alternative to conventional adsorbents such as activated carbon. Additionally, the pyrolysis temperature, gas, properties of feedstocks, and procedure of pyrolysis, can strongly affect the physical characteristics of biochar.[37] Among them, the contents of lignin of biomass directly affect the physicochemical properties of biochar and its yield.

2.2 Biochar application in As-contaminated water and soil

As a low-cost and easily-obtained sorbent, biochar can be used as sorbent and soil amendment, especially for PTEs, for example, lead (Pb), copper (Cu), cadmium (Cd) and As.[43] The main mechanisms of PTEs-immobilization/removal by biochar involve its high alkalinity, cation exchange capacity (CEC), and rich surface functional groups which can form surface complexes with the PTEs, which significantly reduce mobility, leachability, and bioavailability of PTEs in soil.[25,44–46]

Different from the PTEs which are normally metal ions, the surface complexation effects with the surface functionally groups of biochar is one the most important reasons for immobilization of As in soil.[47] The As affinity to biochar is governed by the pyrolysis conditions, including feedstock, temperature, retention time.[18] Previous studies also supported that unmodified biochar produced at lower temperatures (300–500 °C) have better removal performance of As in water as they have more O-containing functional groups.[6] The high CEC of biochar can also increase the effects of electrostatic adsorption.

However, there are still some drawbacks for using biochar in As immobilization in soil. Negligible or adverse effects, for example promoted As concentration in the porewater, were found for As immobilization in soil.[48,49] Some studies suggested that biochar contains large amounts of dissolved organic carbon (DOC) which can be released into soil and complex with As, leading to an elevated As mobility in the porewater of rhizosphere.[50] Besides, enhanced alkalinity caused by the addition of biochar in the soil-water system can lower the surface positive charge on the biochar surface. The reduction of As^V to As^{III} on the surface of biochar can be also enhanced by the functional groups which act as electron donors.[18,51] As a result, As mobility can be promoted in soil, which might cause food safety problems.

The variation of redox conditions in natural environment, for example, paddy soil, can significantly affect mobility and speciation of As.[52] Additionally, previous studies showed that the biochar immobilization effects of the PTEs, especially for As, are remarkably affected by the change of redox conditions.[34,53] Beiyuan et al.[34] found that unmodified pine sawdust produced at 550 °C increased the mobility and phytoavailability which were evaluated by SPLP (synthetic precipitation leaching procedure, US EPA Method 1312) and simulated rhizosphere exudates of As under anaerobic (−300 to −100 mV) and slightly aerobic (100 mV) conditions. Additionally, the adding of biochar can enrich the abundance of Fe-reducing bacteria and lead to reductive dissolution of Fe (oxyhydr)oxides, leading to release of As in the rhizosphere.[50] Therefore, it is of great importance to modify biochar to have higher affinity to As and the immobilization effects are stable to altered natural conditions.

2.3 Limitations of biochar in its use of As-contaminated soil

Though it has been proved many times that biochar can be used as suitable stabilizers for PTEs in soil, some studies found that there are still some problems for As–contaminated soil. Most of the pristine biochar has a relatively low capacity of adsorption of As^V or As^{III}, because its surface is anionic charge during the production. Besides, the rich organic matter with quinoid functional groups in the surface of biochar can promote the reduction of As^V to As^{III} which has higher toxicity. Some studies suggested that biochar has a certain effect on the reduction of As^V on account of both biological and non-biological conditions which results in increasing As toxicity.[54] Furthermore, in general, As has low mobility in acid soil than in alkaline soil. The application of biochar normally can lead to an increase in soil pH, which can cause the desorption of As^V and As^{III}, and even reduce the possibility to form As precipitates.[55] And in chemical properties the phosphate is similar to the As meaning that they may compete with each other for absorption sites, and if the solubility of the phosphorus is higher, As may be less adsorbed to the surface of the biochar.[56]

There are still some limitations of biochar can be found besides its limited effects on As immobilization in soil, with its extensive development and application.[56] For instance, the low surface area, poor pore structure, and scant surface functional groups are limited by the original properties of feedstock during the pyrolysis, which can be improved by further modification.

2.4 Modification of biochar for As

Currently, the modification can be divided into chemical and physical methods to improve the immobilization capacity of contaminants by biochar. Additionally, the application of the modified biochar with new properties and characterization can reduce some possible problems caused by the biochar. The chemical modification is mixing chemicals into the feedstock before the pyrolysis or mixing them with the produced biochar before second pyrolysis, which can improve the surface area and pore volume, enhance the pore structure/distribution, and enrich the functional groups on its surface to better immobilization effects.[18] The chemical modification is the most common one, which includes acid-based modification, oxidation modification, and reduction modification.

The higher surface area and pore volume can improve As immobilization by enhancing the change of activated points. However, some works found that materials with surface area and pore volume can still have low adsorption capacity on As, which indicates that the pore distribution/structure and surface functional groups may also play crucial roles in As immobilization.[57] One of the common methods of chemical modification is to amend the biochar with minerals to enhance the adsorption of As on the surface,[36] as summarized in Fig. 1. For example, both As^V to As^{III} can have strong affinity or complexation ability on Fe oxides in soil[58], thus it is a promising way to load Fe oxides in biochar, leading to a reduction of As mobility in soil. Besides, Mn oxides also have strong complexation capacities with As in natural soil environment,[59] which has been used for biochar modification as well. In addition, the use of manganese oxides to modify biochar can make As^{III} and As^V in the soil form surface complexations with many different oxides, especially Fe, Mn, and Al oxides, and promote the oxidation of ferrous iron. Acid and base modification can change the physical structure of the feedstock biomass or biochar. The acid treatment removes the residual metals and impurity on the surface of biochar and makes more carboxyl groups to improve the adsorption of PTEs. By contrast, the base treatment can remove the ash content and enrich hydroxyl groups in the surface of biochar.[60]

The physical modification can form higher amounts of micro/mesopore structure and improve the pore structure, resulting in a higher surface area of the pristine biochar, which contributes to an enhanced adsorption capacity of the As.[61] Besides, it includes using low-temperature production to make the biochar contain large amounts of O-containing functional groups, for example, carboxyl and hydroxyl groups, which can improve the As immobilization effects by enhancing the effects of complexation with the O-containing

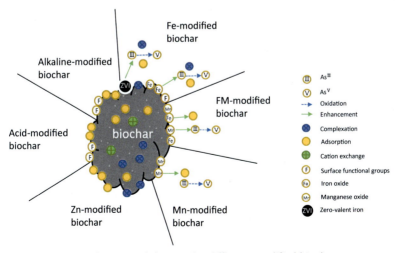

Fig. 1 Mechanisms of As immobilization by different modified biochars.

functional groups, electrostatic adsorption, and precipitation.[62] Some works found that applying biochar with higher lignin contents can greatly reduce the As mobility in soil, which might be due to the higher contents of aromatic component of biochar.[63]

3. Effect of Fe/Mn-based modified biochar
3.1 Effect of Fe-modified biochar

As mentioned above, Fe minerals in soil are important to the adsorption behaviors of As under natural conditions. Iron exists in soil as Fe^{II} and Fe^{III} in various minerals, which mostly are iron oxide minerals, such as goethite (α-FeOOH), hematite (α-Fe$_2$O$_3$), and magnetite (Fe$_3$O$_4$). These Fe-containing minerals are thermodynamically stable cannot be changed easily under natural conditions. Besides, there are some other Fe-containing minerals that are considered as intermediates, such as hydrated iron (Fe$_5$HO$_8 \cdot$4H$_2$O), maghemite (γ-Fe$_2$O$_3$).[10] For example, previous studies showed that at a low pH value of 5–6, As^V is more favorable for the adsorption of amorphous iron oxide and goethite than As^{III}, and when the pH value is higher than 7–8, As^{III} tends to be adsorbed on amorphous iron oxide and goethite than As^V.[64] Therefore, modification by embedding Fe-containing minerals into biochar could largely enhance the adsorption/immobilization performance of As, which could lead to higher adsorption points for As and the oxidation of Fe-containing oxides can affect the oxidation state of As resulting in a less toxic form, i.e., As^V.[59] Additionally, the addition of Fe-modified biochar can affect the surface charge of soil colloidal crystal lattice and promote the generation of OH^- which can replace As to form undissolved minerals to improve the As stabilization in soil (Table 1).[65]

Zhu and Zhao[66] studied the immobilization effects of As-contaminated soil by goethite-loaded biochar (GMB) which is modified by soaking the wheat straw biochar produced at 600 °C in solution of KOH and Fe(NO$_3$)$_3$. According to the results of the Langmuir model, the adsorption capacity of GMB for As^{III} solution is 65.30 mg g^{-1}. In addition to the significant adsorption behavior of As in the solution phase, the immobilization effects in As-contaminated soil are also remarkable. After the GMB treatment, the available As concentration extracted by NaHCO$_3$ in the soil can be reduced by 58.7%. The addition of KOH at the early stage of GMB preparation can increase micropores on the biochar surface, which could promote the adsorption of As by diffusion. The loading of goethite strengthened the amount of −OH functional groups on the biochar surface, which can facilitate the oxidation of As^{III} to As^V, reducing the toxicity of As.

Table 1 Physicochemical properties of some Fe/Mn-modified biochars.

Feedstock	Temperature	Modification methods	BET Surface area $m^2\,g^{-1}$	Pore volume $cm^3\,g^{-1}$	Pore Size nm	C %	N %	H %	O %	Fe %	Mn %	pH	pH_{ZPC}	References
Fe-based modified biochar														
Rice straw	500	Two-step modification; $FeCl_3$; modification temperature: 80 °C	23.37	0.054	3410	46.1	2.23	2.86	16.2	1.97		9.97	9.94	Zhang and Fan[81]
		Two-step modification; $FeCl_3$; modification temperature: 80 °C	26.45	0.047	7.35	43.8	2.27	2.79	15.9	4.87		7.93	5.73	
		Two-step modification; $FeCl_3$; modification temperature: 80 °C	7.31	0.017	9.44	32.5	1.72	2.58	15.8	5.26		1.90	3.16	
		One-step modification; $FeCl_3$	14.63	0.039	18.63	45.5	2.23	2.06	11.2	3.03		9.35	10.23	
		One-step modification; $FeCl_3$	5.98	0.024	24.51	44.2	2.32	2.10	12.0	8.12		2.11	3.42	
		One-step modification; $FeCl_3$	5.73	0.022	26.03	44.2	2.32	2.08	13.8	8.25		2.02	3.14	

Continued

Table 1 Physicochemical properties of some Fe/Mn-modified biochars.—cont'd

Feedstock	Temperature	Modification methods	Biochar Properties											References
			BET Surface area $m^2\,g^{-1}$	Pore volume $cm^3\,g^{-1}$	Pore Size nm	C %	N %	H %	O %	Fe %	Mn %	pH	pH$_{ZPC}$	
Cotton fiber	800	Two-step modification; iron oxide; modification temperature: 120 °C	8.68			23.17			69.71	8.6				Wei and Wei[68]
Palm thread	300	Two-step modification; ZVI				>50				2.16		6.0		Pan and Liu[36]
Oil palm fibers	700	One-step modification; ZVI	241.6			86.7		1.8	3.2	5		9.3		Qiao and Liu[25]
Rice straw	300	Two-step modification; FeSO$_4$; modification temperature: 30 °C	4.44			37.4	0.87	3.94	34.6					Wu and Cui[65]
		One-step modification; FeCl$_3$	3.27			40.9	0.960	4.37	30.8					
		One-step modification; ZVI	3.18			38.5	0.76	3.42	23.6					
Wheat straw	600	Two-step modification; Goethite; modification temperature: 70 °C	276.24	0.1911		45.74			19.19	15.24				Zhu and Zhao[66]

Miscanthus	500	Two-step modification; $Fe(NO_3)_3$; modification temperature: 120°C	123.8	0.07	4.00	66.32	0.19	4.04	13.55					Kim and Song[100]
Hickory chips	600	Two-step modification; $Fe(NO_3)_3$; modification temperature: 100–120°C	16.0			68.8	1.83	2.03		3.88				Hu and Ding[101]
Corn straw	600	One-step modification; $FeCl_3$	4.81	0.012		38.08	1.01	1.44	27.76	24.17	0.01	2.55	6.93	Fan and Xu[102]
Sawdust	800	One-step modification; ZVI	274			45.3			14.2	40.3				Fan and Chen[103]
Guadua chacoensis	700	Two-step modification; KOH; modification temperature: 60°C	1240		12.1	40.98	0.61	2.03	45.99	0.08			8.0	Alchouron and Navarathna[104]
		Two-step modification; Fe_3O_4; modification temperature: 25–60°C	28.9		9.7	16.09	0.57	0.86	15.25	17.9			4.0	
		Two-step modification; Fe_3O_4; modification temperature: 25–60°C	482.4		13.4	10.62	0.38	0.57	41.23	18.7			10.9	
Pinewood	600	Two-step modification; ZVI; modification temperature: 90°C	211.7			72.16	0.37			10.5		4.1		Wang and Gao[105]

Continued

Table 1 Physicochemical properties of some Fe/Mn-modified biochars.—cont'd

Feedstock	Temperature	Modification methods	Biochar Properties												References
			BET Surface area	Pore volume	Pore Size	C	N	H	O	Fe	Mn	pH	pH$_{ZPC}$	
			$m^2\,g^{-1}$	$cm^3\,g^{-1}$	nm	%	%	%	%	%	%			
Coffee ground	700	Two-step modification; FeCl$_3$; modification temperature: 80 °C	8.3	0.018	8.66							0.24		Cho and Yoon[106]
Corn husk	600	One-step modification; Fe(NO$_3$)$_3$	208.0			67.3	1.83	2.38					9.60	Lin and Li[59]
		One-step modification; FeSO$_4$	7.53			53.8	1.41	2.22					3.17	
Rice husk	500	Two-step modification; FeSO4; modification temperature: 70 °C	82.77	0.102	4.903	38.60	0.30					8.353	3.15	Guo and Yan[107]
Dried ground corn straw	600	Two-step modification; KMnO$_4$,; modification temperature: 600 °C	3.18			73.00	0.72	0.33	10.90		7.41	10.75		Yu and Zhou[72]
Loblolly pine wood	600	Two-step modification; MnCl$_2$·4H$_2$O; modification temperature: 600 °C	282.8			54.90	0.41				10.90	8		Wang and Gao[70]

Feedstock	Pyrolysis temperature (°C)	Modification												Reference
Dry stem of *Sesbania bispinosa* plants	450	Two-step modification; MnCl$_2$; modification temperature: 350 °C	NA	NA	NA	NA	NA		NA	NA	NA	NA	NA	Imran and Iqbal[75]
Grape stalks	600	Two-step modification; KMnO$_4$; modification temperature: laboratory temperature	44.0									8.47	7.88	Trakal and Michalkova[71]
Loblolly pine wood	600	Two-step modification; KMnO$_4$; modification temperature: 80 °C	67.4			61.54	0.25	1.85	27.65		8.14			Wang and Gao[108]
Rice husk	700	Two-step modification; KMnO$_4$; modification temperature: 80 °C	71.0	0.11										Cuong and Wu[109]
Air-dried rice hulls	600	Two-step modification; KMnO$_4$; modification temperature: 80 °C	116.3	0.18	6.10						5.83			Kumarathilaka and Bundschuh[76]
Rice husk	600	Two-step modification; KMnO$_4$; modification temperature: 60 °C	42.9	0.05		40.8	0.46				11.5	6.72	6.4	Wang and Chen[79]
Corn straw	600	Two-step modification; KMnO$_4$; modification temperature: 120 °C	3.18			73.0	0.72	0.33	10.9		7.41	10.8		Yu and Qiu[33]

An enhanced reduction of Fe^{III} to Fe^{II} on the biochar surface can further support it. Similarly, Wang and Gao[67] found that the adsorption capacity of As^V by hematite (γ-Fe_2O_3) modified biochar (HPB, $429\,mg\,kg^{-1}$) was remarkably improved than the unmodified biochar ($265\,mg\,kg^{-1}$). The study also found that enhanced adsorption on the γ-Fe_2O_3 particles and complexations via surface functional groups are the possible reasons for the improved As^V removal.

Recently, Wei and Wei[68] used cotton fiber to prepare biochar fibers as supporting materials for Fe oxides which is considered can be useful for removing As. Vertical iron oxide nanoneedle array-decorated biochar fibers (Fe-NN/BFs) were finally produced at $800\,°C$ with $FeCl_3$ and Na_2SO_4. The maximum adsorption capacity of Fe-NN/BFs for As^V and As^{III} at pH 7.0 are 93.94 and $70.22\,mg\,g^{-1}$, respectively. The main mechanisms for removing As^V and As^{III} are mainly attributed to chemical adsorption and surface complexation, respectively. The results also showed that an increase in pH can enhance the complexation of As^{III} and FeOOH. Besides, the Fe-NN/BFs can lead to oxidation of As^{III} to As^V with a reduction of toxicity finally.

Besides, there are some other Fe-containing materials which have been suggested, such as zero-valent iron (Fe^0, ZVI) which has rich reactive surface sites and specific surface are for the As immobilization. Yang and Wang[69] investigated the Fe^0-modified biochar on the As adsorption by loading Fe^0 on corn stalk biochar (nZVI-BC). $NaBH_4$ was used to reduce Fe^{3+} to Fe^0 in the modification system. The results showed that the nZVI-BC has a high removal rate of As^{III} in the pH range of 3.0 to 8.0, and the maximum adsorption capacity of As^{III} is $148.5\,mg\,g^{-1}$ at both 2h and 1h. The change of the surface electrical properties of the adsorbent and the adsorbate are the main factors affecting the adsorption of As^{III}. The change of the surface electrical properties of the nZVI-BC and the adsorbate are the main factor affecting the adsorption of As^{III}. pH is the dominant factor for the As adsorption. When pH <9.0, neutral H_3AsO_3 is the main form. As^{III} can combine with the nZVI-BC and form an inner spherical complex with Fe oxides. When pH > 9.0, H_2AsO_3 is the dominant species. In such conditions, electrostatic repulsion leads to a decrease in the net As^{III} adsorption capacity. Besides, alkaline conditions can inhibit the formation of As oxide, resulting in decrease As^{III} adsorption capacity. Therefore, the optimal pH condition for the As^{III} isotherm experiment is pH 4.0.

A recent study compared different types of Fe-based biochar (Fe-oxyhydroxy sulfate, $FeCl_3$, and ZVI, referring as Biochar-FeOS, Biochar-$FeCl_3$, and Biochar-Fe, respectively) which produced at $300\,°C$ with a feedstock of rice straw.[65] All the three modified biochar have limited effects on the soil pH. The results of incubation studies showed that all the modified biochar can significantly reduce As mobility and bioavailability, which was evaluated by the extraction with $NaHCO_3$. Additionally, geochemical distribution of As were altered by the application of Fe-based biochar. The non-specifically sorbed and specifically sorbed As fractions in soil were transformed to relatively stable fractions, such as the amorphous and poorly crystalline, hydrated Fe, Al oxide-bound, and residual fractions. Biochar-FeOS showed the best performance for As immobilization, which could be due to the extra FexOy or FexOHy groups, resulting in an enhanced complexation of –OH and As. BC-FeOS also provides SO_4^{2-} which can facilitate the formation of Fe oxides and result in a better immobilization of As in soil.

3.2 Effect of Mn-modified biochar

Manganese (III/IV) oxides are strong oxidants, which can oxidize many trace metal(loid)s found in nature, including As. Previous studies showed that As^{III} can be oxidized to As^V on the surface of Mn oxides which also has a strong adsorption capacity of As^V.[24] Therefore, some studies suggested to embed Mn oxides on the sufure of biochar to achieve a better immobilization effects on As. The Mn-modified method mainly uses Mn oxides (for example, MnO_2, $KMnO_4$, etc.) or Mn complexes with other metals (for example, Ni) to modify biochar.[70,71] The Mn oxides attached to the biochar improve the oxidizing properties of biochar, which can enhance the oxygen content of the functional groups of the biochar an d oxidize As^{III} which has a higher toxicity than As^V. Besides, the Mn oxides can because of the huge surface area and many functional groups with negative charge, biochar become a cheap absorbent material what could ues to removal many heavy metals in contaminated soil.[72] As exists in the environment mainly as oxygen anions so that maybe hinders the As adsorption of biochar.[73] In the face of biochars disadvantages, Mn oxides are regarded as the outstanding materials applying to removal and oxidation of As[74] since thay get command of As mobility by the anion exchange in contaminated soil.[72] The Mn-modified method mainly uses manganese oxides (for example, MnO_2,[75] $KMnO_4$,[76] etc.) or Mn complexes with other metals (for example, Mn-Ni,[70] Mn-Zn,[77] Mn-Al[78]) to modify biochar.

Mn-modified biochar (MB) was produced by soaking corn stalks in $KMnO_4$ solution then pyrolyzed at $600\,°C$, with a final content of 7.41% Mn.[33] Compared with the pristine biochar (BC), MB has a higher total carbon content, pH value, and ash content; however, its surface area becomes smaller. The application of MBC reduced the mobility of both As^V and As^{III}, which leads to a remarkable reduction of the As contents in rice grain and root compared with the unmodified biochar. Dvca and Pcw[19] successfully synthesized MnO$_2$/biochar composite (MBC) that improved its porosity and surface chemical properties compared with the unmodified biochar, which enhanced As^{III} oxidation and immobilization on MnO$_2$, resulting in an outstanding removal of As^V (94.6%) from simulated groundwater.

There are some more examples of multiple elements were involved in the Mn modifications. For example, Wang and Gao[70] compared the As^V sorption of biochars modified by Ni/Mn and Ni/Mn-LDH). The maximum As^V adsorption capacity of NMMF and NMMB reached 0.549 and 6.52 mg g^{-1}, respectively. The increase was attributed to the additional performance of anion exchange and surface complexation in NMMB. A recent study prepared birnessite-loaded biochar (BRB) which has a high affinity to both As^{III} and As^V (3.543 and 2.412 mg g^{-1}, respectively, calculated by the Langmuir isotherm model).[79] Higher amounts of As^{III} can be removed owing to the oxidation of As^{III} was found with a release of Mn^{II}.

3.3 Effect of Fe-Mn-modified biochar

Fe-Mn binary oxides have also been proved to have high efficiency in removing both As^V and As^{III}[24], as mentioned in Sections 3.1 and 3.2. Therefore, biochar modified with Fe-Mn oxides can enhance the oxidation of As^{III} and adsorption of As to have a better performance of reduction of toxicity and mobility of As in soil.[58] Lin and Qiu[80] compared the unmodified BC which is produced at 620°C, Fe-Mn binary oxide (FMO) and Fe/Mn modified biochar (FMBC) on the As immobilization in soil. After modification, Fe-Mn oxides, Mn$_3$O$_4$, Mn$_2$O$_3$, and few MnO$_2$ mineral components are found in FMBC. The results also support that As adsorption by FMBC is mainly due to the strong affinity of As^V to the iron oxides. The main role of the Mn oxides (Mn$_3$O$_4$ and Mn$_2$O$_3$) is oxidation of As^{III} to As^V, which is associated with higher removal rates of As.

4. Effects of other modified biochar in As-contaminated soil

By modifying biochar, more targeted biochar can be developed, which can further enhance the ability of biochar to improve As contaminated soil.[6] Some of them may have a higher surface area and better pH$_{PZC}$ which could be beneficial for As adsorption.[62] Besides the most important group of Fe-modified,[81] there are some other types of modified biochars, including Mn-modified,[19] Zn-modified,[82] Si-modified,[82,83] rare earth elements-modified, acid-base-modified,[84] and other modified methods (Table 2).

4.1 Rare earth elements-modified biochar

The oxides of rare earth elements, including La, Ce, Bi, etc., have been used in water purification due to their high capacity of anion adsorption.[85] However, direct application of the rare earth elements is not easy to be recycled and disposed, which could lead to high cost and secondary contaminations. Therefore, it could be beneficial to use biochar as loading materials for the rare earth elements, which can ease the above problems and distribute the rare earth elements evenly on the surface of biochar to enhance the adsorption capacity of anions like As.[86]

Zhang and Liu[9] embedded Ce with Fe-Mn modified biochar (FMCBC) by pyrolyzing corn stover at $600\,^{\circ}C$ then mix the produced biochar with $Fe(NO_3)_3$, $KMnO_4$, and $Ce_2(CO_3)_3$, and compared with the Fe-Mn modified biochar. FMCBC improves soil pH, significantly improves soil oxidation-reduction capacity, and reduces bioavailable As. With the application of FMCBC, As can be transformed then immobilized by forming a specific or non-specific binding to amorphous hydrated and crystalline hydrated oxides. The FMCBC increased soil enzyme activities and increased tolerance of soil enzymes to PTEs, thereby reducing the bioavailability and toxicity of As.

Lin and Song[87] studied the loading of La^{3+} on Fe-Mn-modified biochar and suggested that La-O functional groups play an important role in As^{III} adsorption by forming new inner-sphere La-O-As complexation. The Fe-Mn-La-modified biochar has a maximum As^{III} adsorption capacity of $14.9\,mg\,g^{-1}$. Zhu and Yan[88] impregnated biochar with Bi_2O_3 and found that the modification boosts the micropore volume and specific surface area. The results also suggested that the embedded Bi_2O_3 significantly enhanced the adsorption of As^{III} than the enhanced surface area.

4.2 Zn-modified biochar

Zn-modified biochar can improve As removal by three main mechanisms: physical adsorption, complexation reaction, and electrostatic adsorption.[89] The physical adsorption can be enhanced by enrichment of its specific surface area and pore structure of the Zn-modified biochar.[90] Previous studies

Table 2 Physicochemical properties of some other modified biochars.

Feedstock	Temperature °C	Modification methods	BET Surface area mg^2 g^{-1}	Pore volume cm^3 g^{-1}	Pore Size nm	C %	N %	H %	O %	Fe %	Mn %	pH	pH$_{ZPC}$	References
Zn-modified biochar														
Corncob	600	Two-step modification; ZnO; modification temperature: 80 °C	35.0	39									8.09	Hua[110]
Crawfish shell	450	Two-step modification; ZnCl$_2$; modification temperature: NA	134.19	0.11		31.31	22.83	1.26	0.22					Yan and Xue[89]
Cladodes of Opuntia ficus indica	400	One-step modification; ZnCl$_2$	47.47	0.01	1.65	47.70	1.13	3.26	27.15				7.42	Rra and Mcb[111]
Raw pine cone	500	Two-step modification; Zn(NO$_3$)$_2$; modification temperature: 110 °C	11.54	0.028		71.21	0.51	3.03	20.43					Vinh and Za Fa R[82]
Acid/base-modified biochar														
Sewage sludge	350	Two-step modification; KOH; modification temperature: 50 °C	7.9									8.4	3.4	Wongrod and Simon[84]
Straw	350	Two-step modification; KOH; modification temperature: 800 °C	1360.99	0.81										Xiong and Tong[112]

Chicken litter	250	Two-step modification; H_3PO_4; modification temperature: $450\,^{\circ}C$	789			52.2		1.75	28.4	8.37		3.9	Lima and Ro[113]
Grape Seeds-Derived	400	Two-step modification; HNO_3; modification temperature: $60\,^{\circ}C$	348	0.12		69.0	3.1	2.4	14.6	4.7			Friták and Vladimír[114]
Si-modified biochar													
Bamboo	300	Two-step modification; K_2SiO_3; modification temperature: $100\,^{\circ}C$				54.7	0.36	3.84	41.1			8.84	Zama and Reid[83]
Rice husk	700	Two-step modification; $CaSiO_3$; modification temperature: $80\,^{\circ}C$	182.3	0.20	4.32	53.65			16.87			10.41	Herath and Zhao[50]
Ti-modified biochar													
Raw corncobs	550	Two-step modification; butyl titanate; modification temperature: $100\,^{\circ}C$	450.43	0.05	0.46								Mingke and Luo[115]
Bi-modified biochar													
Wheat straw	500	Two-step modification; Bi_2O_3; modification temperature: $105\,^{\circ}C$	190.4	0.10	2.00								Zhu and Yan[88]

showed that well-developed pore structures can improve the physical adsorption behaviors of As. In addition, the complexation reaction can be enriched by the activated sites caused by Zn oxide.[91] Li and Ronghua[92] suggested that Zn-O-As complexes can be formed to have a better immobilization which is caused by the exchange of Zn-OH hydroxyl ligands of the ZnO-modified biochar. Besides, electrostatic adsorption of As^V is also enhanced by the Zn modification which leads to higher positively charged surface and surface zeta potential.[82,92]

Cruz and Mondal[90] impregnated ZnO on corn cob biochar (CC-B) to improve the As^V adsorption capacity. A maximum adsorption capacity of $25.9 \, mg \, g^{-1}$ of As^V was found. The authors also suggested that the CC-B showed better structural properties to cause a better impregnation of ZnO. Similarly, Yan and Xue[89] used crayfish shell biochar modified by ZnO nanoparticles, which has a maximum As^V sorption of $17.2 \, mg \, g^{-1}$ caused by the improved electrostatic attraction.

4.3 Si-modified biochar

Silicon is the second most abundant element in the earth's crust and plays many significant roles in the growth of plants. Salicyclic acid (H_4SiO_4) is the effective plant form of Si in soil. Arsenous acid (H_3AsO_3), which is the main form in plants, has a common uptake pathway with H_4SiO_4.[93] Previous studies revealed that Si has been widely used to reduce the toxicity of As in plants, especially for rice in the As-contaminated paddy soil.[94] Therefore, some researchers suggested embedding Si in biochar to control the bioaccumulation capacity of As in crops and increase the crop yield.[95] In this study, the Si-modified biochar can decrease the As uptake of plants and As mobilization though the As^{III} concentration has been promoted by 64.4% by the biochar, which could be due to the transport pathway of As in spinach (*Spinacia oleracean*) has been occupied by Si provided by the modified biochar. It is also suggested that the higher amount of Si of the modified biochar, the lower amounts of the uptake of H_3AsO_3. Herath, Zhao[50] loaded Si-fertilizer on rice husk biochar and found it can remarkably decrease As concentration in porewater by 62.5%, compared with the control soil. It also significantly reduced As^{III} concentration by 76.0% in the rhizosphere during the tillering stage. The possible mechanisms for the promoted As immobilization are (1) enhanced adsorption on the surface of Si-ferrihydrite complexes that formed by the Si-modified biochar and (2) formed ferrihydrite layer on the biochar surface which can oxidize the promoted microbe mediated As.[50]

4.4 Acid-base-modified biochar

The physical structure and chemical properties of biochar or feedstock of the biochar can be altered by modification of acidic and alkaline solvents, which leads to enhancement of adsorption performance of As.[6]

Acidic solutions remove metal residues and impurities on the surface of biochar and form large amounts of acidic functional groups, such as carboxyl groups, which can improve the As immobilization.[60] Besides, with the activation by the acidic solutions, micropores on the surface of biochar have been well developed and enhance the As adsorption.[96] For example, phosphoric acid has been used in modification for improving the As immobilization by biochar. Phosphorus and As are belong to the Group VA elements and have similar chemical structures and properties. Due to the smaller molecular and higher valence of phosphoric acid, it has higher adsorption capacity on soil than As acids, which leads to isomorphous replacements of phosphate and arsenate which easily occurs in soil.[97,98]

Similar to acidic solutions, alkaline solutions are helpful to develop a higher surface area and enrich the pore volume of the biochar. Besides, alkaline-modified biochar can increase the amount of O-containing functional groups and alkalinity to improve the affinity to As. In addition, it is found that alkaline-modified biochar can improve the oxidation of As^{III} to As^{V} on its surface.[84] In general, Group I alkali solutions, such as KOH and NaOH, are commonly used to activate biochar.[6] Previous studies showed that the impregnation of K ion caused the formation of different K-containing species which leads to expansion and development of existing pores of the biochar and might increase its pore volume. Goswami, Shim[99] have shown that KOH-activated biochar increases the surface area and the diameter of small pores and clears blocked pores.

5. The way forward

Biochar, as a cost-effective and green sorbent and amendment, has been proposed widely for removing/immobilizing As in environmental studies. Modification of biochar is proposed to improve the affinity to As, stability under varied environmental conditions, and surface area and pore structure of its surface by different methods which can be divided into several groups: Fe/Mn-modified biochar and modified biochar by other metals (including Zn, Si, and rare earth elements), and acid/based-modified biochar. After a review of literature, we found that most of the studies were still at the beginning stage of removing As in water but relatively few studies were conducted in soils. The possible reasons are soil conditions are far more complicated than aqueous solutions. Many factors in soil, for example, dissolved organic matter, phosphate, change of pH and redox conditions, microbial activities, and interaction of plant-soil system, can affect the complexation/sorption of As on the biochar surface, which are difficult to be examined in detail in one single research. Moreover, after biochar immobilization, how to collect and separate the biochar from soil particles could be limitations in application. The stability of As immobilization by modified biochar should be further examined in future studies. Rare earth elements are proposed to be used in the embedment of biochar; however, the cost should be considered for application in large scales. We also found that most of the studies focus on the removal of inorganic As, while the studies for the organic As, for example, DMAA and MMAA, are normally limited, though they have certain toxicity. The current studies also suggested that a combination of different modification methods can get better adsorption efficiency, for example, binary of Fe and Mn oxides. Therefore, development of new modification methods for biochar and evaluate its stability for removing/immobilizing As in water and soil are still urgently needed.

References

1. Lin L, Zhou S, Huang Q, Huang Y, Qiu W, Song Z. Capacity and mechanism of arsenic adsorption on red soil supplemented with ferromanganese oxide-biochar composites. *Environ Sci Pollut Res Int* 2018;**25**(20):20116–24.
2. Fitz WJ, Wenzel WW. Arsenic transformations in the soil–rhizosphere–plant system: fundamentals and potential application to phytoremediation. *J Biotechnol* 2002;**99** (3):259–78.
3. Mandal BK, Suzuki K. Arsenic round the world: a review. *Talanta* 2002;**58**(1):201–35.
4. Meinrat O. Andreae. Distribution and speciation of arsenic in natural waters and some marine algae. *Deep-Sea Res* 1978;**25**(4):391–402.
5. Takamatsu T, Aoki H, Yoshida T. Determination of arsenate, arsenite monomethylarsonate and dimethylarsinate in soil polluted with arsenic. *Soil Sci* 1982;**133** (4):239–46.
6. Amen R, Bashir H, Bibi I, Shaheen SM, Niazi NK, Shahid M, et al. A critical review on arsenic removal from water using biochar-based sorbents: the significance of modification and redox reactions. *Chem Eng J* 2020;**396**:125195.
7. Niazi NK, Bibi I, Shahid M, Ok YS, Burton ED, Wang H, et al. Arsenic removal by perilla leaf biochar in aqueous solutions and groundwater: an integrated spectroscopic and microscopic examination. *Environ Pollut* 2018;**232**:31–41.
8. Smedley PL, Kinniburgh DG. A review of the source, behaviour and distribution of arsenic in natural waters. *Appl Geochem* 2002;**17**(5):517–68.
9. Zhang G, Liu X, Gao M, Song Z. Effect of Fe–Mn–Ce modified biochar composite on microbial diversity and properties of arsenic-contaminated paddy soils. *Chemosphere* 2020;**250**:126249.
10. Asere TG, Stevens CV, Du Laing G. Use of (modified) natural adsorbents for arsenic remediation: a review. *Sci Total Environ* 2019;**676**:706–20.

11. Podgorski J, Berg M. Global threat of arsenic in groundwater. *Science* 2020;**368** (6493):845–50.
12. Zhao FJ, Ma Y, Zhu YG, Tang Z, McGrath SP. Soil contamination in China: current status and mitigation strategies. *Environ Sci Technol* 2015;**49**(2):750–9.
13. E.P. Agency. *Arsenic treatment Technologies for Soil, waste, and water*; 2002. p. 9–16.
14. Sarkar A, Paul B. The global menace of arsenic and its conventional remediation - a critical review. *Chemosphere* 2016;**158**:37–49.
15. Zakhar R, Derco J, Čacho F. An overview of main arsenic removal technologies. *Acta Chim Slov* 2018;**11**(2):107–13.
16. Garelick H, Dybowska A, Valsami-Jones E, Priest N. Remediation Technologies for Arsenic Contaminated Drinking Waters (9 pp). *J Soil Sediment* 2005;**5**(3):182–90.
17. Alka S, Shahir S, Ibrahim N, Ndejiko MJ, Manan FA. Arsenic removal technologies and future trends: a Mini review. *J Clean Prod* 2020;**278**(2):123805.
18. Vithanage M, Herath I, Joseph S, Bundschuh J, Bolan N, Ok YS, et al. Interaction of arsenic with biochar in soil and water: a critical review. *Carbon* 2017;**113**:219–30.
19. Dvca B, Pcw A, Lic A, Chha C. Active MnO 2 /biochar composite for efficient As(III) removal: insight into the mechanisms of redox transformation and adsorption. *Water Res* 2021;**188**:116495.
20. Agency EP. *Arsenic treatment Technologies for Soil, waste, and water*; 2002.
21. Srivastava M, Ma LQ, Santos JA. Three new arsenic hyperaccumulating ferns. *Sci Total Environ* 2006;**364**(1–3):24–31.
22. Kim EJ, Yoo JC, Baek K. Arsenic speciation and bioaccessibility in arsenic-contaminated soils: sequential extraction and mineralogical investigation. *Environ Pollut* 2014;**186**:29–35.
23. Niazi NK, Singh B, Shah P. Arsenic speciation and phytoavailability in contaminated soils using a sequential extraction procedure and XANES spectroscopy. *Environ Sci Technol* 2011;**45**(17):7135–42.
24. Ying SC, Kocar BD, Fendorf S. Oxidation and competitive retention of arsenic between iron- and manganese oxides. *Geochim Cosmochim Acta* 2012;**96**:294–303.
25. Qiao JT, Liu TX, Wang XQ, Li FB, Lv YH, Cui JH, et al. Simultaneous alleviation of cadmium and arsenic accumulation in rice by applying zero-valent iron and biochar to contaminated paddy soils. *Chemosphere* 2018;**195**:260–71.
26. Frohne T, Rinklebe J, Diaz-Bone RA. Contamination of floodplain soils along the Wupper River, Germany, with As, Co, Cu, Ni, Sb, and Zn and the impact of pre-definite redox variations on the mobility of these elements. *Soil Sediment Contam* 2014;**23**(7):779–99.
27. Frohne T, Rinklebe J, Diaz-Bone RA, Du Laing G. Controlled variation of redox conditions in a floodplain soil: impact on metal mobilization and biomethylation of arsenic and antimony. *Geoderma* 2011;**160**(3–4):414–24.
28. Melby ES, Mensch AC, Lohse SE, Hu D, Orr G, Murphy CJ, et al. Formation of supported lipid bilayers containing phase-segregated domains and their interaction with gold nanoparticles. *Environ Sci Nano* 2016;**3**(1):45–55.
29. Wan X, Lei M, Chen T. Review on remediation technologies for arsenic-contaminated soil. *Front Environ Sci Eng* 2019;**14**(2).
30. Palansooriya KN, Shaheen SM, Chen SS, Tsang D, Yong SO. Soil amendments for immobilization of potentially toxic elements in contaminated soils: a critical review. *Environ Int* 2020;**134**:105046.
31. Bolan N, Kunhikrishnan A, Thangarajan R, Kumpiene J, Park J, Makino T, et al. Remediation of heavy metal(loid)s contaminated soils- -to mobilize or to immobilize? *J Hazard Mater* 2014;**266**:141–66.
32. Kumpiene J, Lagerkvist A, Maurice C. Stabilization of As, Cr, Cu, Pb and Zn in soil using amendments—a review. *Waste Manag* 2008;**28**(1):215–25.

33. Yu Z, Qiu W, Wang F, Lei M, Wang D, Song Z. Effects of manganese oxide-modified biochar composites on arsenic speciation and accumulation in an indica rice (*Oryza sativa* L.) cultivar. *Chemosphere* 2017;**168**:341–9.

34. Beiyuan J, Awad YM, Beckers F, Tsang DC, Ok YS, Rinklebe J. Mobility and phytoavailability of As and Pb in a contaminated soil using pine sawdust biochar under systematic change of redox conditions. *Chemosphere* 2017;**178**:110–8.

35. Beiyuan J, Li JS, Tsang DCW, Wang L, Poon CS, Li XD, et al. Fate of arsenic before and after chemical-enhanced washing of an arsenic-containing soil in Hong Kong. *Sci Total Environ* 2017. 599–600:679–88.

36. Pan D, Liu C, Yu H, Li F. A paddy field study of arsenic and cadmium pollution control by using iron-modified biochar and silica sol together. *Environ Sci Pollut Res Int* 2019;**26**(24):24979–87.

37. Ippolito JA, Cui L, Kammann C, Wrage-Mönnig N, Estavillo JM, Fuertes-Mendizabal T, et al. Feedstock choice, pyrolysis temperature and type influence biochar characteristics: a comprehensive meta-data analysis review. *Biochar* 2020;**2**(4):421–38.

38. Figueredo N, Costa L, Melo L, Siebeneichler DEA, Tronto J. Characterization of biochars from different sources and evaluation of release of nutrients and contaminants. *Revciêncagron* 2017;**48**(3).

39. Bolan N, Hoang SA, Beiyuan J, Gupta S, Hou D, Karakoti A, et al. Multifunctional applications of biochar beyond carbon storage. *Int Mater Rev* 2021;1–51.

40. Biswas BK, Inoue J-i, et al. Adsorptive removal of As(V) and As(III) from water by a Zr(IV)-loaded orange waste gel. *J Hazard Mater* 2008;**154**(1–3):1066–74.

41. Gupta K, Maity A, Ghosh UC. Manganese associated nanoparticles agglomerate of iron(III) oxide: synthesis, characterization and arsenic(III) sorption behavior with mechanism. *J Hazard Mater* 2010;**184**(1–3):832–42.

42. Song XD, Xue XY, Chen DZ, He PJ, Dai XH. Application of biochar from sewage sludge to plant cultivation: influence of pyrolysis temperature and biochar-to-soil ratio on yield and heavy metal accumulation. *Chemosphere* 2014;**109**:213–20.

43. Lin L, Song Z, Liu X, Khan ZH, Qiu W. Arsenic volatilization in flooded paddy soil by the addition of Fe-Mn-modified biochar composites. *Sci Total Environ* 2019;**674**:327–35.

44. Ahmad M, Rajapaksha AU, Lim JE, Zhang M, Bolan N, Mohan D, et al. Biochar as a sorbent for contaminant management in soil and water: a review. *Chemosphere* 2014;**99**:19–33.

45. Luke B, Marta M. The immobilisation and retention of soluble arsenic, cadmium and zinc by biochar. *Environ Pollut* 2011;**159**(2):474–80.

46. Qiao JT, Li XM, Li FB. Roles of different active metal-reducing bacteria in arsenic release from arsenic-contaminated paddy soil amended with biochar. *J Hazard Mater* 2018;**344**:958–67.

47. Zhang M, Gao B, Varnoosfaderani S, Hebard A, Yao Y, Inyang M. Preparation and characterization of a novel magnetic biochar for arsenic removal. *Bioresour Technol* 2013;**130**:457–62.

48. Beesley L, Marmiroli M. The immobilisation and retention of soluble arsenic, cadmium and zinc by biochar. *Environ Pollut* 2011;**159**(2):474–80.

49. Beesley L, Dickinson N. Carbon and trace element mobility in an urban soil amended with green waste compost. *J Soil Sediment* 2009;**10**(2):215–22.

50. Herath I, Zhao F-J, Bundschuh J, Wang P, Wang J, Ok YS, et al. Microbe mediated immobilization of arsenic in the rice rhizosphere after incorporation of silica impregnated biochar composites. *J Hazard Mater* 2020;**398**:123096.

51. Choppala G, Bolan N, Kunhikrishnan A, Bush R. Differential effect of biochar upon reduction-induced mobility and bioavailability of arsenate and chromate. *Chemosphere* 2016;**144**:374–81.

52. Yuan Y, Bolan N, Prevoteau A, Vithanage M, Biswas JK, Ok YS, et al. Applications of biochar in redox-mediated reactions. *Bioresour Technol* 2017;**246**:271–81.
53. El-Naggar A, Shaheen SM, Ok YS, Rinklebe J. Biochar affects the dissolved and colloidal concentrations of Cd, Cu, Ni, and Zn and their phytoavailability and potential mobility in a mining soil under dynamic redox-conditions. *Sci Total Environ* 2018;**624**:1059–71.
54. Chen Z, Wang Y, Xia D, Jiang X, Fu D, Shen L, et al. Enhanced bioreduction of iron and arsenic in sediment by biochar amendment influencing microbial community composition and dissolved organic matter content and composition. *J Hazard Mater* 2016;**311**:20–9.
55. Hartley W, Dickinson NM, Riby P, Lepp NW. Arsenic mobility in brownfield soils amended with green waste compost or biochar and planted with Miscanthus. *Environ Pollut* 2009;**157**(10):2654–62.
56. Beesley L, Moreno-Jimenez E, Gomez-Eyles JL, Harris E, Robinson B, Sizmur T. A review of biochars' potential role in the remediation, revegetation and restoration of contaminated soils. *Environ Pollut* 2011;**159**(12):3269–82.
57. Agrafioti E, Kalderis D, Diamadopoulos E. Ca and Fe modified biochars as adsorbents of arsenic and chromium in aqueous solutions. *J Environ Manage* 2014;**146**:444–50.
58. Ma L, Cai D, Tu S. Arsenite simultaneous sorption and oxidation by natural ferruginous manganese ores with various ratios of Mn/Fe. *Chem Eng J* 2020;**382**:123040.
59. Lin L, Li Z, Liu X, Qiu W, Song Z. Effects of Fe-Mn modified biochar composite treatment on the properties of As-polluted paddy soil. *Environ Pollut* 2019;**244**:600–7.
60. Alkurdi SSA, Herath I, Bundschuh J, Al-Juboori RA, Vithanage M, Mohan D. Biochar versus bone char for a sustainable inorganic arsenic mitigation in water: what needs to be done in future research? *Environ Int* 2019;**127**:52–69.
61. Mondal S, Aikat K, Halder G. Biosorptive uptake of arsenic(V) by steam activated carbon from mung bean husk: equilibrium, kinetics, thermodynamics and modeling. *Appl Wat Sci* 2017;**7**(8):4479–95.
62. Benis KZ, Damuchali AM, Soltan J, Mcphedran KN. Treatment of aqueous arsenic – a review of biochar modification methods. *Sci Total Environ* 2020;**739**:139750.
63. Kim H-B, Kim J-G, Kim T, Alessi DS, Baek K. Mobility of arsenic in soil amended with biochar derived from biomass with different lignin contents: relationships between lignin content and dissolved organic matter leaching. *Chem Eng J* 2020;**393**:124687.
64. Suvasis D, Hering JG. Comparison of arsenic(V) and arsenic(III) sorption onto iron oxide minerals: implications for arsenic mobility. *Environ Sci Technol* 2003;**37**(18): 4182–9.
65. Wu C, Cui M, Xue S, Li W, Huang L, Jiang X, et al. Remediation of arsenic-contaminated paddy soil by iron-modified biochar. *Environ Sci Pollut Res Int* 2018;**25**(21):20792–801.
66. Zhu S, Zhao J, Zhao N, Yang X, Chen C, Shang J. Goethite modified biochar as a multifunctional amendment for cationic cd(II), anionic as(III), roxarsone, and phosphorus in soil and water. *J Clean Prod* 2020;**247**:119579.
67. Wang S, Gao B, Zimmerman AR, Li Y, Ma L, Harris WG, et al. Removal of arsenic by magnetic biochar prepared from pinewood and natural hematite. *Bioresour Technol* 2015;**175**:391–5.
68. Wei Y, Wei S, Liu C, Chen T, Tang Y, Ma J, et al. Efficient removal of arsenic from groundwater using iron oxide nanoneedle array-decorated biochar fibers with high Fe utilization and fast adsorption kinetics. *Water Res* 2019;**167**:115107.
69. Yang D, Wang L, Li Z, Tang X, He M, Yang S, et al. Simultaneous adsorption of Cd(II)andAs(III)by a novel biochar-supported nanoscale zero-valent iron in aqueous systems. *Sci Total Environ* 2020;**708**:134823.
70. Wang S, Gao B, Li Y. Enhanced arsenic removal by biochar modified with nickel (Ni) and manganese (Mn) oxyhydroxides. *J Ind Eng Chem* 2016;**37**:361–5.

71. Trakal L, Michalkova Z, Beesley L, Vitkova M, Ourednicek P, Barcelo AP, et al. AMOchar: amorphous manganese oxide coating of biochar improves its efficiency at removing metal(loid)s from aqueous solutions. *Sci Total Environ* 2018;**625**:71–8.
72. Yu Z, Zhou L, Huang Y, Song Z, Qiu W. Effects of a manganese oxide-modified biochar composite on adsorption of arsenic in red soil. *J Environ Manage* 2015;**163**:155–62.
73. Joseph SD, Camps-Arbestain M, Lin Y, Munroe P, Chiaa CH. An investigation into the reactions of biochar in soil. *Soil Res* 2010;**48**(7):501–15.
74. Lenoble V, Laclautre C, Serpaud B, Deluchat V. B Ollinger JC. As(V) retention and As(III) simultaneous oxidation and removal on a MnO2-loaded polystyrene resin. *Sci Total Environ* 2004;**326**(1–3):197–207.
75. Imran M, Iqbal MM, Iqbal J, Shah NS, Khan ZUH, Murtaza B, et al. Synthesis, characterization and application of novel MnO and CuO impregnated biochar composites to sequester arsenic (As) from water: Modeling, thermodynamics and reusability. *J Hazard Mater* 2021;**401**:123338.
76. Kumarathilaka P, Bundschuh J, Seneweera S, Ok YS. Rice genotype's responses to arsenic stress and cancer risk: the effects of integrated birnessite-modified rice hull biochar-water management applications. *Sci Total Environ* 2021;**768**:144531.
77. Niu Z, Feng W, Huang H, Wang B, Su S. Green synthesis of a novel Mn–Zn ferrite/biochar composite from waste batteries and pine sawdust for Pb2+ removal. *Chemosphere* 2020;**252**:126529.
78. Peng G, Jiang S, Wang Y, Zhang Q, Cao Y, Sun Y, et al. Synthesis of Mn/Al double oxygen biochar from dewatered sludge for enhancing phosphate removal. *J Clean Prod* 2020;**251**.
79. Wang HY, Chen P, Zhu YG, Cen K, Sun GX. Simultaneous adsorption and immobilization of As and Cd by birnessite-loaded biochar in water and soil. *Environ Sci Pollut Res Int* 2019;**26**(9):8575–84.
80. Lin L, Qiu W, Wang D, Huang Q, Song Z, Chau HW. Arsenic removal in aqueous solution by a novel Fe-Mn modified biochar composite: characterization and mechanism. *Ecotoxicol Environ Saf* 2017;**144**:514–21.
81. Zhang Y, Fan J, Fu M, Ok YS, Hou Y, Cai C. Adsorption antagonism and synergy of arsenate(V) and cadmium(II) onto Fe-modified rice straw biochars. *Environ Geochem Health* 2019;**41**(4):1755–66.
82. Vinh NV, Za Fa RM, Behera SK, Park HS. Arsenic(III) removal from aqueous solution by raw and zinc-loaded pine cone biochar: equilibrium, kinetics, and thermodynamics studies. *Int J Environ Sci Technol* 2015;**12**(4):1283–94.
83. Zama E, Reid BJ, Sun GX, Yuan HY, Li XM, Zhu YG. Silicon (Si) biochar for the mitigation of arsenic (As) bioaccumulation in spinach (Spinacia oleracean) and improvement in the plant growth. *J Clean Prod* 2018;**189**(10):386–95.
84. Wongrod S, Simon S, Hullebusch EV, Lens P, Guibaud G. Changes of sewage sludge digestate-derived biochar properties after chemical treatments and influence on As(III and V) and Cd(II) sorption. *Int Biodeter Biodegr* 2018;**135**:96–102.
85. Wang Z, Shen D, Shen F, Li T. Phosphate adsorption on lanthanum loaded biochar. *Chemosphere* 2016;**150**:1–7.
86. Biswas BK, Inoue K, Ghimire KN, Kawakita H, Ohto K, Harada H. Effective removal of arsenic with lanthanum(III)- and cerium(III)-loaded Orange waste gels. *Sep Sci Technol* 2008;**43**(8):2144–65.
87. Lin L, Song Z, Khan ZH, Liu X, Qiu W. Enhanced As(III) removal from aqueous solution by Fe-Mn-La-impregnated biochar composites. *Sci Total Environ* 2019;**686**:1185–93.
88. Zhu N, Yan T, Qiao J, Cao H. Adsorption of arsenic, phosphorus and chromium by bismuth impregnated biochar: Adsorption mechanism and depleted adsorbent utilization. *Chemosphere* 2016;**164**:32–40.

89. Yan J, Xue Y, Long L, Zeng Y, Hu X. Adsorptive removal of As(V) by crawfish shell biochar: batch and column tests. *Environ Sci Pollut Res* 2018;**25**:34674–83.
90. Cruz G, Mondal D, Rimaycuna J, Soukup K, Lang J. Agrowaste derived biochars impregnated with ZnO for removal of arsenic and lead in water. *J Environ Chem Eng* 2020;**8**(3):103800.
91. Haddad MY, Alharbi HF. Enhancement of heavy metal ion adsorption using electrospun polyacrylonitrile nanofibers loaded with ZnO nanoparticles. *J Appl Polym Sci* 2019;**136**(11).
92. Ronghua L, Jim W, Gaston J, et al. An overview of carbothermal synthesis of metal-biochar composites for the removal of oxyanion contaminants from aqueous solution. *Carbon* 2018;**129**:674–87.
93. Tubana BS, Babu T, Datnoff LE. A review of silicon in soils and plants and its role in US agriculture: history and future perspectives. *Soil Sci* 2016;**1**.
94. Seyfferth AL, Fendorf S. Silicate mineral impacts on the uptake and storage of arsenic and plant nutrients in Rice (Oryza sativa L.). *Environ Sci Technol* 2012;**46**(24):13176–83.
95. Zama EF, Reid BJ, Arp H, Sun GX, Yuan HY, Zhu YG. Advances in research on the use of biochar in soil for remediation: a review. *J Soil Sediment* 2018;**18**(7):2433–50.
96. Sizmur T, Fresno T, Akgül G, Frost H, Jiménez E. Biochar modification to enhance sorption of inorganics from water. *Bioresour Technol* 2017;**246**:34–47.
97. Violante A, Pigna M. Competitive sorption of arsenate and phosphate on different clay minerals and soils. *Soil Sci Soc Am J* 2002;**66**(6):1788–96.
98. Sun H, Hockaday WC, Masiello CA, Zygourakis K. Multiple controls on the chemical and physical structure of biochars. *Ind Eng Chem Res* 2012;**51**(9):3587–97.
99. Goswami R, Shim J, Deka S, Kumari D, Kataki R, Kumar M. Characterization of cadmium removal from aqueous solution by biochar produced from Ipomoea fistulosa at different pyrolytic temperatures. *Ecol Eng* 2016;**97**:444–51.
100. Kim J, Song J, Lee S-M, Jung J. Application of iron-modified biochar for arsenite removal and toxicity reduction. *J Ind Eng Chem* 2019;**80**:17–22.
101. Hu X, Ding Z, Zimmerman AR, Wang S, Gao B. Batch and column sorption of arsenic onto iron-impregnated biochar synthesized through hydrolysis. *Water Res* 2015;**68**:206–16.
102. Fan J, Xu X, Ni Q, Lin Q, Fang J, Chen Q, et al. Enhanced as (V) removal from aqueous solution by biochar prepared from Iron-impregnated corn straw. *J Chem* 2018;**2018**:1–8.
103. Fan J, Chen X, Xu Z, Xu X, Zhao L, Qiu H, et al. One-pot synthesis of nZVI-embedded biochar for remediation of two mining arsenic-contaminated soils: arsenic immobilization associated with iron transformation. *J Hazard Mater* 2020;**398**:122901.
104. Alchouron J, Navarathna C, Chludil HD, Dewage NB, Perez F, Hassan EB, et al. Assessing South American Guadua chacoensis bamboo biochar and Fe3O4 nanoparticle dispersed analogues for aqueous arsenic(V) remediation. *Sci Total Environ* 2020;**706**:135943.
105. Wang S, Gao B, Li Y, Creamer AE, He F. Adsorptive removal of arsenate from aqueous solutions by biochar supported zero-valent iron nanocomposite: Batch and continuous flow tests. *J Hazard Mater* 2017;**322**(Pt A):172–81.
106. Cho DW, Yoon K, Kwon EE, Biswas JK, Song H. Fabrication of magnetic biochar as a treatment medium for as(V) via pyrolysis of FeCl3-pretreated spent coffee ground. *Environ Pollut* 2017;**229**:942–9.
107. Guo J, Yan C, Luo Z, Fang H, Hu S, Cao Y. Synthesis of a novel ternary HA/Fe-Mn oxides-loaded biochar composite and its application in cadmium(II) and arsenic(V) adsorption. *J Environ Sci (China)* 2019;**85**:168–76.
108. Wang S, Gao B, Li Y, Mosa A, Zimmerman AR, Ma LQ, et al. Manganese oxide-modified biochars: preparation, characterization, and sorption of arsenate and lead. *Bioresour Technol* 2015;**181**:13–7.

109. Cuong DV, Wu PC, Chen LI, Hou CH. Active MnO2/biochar composite for efficient as(III) removal: insight into the mechanisms of redox transformation and adsorption. *Water Res* 2021;**188**:116495.
110. Hua J. Synthesis and characterization of gold nanoparticles (AuNPs) and ZnO decorated zirconia as a potential adsorbent for enhanced arsenic removal from aqueous solution. *J Mol Struct* 2020;**1228**(6):129482.
111. Rra B, Mcb C, Rvb C, et al. Preparation of a new adsorbent for the removal of arsenic and its simulation with artificial neural network-based adsorption models. *J Environ Chem Eng* 2020;**8**(4):103928.
112. Xiong Y, Tong Q, Shan W, Xing Z, Wang Y, Wen S, et al. Arsenic transformation and adsorption by iron hydroxide/manganese dioxide doped straw activated carbon. *Appl Surf Sci* 2017;**416**:618–27.
113. Lima IM, Ro KS, Reddy GB, Boykin DL, Klasson KT. Efficacy of chicken litter and wood biochars and their activated counterparts in heavy metal clean up from wastewater. *Agri* 2015;**5**(3):806–25.
114. Friták V, Moreno-Jimenéz E, Fresno T, et al. Effect of physical and chemical activation on arsenic sorption separation by grape seeds-derived biochar. *Separations* 2018;**5**(4):59.
115. Luo M, Lin H, Bing Y, et al. Efficient simultaneous removal of cadmium and arsenic in aqueous solution by titanium-modified ultrasonic biochar. *Bioresour Technol* 2019;**284**:333–9.

CHAPTER SIX

Biochar for modification of manure properties

Sören Thiele-Bruhn[a,*] and Anastasiah N. Ngigi[b]

[a]Soil Science, University of Trier, Trier, Germany
[b]Department of Chemistry, Multimedia University of Kenya, Nairobi, Kenya
*Corresponding author: e-mail address: thiele@uni-trier.de

Contents

1. Introduction	138
2. Biochar as a feed supplement and amendment to bedding material	140
3. Biochar as a bulking agent for manure handling, storage and further use	142
4. Biochar for improving manure composting	144
4.1 Composting phases	145
4.2 Effects on microbial community	146
4.3 Chemical properties of composted organic matter and nutrient status	146
4.4 Gaseous losses	147
5. Biochar effects on anaerobic digestion and biogas formation	148
6. Nutrient availability and delivery in soil of biochar amended manure	150
7. Impact of biochar on the fate and effects of micropollutants in manure	157
8. Summary and conclusion	161
References	162

Abstract

Available current literature is reviewed on the use of biochar as an additive to manure or alike organic waste materials. The diverse manure substrates are valuable organic fertilizers and can be further processed through composting and anaerobic digestion. However, physical properties, chemical composition and microbial population are hardly optimal but require adjustment and modification. To this end, biochar is a well-suited amendment. Different uses of biochar are explained. These are the use as feed additive for livestock and as an amendment to bedding material. Additionally, biochar is used as a bulking agent for manure handling, storage and further use, to modify manure composting as well as anaerobic digestion for biogas formation. Not last, the properties of biochar to retain nutrients in manure, thus enabling the subsequent use as slow release fertilizer, and for the immobilization of contaminants and pathogenic organisms are presented.

Keywords: Biochar-manure mixture, Digestion, Biogas, Bulking agent, Composting, Detoxification, Feed additive, Greenhouse gases, Pathogens, Slow release fertilizer

---

Advances in Chemical Pollution, Environmental Management and Protection, Volume 7
ISSN 2468-9289
https://doi.org/10.1016/bs.apmp.2021.08.006

Copyright © 2021 Elsevier Inc.
All rights reserved.

1. Introduction

Manure is an inherent by-product of all livestock husbandry. The vast majority of manure is produced by cattle, pigs and poultry[1–3] that are kept either on pastureland or in confined indoor or outdoor feeding operations. In confined operations, manure is collected and stored before it is further used. The estimated worldwide annual production amounts to several billion Mg per year.[1–3] It is to be noted that the focus of this literature research is on the treatment of manure. However, most of the following information also applies to sewage sludge and (un)treated wastewater.

Manure is a complex mixture of a multitude of particulate, suspended, and dissolved organic and inorganic substances with water.[4,5] It comprises all the excreta, i.e., feces and urine (or uric acid in case of poultry), mixed with other constituents in different amounts. The latter can be bedding materials, residues of fodder, hair, drinking as well as cleaning and rainwater, soil and other materials.[1,6] Consequently, the dry mass content of typical manures can vary from 5 to 35% (extremes from <1% to 80%).[6,7] Depending on the portions of water and solids (especially bedding material), different types of manure are distinguished (rough indication of dry mass content in parentheses):

   (i) farmyard or bedding manure (20–80%), stabilized by high proportions of bedding material, especially straw, wood shavings and sawdust;
  (ii) solid manure were feces are separately collected (25–80%);
 (iii) semi-liquid manure or slurry as mixtures of feces and urine in varying ratios (2–10%);
  (iv) liquid manure with different amounts of additional water, or the separately collected urine (<2%).[6]

The individual manure composition and properties depend on the animal species, age level, diet, husbandry conditions, health status, as well as the techniques and conditions of manure treatment and storage. In case of confined indoor operations, manure is collected in cellars underneath slatted floors, in tanks or lagoons that can be protected from weather through buildings, roofs and covers or they are left open and exposed to the weather.[8]

Manure is not a waste product but a valuable substrate.[9] Mostly, it is further used as an organic fertilizer in agriculture, since it contains substantial amounts of organic matter as well as macro- and micronutrients, in particular N, P and K.[2,7,10] Regular application of manure to soil not only increases the nutrient level and availability but also has positive effects on other soil properties such as organic matter content, soil pH, cation exchange capacity, soil structure, porosity and related soil properties such as water holding capacity.[2,11] Additionally, the application of organic amendments changes the composition of the soil organic matter and leads to the enrichment of specific compound classes in the different particle size fractions of soils.[12,13] Hence, the application of manure to agricultural land is common practice in agricultural production systems, and has been used for millennia since the Neolithic revolution.[14] An alternative (intermediate) use of manure is the utilization for biogas (methane) production through anaerobic fermentation.[15,16] The resulting fermented biogas digestate can also be used as soil fertilizer.[17,18]

However, the land application of manure and biogas digestate or composted manure may also have negative effects, leading to non–point-source pollution of agricultural soils. Adverse environmental impacts are especially related to overfertilization with soluble and particulate P, and losses of N through leaching and gaseous efflux.[19] This is due, among other things, to imbalances between the nutrient content in manure and the nutrient demand of crop plants.[10] Additionally, potentially toxic elements, organic micro-pollutants, especially veterinary medicine drugs, and pathogenic microorganisms can be contained in manure and are spread onto agricultural soil when used as fertilizer.[2,20,21] Using manure for biogas fermentation and composting is also associated with problems such as the unintentional release of greenhouse gases and nutrient losses. Additionally, reduced air quality often occurs by emission of volatile organic compounds during manure collection, storage, treatment and in consequence of spreading manure onto soil.[19] All these problems aggravate, when manure utilization practices are first of all targeted to dispose of excessive loads of manure in a way that best management practices such as maintaining nutrient balances on the field and farm level are neglected.[9]

Biochar can be used as an admixture to facilitate positive properties of manure and to mitigate the outlined problems. Biochar with its known properties that depend on feedstock and preparation technique[22,23] affects the chemical, physical and biological properties of manure. To that end, biochar can be added at any stage of the waste stream (Fig. 1).

Usage of biochar reaches from applying it as a feed supplement, to modifying bedding material and manure for storage, biogas fermentation and composting, up to the addition to soil. Consequently, biochar fed to livestock will cascade through the whole usage chain until reaching agricultural soil.[24] The uses of biochar in the waste stream are presented in the following, including the use of biochar amended manure as slow release fertilizer for soil.

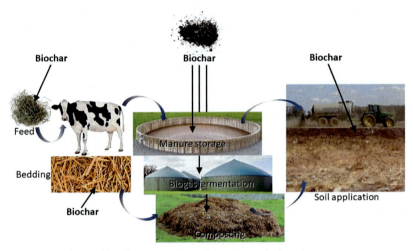

Fig. 1 Usage chain of biochar as an amendment in agricultural manure management.

Information on biochar alone as a soil amendment can be found in other sections of this book. In addition, the properties of biochar prepared from manure as a feedstock are not within the scope of this chapter. It is obvious, though, that good effects can be obtained from the various applications of biochar in combination with manure only when good quality biochar, pyrolysis techniques and feedstock materials, respectively, are used. Contaminated, toxic feedstock will yield contaminated and toxic biochar. Biochar production is not a way of hazardous waste elimination. To ensure good quality of biochar, several guidelines and certificates already exist such as the European Biochar Certificate (EBC) or by the International Biochar Initiative (IBI) guidelines. According to these, Hilber et al.[25] summarized thresholds for inorganic and organic contaminants recommended for biochar. Although the body of evidence on the topic is quickly increasing, still the level of understanding is very much scattered among numerous individual studies, testing specific combinations and quantities of biochar materials and manure substrates. Hence, it is difficult if not impossible to identify optimal combinations for the specific purposes, but rather general information can be derived and synthesized.

2. Biochar as a feed supplement and amendment to bedding material

Previous stages in the cascading usage chain even before the addition of biochar to manure are the addition to feed and bedding material. The use of biochar as a feed additive to livestock animals appears as an innovative alternative use in agriculture.[26] The state of the art on this topic was comprehensively described in an excellent review by Schmidt et al.[24] Substantial research on the topic as well as practical uses in agriculture were done since the last two decades. All this showed that the use of biochar as feed supplement is suited to improve animal health, increase nutrient intake efficiency, lower bioavailability of toxic compounds, and enhance productivity in animal husbandry.[24] Additionally, the positive effects of biochar, e.g., in nutrient and toxicant binding, propagate through the usage chain. For example, nitrogen-enriched biochar can be used as slow release fertilizer in soil.[27]

Literature suggests biochar shows no negative effects on feed quality. For example, when biochar was added to grass for silaging, no negative effects on silage properties, i.e., pH, nitrogen content and species, and content of valuable organic acids (e.g., acetic acid, L(+)-lactic acid) were observed.[28] On the other hand, mycotoxin formation can be reduced, and contaminants such as plant or fungal toxins and pesticides are stronger immobilized.[28] In vitro rumen experiments, mimicking food digestion by ruminants, as well as feeding experiments with ruminants showed no negative effects on rumen chemistry and digestive processes.[28,29] The latter also applies to other livestock animal species such as pigs, poultry, goats and sheep.[30,31]

Instead, numerous positive effects on animal health and growth were reported such as increased feed efficiency and weight gain, improvement of meat and milk quality as well as production.[24,32,33] Within the digestive tract system, biochar acts as an adsorbent for proteins, enzymes, endotoxins and is colonized by gram-negative bacteria, which become thus immobilized.[34] Consequently, unwanted bacteria such as fecal *E. coli* are significantly reduced in manure from livestock fed with biochar, while beneficial bacteria such as *Lactobacillus* proliferate within the digestive tract system and after excretion.[35] The effect on external toxins as well as endotoxins was termed as "enteral dialysis" property of biochar.[34] However, also contradicting results were reported on that latter topic.[36] The methane production in the rumen reacts differently on biochar addition, depending on ruminant species, diet and further food additives.[37] Hence, reported effects varied from no effect up to strong, significant decline by up to -29% compared to a control.[28,29,37] The strongest declines reported would significantly contribute to a reduction of greenhouse gas release from cattle feeding operations.

As an amendment to bedding material, biochar can mitigate negative effects such as loss of nutrients, greenhouse gases (especially NH_3, H_2S and CH_4) and odors. It helps to reduce the formation and spreading of pathogenic organisms.[38] When biochar was combined with straw or saw dust bedding at 5–10% (v/v) hoof diseases were clearly reduced.[39]

It can be assumed that this is due to the high sorptive capacity and specific surface area of biochar. This leads on one hand to a stronger immobilization of compounds and on the other hand to the preferred microbial colonization by beneficial microorganisms, resulting in faster breakdown of low molecular weight chemicals. However, although commercial biochar products are already available as bedding additives there is still much research left on that topic.[40] Much more research was done on the use of biochar for manure composting, though. Respective findings on the effects of biochar can be found in Section 4.

Experiments and practical applications have been conducted with biochar of different feedstock material, preferably plant derived material such as corn stover and maize cobs, rice husk, bamboo and wood; added amounts typically ranged from 0.5% to 8% and went up to 20% (w/w).[24,28] Making use of the strikingly positive effects of biochar addition to feed fully depend on the availability of non-toxic and edible high-quality biochar. To this end, a biochar certification standard for animal feed was introduced.[41]

3. Biochar as a bulking agent for manure handling, storage and further use

Manure, even solid manure is characterized by high moisture content and high amounts of fine material (<2 mm). Hence, solid manure tends to compaction and poor aeration.[42] This may impede aerobic composting of manure. When liquid manure and slurry, respectively, are stored, denser solid material tends to settle while lighter particulate material forms a floating layer during storage time.[43] This leads to undesirable inhomogeneity, when manure is to be removed for further use, e.g., pumped up for subsequent spreading on soil as a fertilizer.

Biochar is a suitable bulking agent to improve the granulation of manure.[44,45] This is probably most of all due to the sequestration of manure derived organic matter in the presence of biochar, forming stable complexes[46,47] and by this small aggregates. Research has been done, testing biochar as bulking agent in comparison with different alternative materials such as sawdust, green waste, straw, woodchips, and clay.[42] Prasai et al.[48] compared the effect of biochar and of clay minerals (bentonite, zeolite). They found that woody green waste biochar applied at contents of 1% to 4% did not change bulk density but altered water content, available water content (declining with the amount of biochar used), and improved mesoporosity of manure from egg-laying birds. The latter effect was due to the porosity of the biochar used. Additionally, the size class distribution of granules increased from largely 1–2 mm in control to mostly 2–4 mm in the presence of biochar, while the fractions of smaller granules (<1 mm) and larger granules (>4 mm) declined. The size classes from 1 to 4 mm are preferred for manure use in agricultural fertilizer spreaders.[48] On the other hand, the formation of too large clumps (>70 mm) is inhibited due to the non-sticking property of biochar.[49] Furthermore, the crushing strength (in Newton) of granules substantially increased with biochar addition from 19.7 N (0% biochar) to 80.4 N (4% biochar).[48] However, the exact effects on granulation and granule stability vary with the manure type.[48] Hence, it is required to find optimal combinations of biochar type, biochar amount and the manure type with its specific properties. Contrasting effects were reported for the effect of biochar on soil aggregate stability.[50] Obviously, aggregate strength of biochar-associates lies between that of manure granules and soil aggregates. Interestingly, the mentioned positive effects of biochar on mechanical properties of manure even persist, when it is previously applied as feed additive and mixes with feces already in the digestive tract system. In the aforementioned studies of Prasai et al.[48,51] biochar was fed to poultry as feed supplement before testing the effects on manure properties during storage.

Biochar cannot only be added to rather solid manure used for composting but also to slurry and liquid manure. Due to its high micro- and mesoporosity, and given its low density[52] it does not settle in liquid manure and slurry. Consequently, it can also improve the homogeneity and pumpability of slurry and liquid manure. The effect of biochar lasts throughout manure storage, composting, etc., since it is sufficiently recalcitrant against decomposition.[44] Subsequently used as soil amendment, biochar–manure mixtures are better suited than biochar alone to preserve soil organic carbon.[53]

Furthermore, stored manure is a major source of gaseous emissions.[54,55] Biochar has the potential to reduce the release of gaseous emissions from manure and thus of volatile organic chemicals (VOCs), greenhouse gases and odor. Corresponding results were reported for manure composting (see Section 4). Adding biochar by surface application to stored swine manure significantly reduces emissions of NH_3 by 12.7% to 22.6%.[54] However, Holly and Larson[43] found no significant effect on NH_3 emissions although biochar increased the amount of ammoniacal N remaining in manure. Concomitantly, the release of CH_4 increased while emissions of other gases, i.e., CO_2, N_2O, H_2S and various VOCs (such as indole, skatole, p-cresol, dimethyl-disulfide) were not significantly affected.[54] The increase in CH_4 coincides with findings on CH_4 production upon anaerobic digestion for biogas formation (see Section 5). The insignificant effect of biochar on other gases compared to the unamended control contradicts with the positive findings for manure composting, where clear reductions of N-gases, greenhouse gases and VOCs were found.[56,57] In case of stored liquid manure, this lack of a further positive effect could be due to the formation of surface crusts with a thickness of cm to dm that may reduce gaseous emissions from stored liquid manure. Such crusts do not only act as a cover but even more as a biofilter, because they are a habitat for complex microbial communities that actively degrade gaseous compounds such as CH_4.[58] The admixture of additives such as biochar requires disturbing such surface crusts,[43] which significantly increases gas emissions.[59] Consequently, it is advised to add biochar before it is introduced into storage reservoirs and the formation of surface crusts starts.

4. Biochar for improving manure composting

Composting of manure, especially farmyard and solid manure, is a traditionally used, biooxidative process, resulting in a stabilized and sanitized product.[42] Biochemical transformation processes lead on one hand to partial mineralization, especially of low molecular weight organic compounds, and to humification of organic matter on the other hand.[60] Phytotoxic compounds, weeds and pathogenic organisms are largely depleted, and volume and moisture of the composted manure are reduced.[42,61] This results in a stabilized, easy to store and to handle product that can be used as a valuable fertilizer and substrate to improve soil quality.[10] Composting requires sufficient aeration of the substrate; it goes along with a release of CO_2 but also of unwanted volatile products such as NH_3, methane, and odor, especially under insufficient oxygen supply.[62] The thermophilic process follows a characteristic temperature profile, starting with a mesophilic phase of a few days, followed by a temperature increase in the thermophilic phase before subsequent cooling and the final maturation or curing phase (Fig. 2). This goes along with subsequent changes in the microbial community composition and activity, producing heat as waste product.[63]

However, animal manure is rather inadequate for composting. This is due to high moisture content, low porosity and tendency to compact, high N-content and occasionally high pH.[42] Successful composting of manure with up to 79% water content was reported but rather contents of <65%

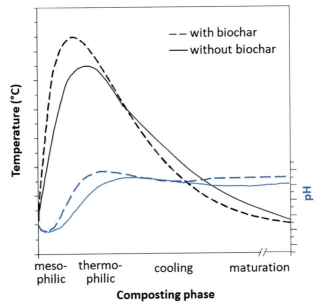

Fig. 2 Effects of biochar addition on the profiles of temperature and pH during composting of manure. Scales for temperature and pH are meant to give a relative rather than a quantitative orientation.

are recommended, requiring water removal from manure.[64] Often, additives such as sawdust, wood shavings, straw, and other crop plant waste products such as cotton waste or bagasse are admixed in order to improve the dry mass and structure of manure as well as the nutrient content and pH.[42] Alternatively, this can be reached through admixtures of biochar to manure, acting as a bulking agent[44,56] (also see Section 3). Additions of biochar—typically in a recommended range of 2% to 10% (up to 20% w/w)—result in improved moisture status, bulk density and air porosity of manure as composting substrate.[56,65–67] Not last, pellets, produced from biochar amended compost have improved mechanical properties.[68] However, also contrasting effects of biochar, e.g., on the air porosity have been reported.[61]

4.1 Composting phases

An optimized porosity of the manure leads to an improved gas-exchange and oxygen supply of the compost pile, which is reflected by changes in the microbial community composition and activity and a substantially stronger breakdown of organic matter.[49] The course and characteristics of the composting phases are altered (Fig. 2). The thermophilic phase starts with a steeper increase in temperature.[69–71] In most studies this went along with a more pronounced dehydration and prolonged thermophilic phase.[70,72,73] Yet, also a reduction of the thermophilic phase was reported.[74] Maximum temperatures reached are by about 5% higher compared to composts without biochar.[66,75] The stronger degradation of organic matter is based on a significantly increased microbial respiration and CO_2 emission.[44,74] The maturation is improved and reached in a shorter period of time. An extreme result was reported by Zhang and Sun,[67] who received mature compost within 24 d with 20% biochar addition compared to 90 to 270 d without biochar.

4.2 Effects on microbial community

The altered composting phases are accompanied and caused by a changed microbial community composition, growth and activities.[76,77] Fungal as well as bacterial community members are altered.[78,79] The structural diversity of fungi and bacteria is increased,[76,80] whereby the abundance of aerobic heterotrophs inclines.[67] Jiang et al.[81] found especially *Aspergillus* and *Myriococcum* promoted during the thermophilic and cooling phase. Among the bacteria, the Actinobacteria phylum was significantly increased by biochar addition.[79] Correspondingly, microbial functions such as the amino acid, carbohydrate and energy metabolism, ammonification and nitrification are substantially increased.[79] This is reflected by enhanced activities of microbial enzymes like urease, phosphatase, phenoloxidase and dehydrogenase.[67,80] The effects on microbiota vary with the dose of applied biochar.[76,78] Changes were well correlated with changes in the structural and physicochemical properties of the composting material caused by the biochar-amendment.[65,81]

4.3 Chemical properties of composted organic matter and nutrient status

The stronger breakdown and transformation of the organic matter during biochar amended composting leads to stronger humification.[82,83] The end product is characterized by a higher polymerization degree and higher portion of stabilized organic matter ('humic acids').[45,67,84] On the other hand, the fraction of more labile organic matter ('fulvic acids', water extractable carbon) declines.[80] Using analytical techniques such as FTIR spectroscopy it was shown that organic matter in mature compost with biochar addition has a higher O-alkyl C/alkyl C ratio and aromaticity, polymerization degree, and thermostable fraction.[85,86] The higher complexity of the compost organic matter results in higher cation exchange capacity.[67] However, Hagemann et al.[87] are more skeptical about the effects of biochar addition on the organic matter composition in mature compost. Testing biochar from different feedstock, they found no altered carbon speciation by analysis with FTIR, ^{13}C NMR and UV-spectroscopy. An increased C/N ratio and total organic carbon content of the mature compost can be fully due to the addition of biochar with its hardly degradable carbonaceous material with high C/N ratio.[87,88]

The retention of nutrients in the composting material and final compost is increased by the addition of biochar[67](see also Section 6). In this context, especially effects on N-retention during composting were investigated. Going along with reduced gaseous losses of volatile N-forms (see next paragraph), more N is preserved in biochar amended compost.[61,89] Losses of N are reduced[69,90] and more N is either stabilized in N-containing organic compounds[83] or converted to NH_4-N and subsequently to NO_3-N.[49,75] The latter, soluble N-species are better retained in the compost with biochar admixture due to the reduced available water content and higher sorption capacity, so that leaching losses are reduced.[87,88] Additionally, the contents of other plant nutrients are increased in the mature compost. For example, plant-available contents of PO_4^{3-}, K^+ and Ca^{2+} increased by 5.6–7.4%, 14.2–58.6%, and 0–12.5%, respectively, when pig manure was mixed with 5%, 10% and 15% of biochar.[91] In addition, the plant available contents of Zn and Cu increased, when pig manure was composted with an admixture of 10% biochar.[71] Effects of biochar addition to manure on pH are contradictory; increased as well as lowered pH–values during the composting period were reported.[75,91,92] The effect on pH presumably depends on the particular combination of biochar and manure with their specific pH values.

4.4 Gaseous losses

Although the emission of CO_2 increases with biochar addition to composting material, this is not the case for other volatile compounds. Emissions of ammonia (NH_3) and greenhouse gases such as N_2O and CH_4 were consistently reduced in all studies.[75,77,92] Adding 20% biochar to compost reduced emissions of N_2O and CH_4 by as much as 59.8% and 54.9%, respectively.[66,89] Also, losses of volatile organic chemicals (VOC) are reduced,[90] which includes N-volatile compounds, originating from microbial transformation of N-containing compounds, and of oxygenated volatile compounds, i.e., ketones, phenols and organic acids, while no effect was observed for other VOC families such as aliphatics, aromatics and terpenes.[57] The reduced volatile losses are due to the enhanced microbial turnover. Effects on N_2O emission rates correlated with the abundance of *nos*Z, *nir*K, and *nir*S genes,[92] coding for microbial N-transformation functions. Higher activities of methanotrophs such as *Methylococcaceae* resulted in reduced losses of CH_4.[93] Biochar amendment altered the abundance of denitrifying bacteria significantly; less N_2O-producing and more N_2O-consuming bacteria were present in the tested mixture of pig manure with biochar and additional woody bulking material and this significantly lowered N_2O emissions in the maturation phase.[92]

Most of the recent research work on biochar as an amendment to manure was done on composting. Although general trends become visible, there are also various reports that contradict one or the other finding. Hence, further research on that topic is required. This especially applies to systematic tests with different biochar material. The few existing studies indicate that effects vary with the surface area, porosity and chemical surface properties, including the sorption capacity of biochar.[75,93] However, the very case specific studies do not enable to derive generally applicable parameters, characteristic values, and recommendations in order to predefine best possible combinations of manure type, biochar properties and mixing ratios.

5. Biochar effects on anaerobic digestion and biogas formation

Using manure for anaerobic digestion, in order to produce biogas (methane, CH_4) is a cost-effective and ecofriendly option.[94,95] Apart from that major issue, anaerobic digestion mitigates negative properties of manure that may set the environment at risk, e.g., when manure is released as soil fertilizer. These are losses of nutrients and volatile compounds during storage and after addition to soil, the abundance of pathogenic organisms in manure, the contamination with organic contaminants such as hydrophobic organic chemicals and pharmaceuticals as well as heavy metals.[94,96] Typically, the digested slurry or digestate obtained after biogas formation contains less pathogens and organic contaminants, while nutrients and metals become more stabilized.[94]

The general process of anaerobic digestion and biogas formation proceeds in three to four steps that are driven by microbiota (Fig. 3). Hydrolytic bacteria degrade complex organic matter into soluble building blocks such as sugars, amino acids, and fatty acids.[96] Next steps are the formation of acids (methanogenic compounds) by acidogenic bacteria that are directly further converted to methane and CO_2 in the final methanogenesis step or compounds such as propionate, alcohols and aldehydes (non-methanogenic compounds) are first converted to methanogenic compounds before methane

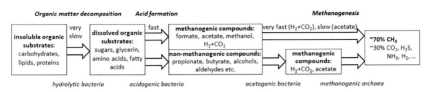

Fig. 3 Phases of anaerobic digestion of organic substrates and biogas formation according to.[97]

formation through the activity of acetogenic bacteria.[98] Methanogenesis is performed by archaea[96] and leads to the formation of about 70% methane and 30% CO_2 plus minor amounts of other, unwanted gases such as NH_3 and H_2S. The final digestate is typically characterized by an increased pH (about 1 pH unit), higher content of total nitrogen and NH_4-N as well as of other nutrients such as K, Mg and Ca (factor of about 1.2), while the C/N ratio is lowered (by about 35%).[94]

Digesting manure has the outlined advantages, plus the fact that especially ruminant manure already contains methanogenic bacteria as a natural inoculum.[98] Consequently, it is a technique with long tradition, especially in Asian countries.[96] However, digesting manure is not favored. This is because livestock manure has a low energy content, a too small C/N ratio and low dry matter or particulate material content. The latter is important for colonization by and growth of the relevant microbial consortium.[96] Consequently, most systems for manure digestion are operated by using organic- and energy-rich co-substrates; the process is called co-digestion.[98] Co-substrates are many kinds of organic waste products such as crop substrates and residues, e.g., maize silage, palm fiber, corn stover and husk, straw, coffee ground, kitchen waste, bedding material, etc.[96]

Additions of biochar to the various types of manure are suited to substantially improve the digestion process. First of all, biochar can increase the CH_4 yield and specific cumulative methane production, respectively.[99–102] Reported optimum increases range from 32% to 69%.[99,100] Additionally, the CH_4 content of biogas is increased due to the reduction in parallel of undesirable gases such as CO_2, H_2S and N_2O.[101,103,104] For example, a reduction by as much as 78% was reported for H_2S.[105] These effects on the digestion process are most of all caused by the facts that biochar acts as a sorbent for solutes and at the same time it is colonized by beneficial microorganisms.[106] Colonization with microorganisms was shown by scanning electron microscopy with bacillus and coccus as most abundant genera.[101] Among the archaea, which are most relevant for biogas formation, the genus of Methanosarcina with broad substrate utilization dominated in the presence of biochar, while strictly aceticlastic Methanosaeta prevailed when no biochar was added to manure.[102]

Biochar with its sorptive properties and alkaline pH acts as a buffer, alleviating the accumulation of propionic acid.[102] Toxic inorganic and organic chemicals such as pharmaceutical antibiotics that may hamper the microbial biogas formation are immobilized by sorption to the biochar surfaces.[106,107] Additionally, the control of gaseous emissions is at least partly explained by sorption.[104] For example, sorption of H_2S and not of its sulfate containing precursor molecules is the dominant process reducing H_2S emissions.[105] Furthermore, reaction rates are increased in the presence of the reactive and biologically active surfaces, the breakdown of macromolecular substrates is accelerated and solutes such as salts and NH_4-N are immobilized.[100]

In various studies, it was shown that the positive effect of biochar on biogas formation was similar or even superior to that of clay minerals or loess.[103,108] Best results were obtained at an amount added of 5% (w/w) biochar, whereby relative quantities of 0.6% to 20% were tested and shown to have a positive effect.[101,105,109] Additionally, using biochar produced at higher pyrolysis temperature (550 °C), resulting in higher surface area,[110] yielded the best effect on CH_4 production, while all types of biochar produced from different plant-derived feedstock material worked evenly well.[99,100] It can be concluded that biochar is especially suited to improve the productivity of digestive biogas formation and at the same time to lower possible adverse effects on the environment that result from losses of volatiles.[1]

6. Nutrient availability and delivery in soil of biochar amended manure

Manure is a relevant organic fertilizer.[10] Irrespective of the specific source and composition of manure, it contains substantial amounts of essential plant nutrients, in particular N in the form of NH_4-N, NO_3-N and urea, as well as P, S, Ca, Mg, B, Cu, Fe, Mn, Mo, and Zn.[2,111,112] Large portions of the major nutrient elements N, P and S are contained in manure in organic form. For example, 8–68% of total P are present in manure as organic P (variations depending on animal species and age, feed, husbandry conditions, etc.).[1] Nutrient contents, chemical binding form and subsequent plant availability after manure addition to soil depend, however, on the animal species, age, diet and husbandry.[113] Typical nutrient contents in animal manures are listed in Table 1. Several further benefits are obtained that may indirectly promote nutrient availability to plants when manure is applied as

Table 1 Contents of selected nutrients in manure from cattle, poultry (chicken) and pigs.

Type	DM[a]	OC	N$_{total}$	NH$_4$-N	P	K	Mg	Ca	S	Cu	Mn	Zn	B	Mo
		%	g kg^{-1}							mg kg^{-1}				
Cattle														
F[b]	32.3	32.3	9.7	1.7	2.1	9.3	0.23	0.63	0.58	4	45	50	5	0.2
S[c]	6.3	25.8	24.3	12.7	3.3	21.9	0.60	1.4	1.5	5	20	15	3	0.1
L[d]	2.0	7.5	2.35	2.0	0.07	5.2	0.06		0.09					
Poultry														
Dry[e]	50	37.5	20	7.5	8.4	10.8	3.0	30.0	0.80	27	45	185	14	0.7
S	14	47.5	8.1	4.8	2.8	3.6	1.1	10.7	0.32	8.4	49	35	4.2	0.42
Pigs														
F	23	37.5	5.9	1.5	1.6	2.1	1.2	2.9	0.24	20	105	440	6	0.4
S	4.5	12.5	3.9	2.6	1.2	2.4	0.54	1.6	0.16	5	12.5	15	1.5	0.05
Biogas														
Digestate	6.0	41.1	18.3	2.6	8.0	21.0	3.1	16.1	0.87	0.05	0.3	0.3		0.05

[a]DM = dry mass.
[b]F = farmyard manure.
[c]S = slurry.
[d]L = liquid manure.
[e]Dry = dry chicken feces.
Data compiled from[6,7,9,114–116]

an organic amendment to soil. In addition to nutrient contents, manure improves important soil functions by increasing the content of soil organic matter, enhancing microbial (fungal and bacterial) biomass and activity,[117] increasing soil pH and electrical conductivity (EC).[118] Furthermore, manure improves soil physical properties by promoting soil aggregation, reducing bulk density, and increasing porosity and water percolation rate.[119] A large body of studies have demonstrated the aforementioned benefits of manure in soils and reported a significant increase in the contents of total and plant available nutrients such as N, P, K, Ca, and Mg as well as of soil pH upon land application of organic manure, with clear positive effects on plant growth.[2,117]

However, potential negative effects of land application of animal manures are also well known. This includes an excess of nutrient inputs especially of N and P exceeding plant uptake. Provided that manure application was done following fertilizer recommendations, such an excess can be due (i) to an imbalance between nutrients contained in manure and nutrients used by plants,[10,19] and (ii) a rapid release of N, P and S after organic matter mineralization, e.g., under warm and moist pedoclimate preceding plant development and nutrient requirements.[120] Thereby, equal degradability and thus mineralization rates were determined for liquid and solid manures.[120] Such an untimely release of nutrients may lead to losses with leaching water and surface runoff. This especially applies to highly mobile ions such as NO_3-N but also nutrients that are bound in or to manure derived dissolved organic matter may be leached. About 50% of the organic matter from slurry is present in the liquid phase[121] and dissolved and colloidal organic matter constitutes a substantial fraction of manure.[122] Furthermore, gaseous losses can be significant when NH_4-N is chemically converted to NH_3 and volatilizes, which especially happens in alkaline soil,[123] or when NO_3-N is microbially reduced to N_2O or N_2, which effect is even stronger in solid compared to liquid manure.[124]

There is a growing interest in amending manure with biochar to mitigate the disadvantages of fertilizing soils with manure. While manure derived organic matter is rather quickly turned over and mineralized, respectively, biochar is recalcitrant to degradation with very slow turnover times of centuries to millennia.[125,126] On the other hand, biochar contains and can release some nutrient elements but the content is rather low and release rates are very slow because of the recalcitrance of biochar to degradation, thus making biochar inadequate as a sole nutrient provider.[127,128] Hence, it is advantageous to mix both substrates before addition to soil, which adds to the favorable characteristics of manure-biochar mixtures in the preceding stages of the waste stream (see previous sections).

It was shown that biochar can take up significant amounts of nutrients (see also Sections 3–5). For example, when hardwood biochar was mixed with liquid dairy manure (1% w/w) and shaken for 24 h, an uptake of N and P of $5.3 \, mg \, g^{-1} \, NH_4^+$ and $0.24 \, mg \, g^{-1} \, PO_3^{4-}$ was obtained.[129] The use of such biochar-manure mixtures as fertilizer results in lowered turnover rates. Biochar affects the mineralization rate of labile organic matter sources such as manure and concomitantly affects nutrient release.[130] Consequently, the release of major nutrients such as NH_4-N, NO_3-N, dissolved organic N and phosphate is much reduced.[111,131] For instance, sequential soil extraction using 2 M KCl released in the first step 70% of total nitrate from loamy soil, which was reduced to 58% in the presence of biochar, although the amount of total extractable NO_3-N was doubled due to the N content of the biochar-manure mixture.[132] Hence, biochar-manure mixtures are termed as "slow release fertilizers".[132,133] That effect is not limited to N and P but also to other nutrients, e.g., alkali and earth-alkali elements such as K, Ca and Mg.[111] A slow nutrient release results in lower levels of available and easily extractable nutrient contents in soil, yet the whole storage of plant available nutrients is increased due to the fertilization effect and thus sufficient for plant growth.[130,134,135] The effect of slow nutrient release is not identical for all nutrients, though, but strongest for moderately released nutrients such a NH_4-N. A much lower effect was reported for very mobile nutrients such as NO_3-N and slow desorbing nutrients such as phosphate.[133]

The effect of biochar on nutrient release is particularly due to complex formation and sorption of manure derived organic matter (negative priming effect) and nutrients on biochar and subsequent interaction with soil constituents.[135] Second, it is caused by effects on the microbial community composition in manure and successively in soil.[130] Biochar stabilizes the rather labile manure derived organic matter through the formation of organo-mineral complexes.[136,137] A coating is formed on the outer and inner surfaces of biochar particles that add further properties such as hydrophilicity, redox-active moieties, and additional mesoporosity, which enhances biochar water interactions and thus further increases nutrient retention.[132] Nutrients are either retained as constituents of the complexed organic matter or as ions through sorption to the large specific surface of biochar.[138] The organic coating on co-composted biochar particles serves as an additional reservoir for nitrate.[132,139,140] The release of dissolved organic carbon correlated positively ($R^2 > 0.99$) to the release of organic N from co-composted biochars.[131] Furthermore, biochar-manure mixtures alter the composition of microbial communities, thus, mediating and enhancing soil microbial processes affecting soil nutrient cycles.[131]

Ippolito et al.[130] reported an increase in gram-positive bacteria and a decrease in gram-negative bacteria. Typically, gram-positive bacteria dominate in the presence of easily decomposable substrate performing fast turnover processes, while gram-negative bacteria are present, when substrate is less available and turned over more slowly.[141,142] Furthermore, biochar-manure mixtures as soil amendment activate bioturbation by dung and soil living faunal species, thus improving soil physical properties.[26]

An improved nutrient use efficiency in the presence of biochar-manure mixtures in soil goes together with reduced losses.[128] Leaching losses, especially of NO_3-N but also of NH_4-N are significantly reduced (in part to less than 50% of losses from control soil), even when additional N was added to soil with the biochar-manure mixture.[143,144] Obviously, nitrate leaching is prevented through the slow release of nitrate from biochar.[132] Furthermore, the improved formation of stable aggregates in the presence of biochar increases soil water storage and reduces nutrient leaching.[145] This goes together with the effects on microbiota and microbial processes. Less nitrification (75% reduction compared to control), ammonification (229%) and cumulative CO_2 release (68%) were found in soil amended with a biochar-manure mixture that was even more effective than biochar alone.[146] Accordingly it was found that gaseous losses from soil, e.g., of N_2O, were less in the presence of biochar-manure mixtures.[146,147]

In numerous studies, positive effects were reported of co-composted biochar and biochar-manure mixtures on soil characteristics and plant growth. In an agronomic evaluation across different cropping systems, compost and biochar-blended compost were most effective in enhancing soil fertility. This was related to increased nutrient cycling (25% mean increase in extractable organic C and 44% increase in extractable N), element availability (26% increase in available K), soil microbial activity (26% increase in soil respiration and two to fourfold enhancement of denitrifying activity), and by this enhanced key crop (grapes and tomatoes) quality parameters.[57] Maize grown in soil amended with digestate-enriched biochar showed a significantly higher biomass yield compared to the control and non-enriched biochar treatments.[148] In a pot experiment, biomass of maize plants increased by 243% with application of co-composted biochar compared to mineral NPK, also showing better growth effects than compost alone.[111] These examples demonstrate that the effect of biochar-manure mixtures on plant performance goes beyond the sole nutrient effect. Further examples for effects of biochar-manure mixtures on plant growth and yield are listed in Table 2. Reviews by Fischer and Glaser,[153] Agegnehu et al.[154] and Wu et al.[155] provide further information on the use of co-composted biochars in soils.

Table 2 Selected examples for responses of soils and crop plants to treatments with different biochar-manure mixtures and nutrient enriched biochars.

Biochar, manure mix and application rate	Crop type, soil, type of experiment	Soil and crop response	References
Blend of 1:1 composted duck manure with hard and soft woods biochar (dry weight), 0.6 and 2.4 kgm^{-2} with and without standard NPK fertilizer; compost at 1.2 kgm^{-2}	Cabbage, sandy loam, field experiments >3 years	Increased pH (>0.5) of soil (0–15 cm), higher cation-exchange capacity (CEC), elevated levels of some trace metals and exchangeable cations (K, Ca, and Mg) in biochar treatments compared to untreated soil. Highest crop yields in treatments with NPK fertilizer	149
Co-composts of biochar (BC) and farmyard manure (FYM) at various ratios (FM$_{100}$:BC$_0$, FM$_{75}$:BC$_{25}$, FM$_{50}$:BC$_{50}$, FM$_{25}$:BC$_{75}$, FM$_0$:BC$_{100}$), added to soil at 2% (w/w).	Wheat, alkaline soil, pot trial testing plant growth, incubation experiment testing carbon mineralization	Stabilization of native soil carbon with increase in BC proportion in composts, negative priming effect; enhanced grain yield, increased total N, P, and K contents; significantly higher N and K concentrations in wheat plants grown in soil with compost; best effect on grain yield in treatments with BC portion > compost portion	150
Co-composted biochar (wood chips; 80% coniferous, 20% deciduous wood) mix with animal manures, straw, rock powder, soil and mature compost at 2% (w/w)	Quinoa, mixture of sandy loam, sand and gravel, potted greenhouse experiments	Increase of biomass yield of up to 305%; untreated biochar decreased biomass to 60% of control, 10–20% increase in water holding capacity	131
Corn cob biochar and fig tree pruning's wood biochar mixed with digestate, added to soil with 10 and 20 t C ha^{-1}	Maize, greenhouse experiments with farmland soils, completely randomized block design	Increase in soil organic matter (232–514%) and macronutrients (110–230%); significantly higher biomass yield compared to the control and non-enriched biochar; slightly lower yields (20–25%) than from chemical fertilizer treatments	148

Continued

Table 2 Selected examples for responses of soils and crop plants to treatments with different biochar-manure mixtures and nutrient enriched biochars.—cont'd

Biochar, manure mix and application rate	Crop type, soil, type of experiment	Soil and crop response	References
Oak biochar with organic residues (sheep manure, compost from olive waste, biowaste and green waste)	Different sites (countries) experiments with olive groves, tomatoes, vineyards	Increased soil organic C (11–35%), pH (0.15–0.50 units) in acidic soils. Increased nutrient cycling (25% mean increase in extractable organic C and 44% increase in extractable N), element availability (26% increase in available K), and soil microbial activity (26% increase in soil respiration and two to fourfold enhancement of denitrifying activity)	57
Co-compost of rice hull biochar with chicken manure, 5%	120-d microcosm experiment, different field soils	Significantly enhanced soil total C and N, inorganic and KCl-extractable organic N, microbial biomass C and N, cellulase enzyme activity, abundance of N_2O-producing bacteria and fungi. Significant reduction of soil CO_2 and N_2O emissions by 35% and 27%, respectively, compared to manure treatment	151
Cotton straw biochar with chicken manure, 0%, 1%, 3% of biochar and 0%, 3%, 6% manure	Cotton, Gray desert soil (loess-like sediments)	Significantly affected root morphology and physiology by improving soil nutrients; morphological and physiological parameters peaked with 6% organic manure combined with 1% biochar	152
Co-composted bokashi-biochar ($60\,t\,ha^{-1}$)	Maize, silty loam soil, potted plant trials	Increased biomass by 243% compared to mineral NPK, Improved soil available nutrients (available P, K +, Ca^{2+} and Mg^{2+}) leading to better plant growth	111

However, there are also reports on insufficient nutrient effects of biochar and respective mixtures.[156] Findings vary from no effects[149,157,158] to a decrease in plant yield and biomass.[148,159] For example, in a greenhouse experiment with potato plants, total potato biomass and tuber yields were lower by 30% and 27% in biochar and biochar-manure mixture (enriched with anaerobic digested dairy manure) treatments compared to mineral fertilizer treatments.[159] A slightly lower maize yield (20–25%, compared to control with mineral fertilizer) was reported for biogas digestate-enriched biochar and attributed to slower mineralization and retarded release of the adsorbed nutrients.[148] Nonetheless, such effects are relevant on a short-term, while on a long-term scale benefits for nutrient availability are expected to prevail, with biochar maintaining a gradual release of micronutrients.[148,149]

7. Impact of biochar on the fate and effects of micropollutants in manure

Manure as a fertilizer is a valuable source of nutrients in soils. However, it may also contain micropollutants and toxic agents, i.e., potentially toxic elements (PTEs), organic contaminants, pathogens and pathogenic agents.[160,161] The organic pollutants include a wide range of compounds such as pharmaceutical drugs, hormones, pesticides, detergents, personal care products, cleaning agents. These contaminants may adversely affect soil organisms, and thus affect soil fertility and functioning,[162,163] and they can be transferred to the food chain, thus posing a risk to the health of plants, animals and humans.[160]

Application of manure onto soils may add excess of trace nutrient elements and PTEs, i.e., copper, manganese, selenium, zinc, cobalt, nickel, chromium, arsenic, cadmium, mercury and lead.[2,164,165] Reported PTE contents in manure vary widely (from a few µg kg^{-1} to several hundred mg kg^{-1}) depending on a variety of factors such as animal type and age, type of feed, conditions of husbandry and manure storage.[164,166] Animal waste and sewage sludge contain higher levels of heavy metals than other organic wastes[167,168] and therefore are the single most important sources of PTEs in soil. Especially contents of Zn and Cu are often excessively high in manure because of their use as feed additives.[166,169]

Land fertilization with organic manure can also lead to accumulation of organic pollutants in soils.[170] Among the various organic pollutants, pharmaceuticals and especially antibiotics have received much attention.

After medication of livestock animals, around 50% and up to >90% of the administered parent compound are quickly excreted from the medicated animals within time periods of several hours and days, respectively.[171,172] The excreted antibiotics end up in manure.[173,174] Due to the use of several pharmaceuticals and/or the mixing of manure from various sources in one storage tank or lagoon, studies consistently show that animal waste materials are typically contaminated with more than one antibiotic compound and other contaminants. Hence, mixed contaminations with numerous antibiotics and other contaminants are reported for manure and the receiving soil environment.[175–177] A detailed report by Thiele-Bruhn[178] gives information on antibiotics that are typically used in animal medication, contamination levels in manure, animal waste and in soils. Other reported organic pollutants in manure include personal care products and hormones.[6] However, there is a wealth of information on these contaminants in animal waste material, manure and sewage sludge, knowledge is rather scarce for other organic pollutants. For example, Krzebietke et al.[179] reported that fertilizing soils with farmyard manure increased the content of light and heavy polycyclic aromatic hydrocarbons. Some herbicides may reach manure tanks through wash off from nearby surfaces or may pass through animals' digestive tract and are excreted. Therefore, manure may contain herbicide residues, some of which may remain active even after composting, for example picloram, clopyralid and aminopyralid, and thus hamper plant growth.[180]

In addition to aforementioned micropollutants, animal waste and manure contain microbial pathogens (from both human and animal origin) including viruses, bacteria and parasites. From the viruses contained in manure, several are human pathogenic enteric and respiratory viruses such as rotaviruses, adenoviruses, hepatitis E viruses, parvoviruses, etc.[181] Important pathogenic bacteria species are *Clostridium botulinum*, *Escherichia coli*, *Salmonella* species, *and Brucella abortus* among others. Similarly, parasites (Protozoans and Helminths) are also found in manure and animal waste.[181] Several of these pathogens are zoonotic in nature, which imposes a direct threat to the health of humans and animals.[181,182] Several reviews give detailed insights on the variety of pathogens in animal waste and manure, routes of exposure to humans and their health risks, contributory factors to their survival and persistence in manure and environment.[181,183]

The presence of antibiotic resistant bacteria (ARB) and antibiotic resistance genes (ARGs) in agricultural animal manures is of special global concern. The use of antibiotic contaminated manure in agriculture directly adds antibiotic resistance to soils and the food chain.[184–186] Antibiotics resistance selection and proliferation of antibiotic resistant bacteria already occurs in the digestive systems of medicated animals.[184,187] Excreted antibiotics continue exerting selection pressure in manure, which together with antibiotic resistant bacteria can further increase the abundance of ARGs in manure during storage.[188–190] In addition, manure contains mobile genetic elements that facilitate and accelerate the horizontal transfer of ARGs and manure has unique features that make it a potential hotspot for ARGs dissemination by horizontal gene transfer.[184,185] Consequently, soils fertilized with contaminated manure are characterized by a significantly increased abundance of numerous ARGs.[189–191] Once in the environment, ARB and ARGs can be transmitted to humans through various pathways.[192] The spread and increase of ARGs in contaminated environments is of special concern for human health.[188,192] Hence, it can be concluded that manure storage management is highly relevant for the modulation of antibiotic residues and abundance of ARGs.[193]

As a strategy for manure management, the effects of micropollutants can be alleviated through biochar manure co-composting and biochar-compost biochar-manure mixtures. The potential and use of biochar to decontaminate soils from diverse inorganic and organic pollutants (including ARGs and pathogenic bacteria) has been comprehensively investigated.[194–196] It can be assumed that biochar acts similarly well on contaminants contained in manure, however, the number of reports is scarce. The few existing studies suggest that co-composted biochars have enhanced pollution remediation qualities for manure, sludge (Table 3) and soils than biochar alone.[203] The interaction of biochar and manure during storage, composting and anaerobic digestion, respectively, influences and modifies the physicochemical/biological properties of both substrates (see Sections 3–5). The surface functionalities of biochar are altered upon contact with manure and especially during composting. Additional and/or modified organic surfaces increase the affinity of biochar for organic pollutants.[204] Impacts of biochar on contaminants in organic waste materials include a reduced bioavailability of micropollutants through sorptive immobilization[199,205] and the provision of a suitable habitat for beneficial microorganisms[206,207] among others. Enhanced sorption may occur through electrostatic attraction, H-bonding, ion-exchange, surface complexation and surface precipitation mechanisms of polar micropollutants. Non-polar pollutants are stabilized through pore-filling, partitioning and hydrophobic interaction. As was mentioned, biochar also offers a favorable environment for the proliferation of microorganisms,[206] promoting the biodegradation of organic chemicals, and mitigating harmful effects on other living organisms.[205]

Table 3 Selected examples for effects of biochar on contaminants and antibiotic resistance genes (ARGs) in manure and manure compost.

Biochar	Compost material	Nature of product	Decontamination effect	References
Bamboo charcoal	Pig manure	BC added to manure composting piles	Reduced mobility of Cu and Zn.	197
Biochar from rice straw and spent mushroom substrate	Chicken manure	Biochar added to lab-scale chicken manure composting	Mushroom biochar had higher removal rate of ARGs and pathogenic bacteria. Parallel decline of bio-available heavy metals (Cu, Zn and As).	198
Biochar from sawdust, corn stover and peanut hulls	Pig manure	Biochars spiked to the composting mixture at levels of low (6%), medium (12%) and high (24%) on a dry weight basis	Removal of ARGs in the composting pile more dependent on biochar content than on biochar type. Low-level biochar addition could enhance the elimination of studied ARGs. Medium level biochar addition to the composting piles would increase the risk of ARGs' propagation.	199
Bamboo charcoal	Chicken manure compost	Biochar added to manure during composting	BC significantly decreased bioavailable Cu and Zn. Addition of 0%, 5%, 10%, and 20% BC reduced ARGs in chicken manure by 0.85, 1.05, 1.08 and 1.15 logarithmic units.	200
Bamboo charcoal	Sludge compost	Biochar added to sludge during laboratory scale composting	Decline in DTPA-extractable Cu and Zn contents of composted sludge by 44.4% and 19.3%, respectively.	201
Biochar and activated carbon (AC)	Sewage sludge		Significant decrease of freely dissolved PAHs concentration with addition of AC/biochar to sewage sludge; dose dependent effect ranged from 56 to 95% (AC) and from 0 to 57% (biochars); surface area of AC/biochar and reduction of PAHs significantly related.	202

Among these bacteria are Proteobacteria, Actinobacteria and Bacteroidetes.[208] These detoxifying properties can be used at all stages of the waste stream, making biochar-manure mixtures suitable for many applications, ranging from the detoxification of livestock feed up to the remediation and detoxification of contaminated soils.[203,209]

For example, the dissolved fraction of antibiotics from various structural classes (sulfonamides, fluoroquinolones, fenicols) substantially declined when biochar was added to manure.[107] Thereby, plant derived biochar was superior to biosolids derived biochar. Additionally, the dissipation kinetics were substantially accelerated and the formation of ARGs was significantly lowered.[189] Riaz et al.[208] received similar findings showing a stronger reduction of antibiotics and related ARGs as well as of heavy metals upon composting and anaerobic digestion of chicken manure in the presence of biochar. Li et al.[210] reported better immobilization of Zn and Cu during composting, when compost was amended with biochar produced at a pyrolysis temperature between 450°C and 500°C. Further examples are compiled in Table 3.

However, contrary or no effects have also been reported. Rice straw biochar manure compost was not effective in removal of ARGs and pathogenic bacteria.[198] No effects on the bioavailability of heavy metals after manure composting with biochar were explained by unfavorable surface properties of the specific biochar used.[211] The effects of biochars on the sorbed fraction of the antibiotic oxytetracycline in manure were indifferent presumably because that compound already showed strong sorption to manure alone and changed non-linearly with the added amount of biochar.[107]

The existing knowledge shows that biochar has highest potential for the immobilization and detoxification of contaminants in manure. However, further research should be extended to a wide range of environmental pollutants and pathogenic agents, respectively, to gather chemical, microbiological and toxicological data for a better understanding of the removal mechanisms. As of now, information obtained from only a few studies on few pollutants is not enough to make general conclusions.

8. Summary and conclusion

There are several more uses for biochar than simply the use as a soil additive. For example, biochar can be used at all stages of the waste stream to modify and improve the properties and recycling of manure or similar organic waste materials. The possible uses stem from the application of biochar as feed additive to the utilization of biochar-manure mixtures as slow release fertilizers. Thereby, positive effects of biochar may cascade down the utilization chain. The rapidly increasing body of evidence documents the positive effects of biochar on livestock health and performance when it is used as a feed additive. Mixing biochar with bedding material further improves livestock health by affecting hygiene and climate in the stable. Physical properties of manure as well as the immobilization and thus (partial) detoxification of contaminants and pathogenic organisms and pathogenic agents such as antibiotic resistance genes are increased. Losses of manure constituents such as nutrients through leaching or volatilization are lowered, while transformation processes and product yield and quality in composting and anaerobic digestion for biogas formation are enhanced. All these positive effects of biochar depend on the outstanding surface area and surface properties of biochar, acting as sorption sites for chemicals and habitat for microorganisms. Consequently, effects of biochar depend on physicochemical interactions on the one hand and on biochemical effects on the other, making biochar highly effective in buffering and balancing organic matter, nutrients, contaminants, beneficial as well as pathogenic organisms, etc. However, not only positive effects but also no or even negative effects of biochar have been reported. It very much depends on the proper combination of biochar type (feedstock and preparation technique), amount and purpose of manure treatment to obtain the outlined positive effects. There is even more potential in engineered, tailor-made biochars with designed properties that are optimized for the specific purpose. The state of knowledge, however, is still incomplete. Much research on the topic has been done on manure composting, but systematic research and meta-analyses are missing in order to come from case specific research to an overall systematic understanding. It is obvious, though, that good effects can only be obtained when good quality biochar is used. Contaminated feedstock will result in contaminated biochar and hazardous biochar-manure mixtures. First guidelines have been released to define good quality biochar opening the way to reuse plant waste material as a multi-purpose substrate that is able to improve environmental status and health.

References

1. He Z, Pagliari PH, Waldrip HM. Applied and environmental chemistry of animal manure: a review. *Pedosphere* 2016;**26**(6):779–816.
2. Bolan NS, Adriano DC, Mahimairaja S. Distribution and bioavailability of trace elements in livestock and poultry manure by-products. *Crit Rev Environ Sci Technol* 2004;**34**(3):291–338.

3. FAOSTAT. *Livestock manure [internet].* Food and Agricultural Organization of the United Nations; 2020.
4. Chadwick DR, Chen S. Manures. In: Haygarth PM, Jarvis SC, editors. *Agriculture, hydrology and water quality.* Wallingford, UK: CABI Publishing; 2002. p. 57–82.
5. Leenheer JA, Rostad CE. *Fractionation and characterization of organic matter in wastewater from a swine waste-retention basin. Scientific Investigations Report 2004-5217.* Reston, Virginia: U.S. Department of the Interior, U.S. Geological Survey; 2004. p. 21.
6. Ghirardini A, Grillini V, Verlicchi P. A review of the occurrence of selected micropollutants and microorganisms in different raw and treated manure—environmental risk due to antibiotics after application to soil. *Sci Total Environ* 2020;**707**, 136118.
7. Paulsen HM, Blank B, Schaub D, Aulrich K, Rahmann G. Zusammensetzung, Lagerung und Ausbringung von Wirtschaftsdüngern ökologischer und konventioneller Milchviehbetriebe in Deutschland und die Bedeutung für die Treibhausgasemissionen. *Landbauforschung* 2013;**1**(63):29–36.
8. Samer M. GHG emission from livestock manure and its mitigation strategies. In: Sejian V, Gaughan J, Baumgard L, Prasad C, editors. *Climate change impact on livestock: adaptation and mitigation.* New Delhi, India: Springer; 2015. p. 321–46.
9. Sims JT, Maguire RO. Manure management. In: Hillel D, Rosenzweig C, Powlson D, Scow K, Singer M, Sparks D, editors. *Encyclopedia of soils in the environment.* vol. 2. Dordrecht, NL: Elsevier; 2005. p. 402–10.
10. Edmeades DC. The long-term effects of manures and fertilisers on soil productivity and quality: a review. *Nutr Cycl Agroecosyst* 2003;**66**(2):165–80.
11. Haynes RJ, Naidu R. Influence of lime, fertilizer and manure applications on soil organic matter content and soil physical conditions: a review. *Nutr Cycl Agroecosyst* 1997;**51**(2):123–37.
12. Senesi N, Xing B, Huang PM. In: Senesi N, Xing B, Huang PM, editors. *Biophysico-chemical processes involving natural nonliving organic matter in environmental systems.* Hoboken, NJ: John Wiley & Sons Inc.; 2009. p. 876.
13. Schulten HR, Leinweber P. Influence of long-term fertilization with farmyard manure on soil organic matter: characteristics of particle-size fractions. *Biol Fertil Soils* 1991;**12**(2):81–8.
14. Bogaard A, Fraser R, Heaton THE, Wallace M, Vaiglova P, Charles M, et al. Crop manuring and intensive land management by Europe's first farmers. *Proc Natl Acad Sci U S A* 2013;**110**(31):12589–94.
15. Weiland P. Biogas production: current state and perspectives. *Appl Microbiol Biotechnol* 2010;**85**(4):849–60.
16. Burg V, Bowman G, Haubensak M, Baier U, Thees O. Valorization of an untapped resource: energy and greenhouse gas emissions benefits of converting manure to biogas through anaerobic digestion. *Resour Conserv Recycl* 2018;**136**:53–62.
17. Kall K, Roosmaa Ü, Viiralt R. Assessment of the economic value of cattle slurry and biogas digestate used on grassland. *Agron Res* 2016;**14**(1):54–66.
18. Sogn TA, Dragicevic I, Linjordet R, Krogstad T, Eijsink VGH, Eich-Greatorex S. Recycling of biogas digestates in plant production: NPK fertilizer value and risk of leaching. *Int J Recycl Org Waste Agric* 2018;**7**(1):49–58.
19. Sharpley A, Meisinger JJ, Breeuwsma A, Sims JT, Danie TC, Schepers JS. Impact of animal manure management on ground and surface water quality. In: Hatfield JL, Stewart BA, editors. *Animal waste utilization: effective use of manure as a soil resource.* Chelsea, MI: Ann Arbor Press; 1998. p. 173–242.
20. Reichel R, Rosendahl I, Peeters ETHM, Focks A, Groeneweg J, Bierl R, et al. Effects of slurry from sulfadiazine- (SDZ) and difloxacin- (DIF) medicated pigs on the structural diversity of microorganisms in bulk and rhizosphere soil. *Soil Biol Biochem* 2013;**62**:82–91.

21. Van Elsas JD, Semenov AV, Costa R, Trevors JT. Survival of *Escherichia coli* in the environment: Fundamental and public health aspects. *ISME J* 2011;**5**(2):173–83.
22. Aller MF. Biochar properties: transport, fate, and impact. *Crit Rev Environ Sci Technol* 2016;**46**(14–15):1183–296.
23. Ahmad M, Rajapaksha AU, Lim JE, Zhang M, Bolan N, Mohan D, et al. Biochar as a sorbent for contaminant management in soil and water: a review. *Chemosphere* 2014;**99**:19–23.
24. Schmidt HP, Hagemann N, Draper K, Kammann C. The use of biochar in animal feeding. *PeerJ* 2019;**2019**(7), e7373.
25. Hilber I, Bastos AC, Loureiro S, Soja G, Marsz A, Cornelissen G, et al. The different faces of biochar: contamination risk versus remediation tool. *J Environ Eng Landsc Manag* 2017;**25**(2):86–104.
26. Joseph S, Pow D, Dawson K, Mitchell DRG, Rawal A, Hook J, et al. Feeding biochar to cows: an innovative solution for improving soil fertility and farm productivity. *Pedosphere* 2015;**25**(5):666–79.
27. Oh TK, Shinogi Y, Lee SJ, Choi B. Utilization of biochar impregnated with anaerobically digested slurry as slow-release fertilizer. *J Plant Nutr Soil Sci* 2014;**177**(1):97–103.
28. Calvelo Pereira R, Muetzel S, Camps Arbestain M, Bishop P, Hina K, Hedley M. Assessment of the influence of biochar on rumen and silage fermentation: a laboratory-scale experiment. *Anim Feed Sci Technol* 2014;**196**:22–31.
29. Hansen HH, Storm IMLD, Sell AM. Effect of biochar on in vitro rumen methane production. *Acta Agric Scand Sect A Anim Sci* 2012;**62**(4):305–9.
30. Silivong P, Preston TR. Supplements of water spinach (*Ipomoea aquatica*) and biochar improved feed intake, digestibility, N retention and growth performance of goats fed foliage of *Bauhinia acuminata* as the basal diet. *Livest Res Rural Dev* 2016;**28**(5), 98.
31. Chu GM, Kim JH, Kim HY, Ha JH, Jung MS, Song Y, et al. Effects of bamboo charcoal on the growth performance, blood characteristics and noxious gas emission in fattening pigs. *J Appl Anim Res* 2013;**41**(1):48–55.
32. Gerlach A. Pflanzenkohle in der Rinderhaltung. *Ithaka J* 2012;**1**:80–4.
33. Di Natale F, Gallo M, Nigro R. Adsorbents selection for aflatoxins removal in bovine milks. *J Food Eng* 2009;**95**(1):186–91.
34. Schirrmann U. *Aktivkohle und ihre Wirkung auf Bakterien und deren Toxine im Gastrointestinaltrakt*. Munich, Germany: TU München; 1984. p. 153. Dissertation.
35. Kim KS, Kim Y-H, Park J-C, Yun W, Jang K-I, Yoo D-I, et al. Effect of organic medicinal charcoal supplementation in finishing pig diets. *Korean J Agric Sci* 2017;**44**(1):50–9.
36. Toth JD, Dou Z. Use and impact of biochar and charcoal in animal production systems. In: Guo M, He Z, Uchimiya SM, editors. *Agricultural and environmental applications of biochar: advances and barriers*. Madison: Soil Science Society of America; 2016. p. 199–224.
37. Leng RA, Preston TR, Inthapanya S. Biochar reduces enteric methane and improves growth and feed conversion in local "Yellow" cattle fed cassava root chips and fresh cassava foliage. *Livest Res Rural Dev* 2012;**24**(11), 199.
38. EBC. *European Biochar Certificate —Richtlinien für die nachhaltige Produktion von Pflanzenkohle*. Arbaz, Switzerland: European Biochar Foundation (EBC); 2012. http://www.european-biochar.org/en/download. Version 7of 2nd December 2015, DOI: 10.13140/RG.2.1.4658.7043.
39. O'Toole A, Andersson D, Gerlach A, Glaser B, Kammann CI, Kern J, et al. Current and future applications for biochar. In: Shackley S, Ruysschaert G, Zwart K, Glaser B, editors. *Biochar in European soils and agriculture: science and practice*. Abington: Taylor and Francis; 2016. p. 253–80.

40. Haubold-Rosar M, Heinkele T, Rademacher A, Kern J, Dicke C, Funke A, et al. *Chancen und Risiken des Einsatzes von Biokohle und anderer "veränderter" Biomasse als Bodenhilfsstoffe oder für die C-Sequestrierung in Böden. UBA Texte 04/2016*. Dessau-Roßlau: Umweltbundesamt; 2016. p. 254.
41. EBC. *European Biochar Foundation—Guidelines for EBC-feed certification*; 2018. Available at *http://wwweuropean-biocharorg/biochar/media/doc/ebc-feedpdf*. [Accessed 31 March 2020].
42. Bernal MP, Alburquerque JA, Moral R. Composting of animal manures and chemical criteria for compost maturity assessment. A review. *Bioresour Technol* 2009;**100**(22):5444–53.
43. Holly MA, Larson RA. Effects of manure storage additives on manure composition and greenhouse gas and ammonia emissions. *Trans ASABE* 2017;**60**(2):449–56.
44. Steiner C, Melear N, Harris K, Das K. Biochar as bulking agent for poultry litter composting. *Carbon Manag* 2011;**2**(3):227–30.
45. Dias BO, Silva CA, Higashikawa FS, Roig A, Sánchez-Monedero MA. Use of biochar as bulking agent for the composting of poultry manure: effect on organic matter degradation and humification. *Bioresour Technol* 2010;**101**(4):1239–46.
46. Ngo PT, Rumpel C, Ngo QA, Alexis M, Vargas GV, Mora Gil MDLL, et al. Biological and chemical reactivity and phosphorus forms of buffalo manure compost, vermicompost and their mixture with biochar. *Bioresour Technol* 2013;**148**:401–7.
47. Sukartono UWH, Kusuma Z, Nugroho WH. Soil fertility status, nutrient uptake, and maize (*Zea mays* L.) yield following biochar and cattle manure application on sandy soils of Lombok, Indonesia. *J Trop Agric* 2011;**49**:47–52.
48. Prasai TP, Walsh KB, Midmore DJ, Jones BEH, Bhattarai SP. Manure from biochar, bentonite and zeolite feed supplemented poultry: moisture retention and granulation properties. *J Environ Manag* 2018;**216**:82–8.
49. Sánchez-García M, Alburquerque JA, Sánchez-Monedero MA, Roig A, Cayuela ML. Biochar accelerates organic matter degradation and enhances N mineralisation during composting of poultry manure without a relevant impact on gas emissions. *Bioresour Technol* 2015;**192**:272–9.
50. Sokołowska Z, Szewczuk-Karpisz K, Turski M, Tomczyk A, Cybulak M, Skic K. Effect of wood waste and sunflower husk biochar on tensile strength and porosity of dystric Cambisol artificial aggregates. *Agronomy* 2020;**10**(2), 244.
51. Prasai TP, Walsh KB, Bhattarai SP, Midmore DJ, Van Thi TH, Moore RJ, et al. Biochar, bentonite and zeolite supplemented feeding of layer chickens alters intestinal microbiota and reduces campylobacter load. *PLoS One* 2016;**11**(4), e0154061.
52. Preston CM, Schmidt MWI. Black (pyrogenic) carbon: a synthesis of current knowledge and uncertainties with special consideration of boreal regions. *Biogeosciences* 2006;**3**(4):397–420.
53. Rogovska N, Laird D, Cruse R, Fleming P, Parkin T, Meek D. Impact of biochar on manure carbon stabilization and greenhouse gas emissions. *Soil Sci Soc Am J* 2011;**75**(3):871–9.
54. Maurer DL, Koziel JA, Kalus K, Andersen DS, Opalinski S. Pilot-scale testing of non-activated biochar for swine manure treatment and mitigation of ammonia, hydrogen sulfide, odorous volatile organic compounds (VOCs), and greenhouse gas emissions. *Sustainability* 2017;**9**(6), 929.
55. Sommer SG, Petersen SO, Søgaard HT. Greenhouse gas emission from stored livestock slurry. *J Environ Qual* 2000;**29**(3):744–51.
56. Janczak D, Malińska K, Czekała W, Cáceres R, Lewicki A, Dach J. Biochar to reduce ammonia emissions in gaseous and liquid phase during composting of poultry manure with wheat straw. *Waste Manag* 2017;**66**:36–45.

57. Sánchez-Monedero MA, Sánchez-García M, Alburquerque JA, Cayuela ML. Biochar reduces volatile organic compounds generated during chicken manure composting. *Bioresour Technol* 2019;**288**, 121584.
58. Petersen SO, Ambus P. Methane oxidation in pig and cattle slurry storages, and effects of surface crust moisture and methane availability. *Nutr Cycl Agroecosyst* 2006;**74**(1):1–11.
59. Vanderzaag AC, Gordon RJ, Jamieson RC, Burton DL, Stratton GW. Effects of winter storage conditions and subsequent agitation on gaseous emissions from liquid dairy manure. *Can J Soil Sci* 2010;**90**(1):229–39.
60. Zucconi F, de Bertoldi M. Compost specifications for the production and characterization of compost from municipal solid waste. In: de Bertoldi M, Ferranti MP, L'Hermite P, Zucconi F, editors. *Compost: production, quality and use.* Dordrecht: Elsevier; 1987. p. 30–50.
61. Jain MS, Jambhulkar R, Kalamdhad AS. Biochar amendment for batch composting of nitrogen rich organic waste: effect on degradation kinetics, composting physics and nutritional properties. *Bioresour Technol* 2018;**253**:204–13.
62. Larney FJ, Sullivan DM, Buckley KE, Eghball B. The role of composting in recycling manure nutrients. *Can J Soil Sci* 2006;**86**(4):597–611.
63. Ryckeboer J, Mergaert J, Vaes K, Klammer S, De Clercq D, Coosemans J, et al. A survey of bacteria and fungi occurring during composting and self-heating processes. *Ann Microbiol* 2003;**53**(4):349–410.
64. Imbeah M. Composting piggery waste: a review. *Bioresour Technol* 1998;**63**(3):197–203.
65. Jain MS, Paul S, Kalamdhad AS. Utilization of biochar as an amendment during lignocellulose waste composting: impact on composting physics and realization (probability) amongst physical properties. *Process Saf Environ Prot* 2019;**121**:229–38.
66. Jia X, Wang M, Yuan W, Ju X, Yang B. The influence of biochar addition on chicken manure composting and associated methane and carbon dioxide emissions. *Bioresources* 2016;**11**(2):5255–64.
67. Zhang L, Sun X. Changes in physical, chemical, and microbiological properties during the two-stage co-composting of green waste with spent mushroom compost and biochar. *Bioresour Technol* 2014;**171**(1):274–84.
68. Pampuro N, Bagagiolo G, Priarone PC, Cavallo E. Effects of pelletizing pressure and the addition of woody bulking agents on the physical and mechanical properties of pellets made from composted pig solid fraction. *Powder Technol* 2017;**311**:112–9.
69. Steiner C, Das KC, Melear N, Lakly D. Reducing nitrogen loss during poultry litter composting using biochar. *J Environ Qual* 2010;**39**(4):1236–42.
70. Huang XD, Xue D. Effects of bamboo biochar addition on temperature rising, dehydration and nitrogen loss during pig manure composting. *Chin J Appl Ecol* 2014;**25**(4):1057–62.
71. Mao H, Li R, Huang Y, Wang Z. Effect of additives on forms of Zn and Cu during aerobic composting of pig manure. *Nongye Jixie Xuebao/Trans Chin Soc Agric Eng* 2013;**44**(10). 164-71 + 202.
72. Liu H. Biochar is conducive to reduce thermal loss caused by mechanical turning during swine manure composting. *Bioresour Technol* 2019;**290**, 121810.
73. Qiu X, Zhou G, Zhang J, Wang W. Microbial community responses to biochar addition when a green waste and manure mix are composted: a molecular ecological network analysis. *Bioresour Technol* 2019;**273**:666–71.
74. Czekała W, Malińska K, Cáceres R, Janczak D, Dach J, Lewicki A. Co-composting of poultry manure mixtures amended with biochar—the effect of biochar on temperature and C-CO2 emission. *Bioresour Technol* 2016;**200**:921–7.

75. Chen W, Liao X, Wu Y, Liang JB, Mi J, Huang J, et al. Effects of different types of biochar on methane and ammonia mitigation during layer manure composting. *Waste Manag* 2017;**61**:506–15.
76. Awasthi MK, Duan Y, Liu T, Awasthi SK, Zhang Z. Relevance of biochar to influence the bacterial succession during pig manure composting. *Bioresour Technol* 2020;**304**, 122962.
77. Liu N, Zhou J, Han L, Ma S, Sun X, Huang G. Role and multi-scale characterization of bamboo biochar during poultry manure aerobic composting. *Bioresour Technol* 2017;**241**:190–9.
78. Li J, Bao H, Xing W, Yang J, Liu R, Wang X, et al. Succession of fungal dynamics and their influence on physicochemical parameters during pig manure composting employing with pine leaf biochar. *Bioresour Technol* 2020;**297**, 122377.
79. Zhou G, Xu X, Qiu X, Zhang J. Biochar influences the succession of microbial communities and the metabolic functions during rice straw composting with pig manure. *Bioresour Technol* 2019;**272**:10–8.
80. Jindo K, Suto K, Matsumoto K, García C, Sonoki T, Sanchez-Monedero MA. Chemical and biochemical characterisation of biochar-blended composts prepared from poultry manure. *Bioresour Technol* 2012;**110**:396–404.
81. Jiang X, Deng L, Meng Q, Sun Y, Han Y, Wu X, et al. Fungal community succession under influence of biochar in cow manure composting. *Environ Sci Pollut Res* 2020;**27 (9)**:9658–68.
82. Tu Q, Wu W, Lu H, Sun B, Wang C, Deng H, et al. The effect of biochar and bacterium agent on humification during swine manure composting. In: Xu J, Wu J, He Y, editors. *Functions of natural organic matter in changing environment*. The Netherlands: Springer Dordrecht; 2013. p. 1021–5. 9789400756342 2013.
83. Li B, Ye J, Liu C, Li Y, Weng B, Wang Y. Effects of biochar addition on carbon transformation during composting of pig manure. *Huanjing Kexue Xuebao/Acta Sci Circumst* 2017;**37**(9):3511–8.
84. Jindo K, Sánchez-Monedero MA, Matsumoto K, Sonoki T. The efficiency of a low dose of biochar in enhancing the aromaticity of humic-like substance extracted from poultry manure compost. *Agronomy* 2019;**9**(5), 248.
85. Wang C, Tu Q, Dong D, Strong PJ, Wang H, Sun B, et al. Spectroscopic evidence for biochar amendment promoting humic acid synthesis and intensifying humification during composting. *J Hazard Mater* 2014;**280**:409–16.
86. Jindo K, Sonoki T, Matsumoto K, Canellas L, Roig A, Sanchez-Monedero MA. Influence of biochar addition on the humic substances of composting manures. *Waste Manag* 2016;**49**:545–52.
87. Hagemann N, Subdiaga E, Orsetti S, de la Rosa JM, Knicker H, Schmidt HP, et al. Effect of biochar amendment on compost organic matter composition following aerobic compositing of manure. *Sci Total Environ* 2018;**613–614**:20–9.
88. Khan N, Clark I, Sánchez-Monedero MA, Shea S, Meier S, Bolan N. Maturity indices in co-composting of chicken manure and sawdust with biochar. *Bioresour Technol* 2014;**168**:245–51.
89. Jia X, Wang M, Yuan W, Shah S, Shi W, Meng X, et al. N2O emission and nitrogen transformation in chicken manure and biochar co-composting. *Trans ASABE* 2016;**59** (5):1277–83.
90. Duan Y, Awasthi SK, Liu T, Zhang Z, Awasthi MK. Evaluation of integrated biochar with bacterial consortium on gaseous emissions mitigation and nutrients sequestration during pig manure composting. *Bioresour Technol* 2019;**291**, 121880.
91. Zhang J, Chen G, Sun H, Zhou S, Zou G. Straw biochar hastens organic matter degradation and produces nutrient-rich compost. *Bioresour Technol* 2016;**200**:876–83.

92. Wang C, Lu H, Dong D, Deng H, Strong PJ, Wang H, et al. Insight into the effects of biochar on manure composting: evidence supporting the relationship between N_2O emission and denitrifying community. *Environ Sci Technol* 2013;**47**(13):7341–9.

93. He X, Yin H, Han L, Cui R, Fang C, Huang G. Effects of biochar size and type on gaseous emissions during pig manure/wheat straw aerobic composting: insights into multivariate-microscale characterization and microbial mechanism. *Bioresour Technol* 2019;**271**:375–82.

94. Insam H, Gómez-Brandón M, Ascher J. Manure-based biogas fermentation residues—friend or foe of soil fertility? *Soil Biol Biochem* 2015;**84**:1–14.

95. Nasir IM, Mohd Ghazi TI, Omar R. Anaerobic digestion technology in livestock manure treatment for biogas production: a review. *Eng Life Sci* 2012;**12**(3):258–69.

96. Neshat SA, Mohammadi M, Najafpour GD, Lahijani P. Anaerobic co-digestion of animal manures and lignocellulosic residues as a potent approach for sustainable biogas production. *Renew Sust Energ Rev* 2017;**79**:308–22.

97. Leuchtenberger A. *Grundwissen zur mikrobiellen Biotechnologie*. Stuttgart, Germany: B.G. Teubner; 1998.

98. Tufaner F, Avşar Y. Effects of co-substrate on biogas production from cattle manure: a review. *Int J Environ Sci Technol* 2016;**13**(9):2303–12.

99. Indren M, Birzer CH, Kidd SP, Hall T, Medwell PR. Effects of biochar parent material and microbial pre-loading in biochar-amended high-solids anaerobic digestion. *Bioresour Technol* 2020;**298**, 122457.

100. Pan J, Ma J, Liu X, Zhai L, Ouyang X, Liu H. Effects of different types of biochar on the anaerobic digestion of chicken manure. *Bioresour Technol* 2019;**275**:258–65.

101. Pan JT, Ma JY, Qiu L, Gou XH, Gao TL. The performance of biochar-mediated anaerobic digestion of chicken manure. *Zhongguo Huanjing Kexue/China Environ Sci* 2016;**36**(9):2716–21.

102. Ma J, Pan J, Qiu L, Wang Q, Zhang Z. Biochar triggering multipath methanogenesis and subdued propionic acid accumulation during semi-continuous anaerobic digestion. *Bioresour Technol* 2019;**293**, 122026.

103. Lei M, Cheng Y, Miao N, Zhou J, Chen Z. Effects of mixing loess and other additives with pig manure on ammonia and greenhouse gas emissions during storage. *Huanjing Kexue Xuebao/Acta Sci Circumst* 2019;**39**(12):4132–9.

104. Baltrėnas P, Paliulis D, Kolodynskij V. The experimental study of biogas production when digesting chicken manure with a biochar additive. *Greenhouse Gases Sci Technol* 2019;**9**(4):837–47.

105. Wang H, Larson RA, Runge T. Impacts to hydrogen sulfide concentrations in biogas when poplar wood chips, steam treated wood chips, and biochar are added to manure-based anaerobic digestion systems. *Bioresour Technol Rep* 2019;**7**, 100232.

106. Yue X, Arena U, Chen D, Lei K, Dai X. Anaerobic digestion disposal of sewage sludge pyrolysis liquid in cow dung matrix and the enhancing effect of sewage sludge char. *J Clean Prod* 2019;**235**:801–11.

107. Ngigi AN, Ok YS, Thiele-Bruhn S. Biochar-mediated sorption of antibiotics in pig manure. *J Hazard Mater* 2019;**364**:663–70.

108. Li DN, Zhang KQ, Liang JF, Gao WX, Kong DW, Du LZ. Solid-state anaerobic digestion of pig manure with three kinds of additives. *J Agro-Environ Sci* 2019;**38**(8):1777–85.

109. Pan J, Qiu L, Guo X, Ma J, Gao T. Optimizing process parameters for methane production during biochar-mediated anaerobic digestion of poultry manure. *Nongye Jixie Xuebao/Trans Chin Soc Agric Eng* 2016;**47**(3):167–73.

110. Tomczyk A, Sokołowska Z, Boguta P. Biochar physicochemical properties: pyrolysis temperature and feedstock kind effects. *Rev Environ Sci Biotechnol* 2020;**19**(1):191–215.

111. Pandit NR, Schmidt HP, Mulder J, Hale SE, Husson O, Cornelissen G. Nutrient effect of various composting methods with and without biochar on soil fertility and maize growth. *Arch Agron Soil Sci* 2020;**66**(2):250–65.

112. Matsi T. *Liquid cattle manure application to soil and its effect on crop growth, yield, composition, and on soil properties.* InTech; 2012.

113. Miller JJ, Beasley BW, Drury CF, Zebarth BJ. Available nitrogen and phosphorus in soil amended with fresh or composted cattle manure containing straw or wood-chip bedding. *Can J Soil Sci* 2010;**90**(2):341–54.

114. DBP B, Nabel M, Jablonowski ND. Biogas-digestate as nutrient source for biomass production of Sida hermaphrodita, Zea mays L. and Medicago sativa L. *Energy Procedia* 2014;**59**:120–6.

115. Reichel R, Radl V, Rosendahl I, Albert A, Amelung W, Schloter M, et al. Soil microbial community responses to antibiotic-contaminated manure under different soil moisture regimes. *Appl Microbiol Biotechnol* 2014;**98**(14):6487–95.

116. KTBL. In: (KTBL) KfTuBidLeV, editor. *Faustzahlen für die Landwirtschaft.* 15 ed. Frankfurt/M: Druck- und Verlagshaus Zarbock GmbH Co. KG; 2018.

117. Antonious GF, Turley ET, Dawood MH. Monitoring soil enzymes activity before and after animal manure application. *Agriculture (Switzerland)* 2020;**10**(5), 166.

118. Zhao Y, Yan Z, Qin J, Xiao Z. Effects of long-term cattle manure application on soil properties and soil heavy metals in corn seed production in Northwest China. *Environ Sci Pollut Res* 2014;**21**(12):7586–95.

119. Thangarajan R, Bolan NS, Tian G, Naidu R, Kunhikrishnan A. Role of organic amendment application on greenhouse gas emission from soil. *Sci Total Environ* 2013;**465**:72–96.

120. Rochette P, Angers DA, Chantigny MH, Gagnon B, Bertrand N. In situ mineralization of dairy cattle manures as determined using soil-surface carbon dioxide fluxes. *Soil Sci Soc Am J* 2006;**70**(3):744–52.

121. Japenga J, Harmsen K. Determination of mass balances and ionic balances in animal manure. *Neth J Agric Sci* 1990;**38**:353–67.

122. Aust MO, Thiele-Bruhn S, Eckhardt KU, Leinweber P. Composition of organic matter in particle size fractionated pig slurry. *Bioresour Technol* 2009;**100**(23):5736–43.

123. Webb J, Pain B, Bittman S, Morgan J. The impacts of manure application methods on emissions of ammonia, nitrous oxide and on crop response-a review. *Agric Ecosyst Environ* 2010;**137**(1–2):39–46.

124. Loro PJ, Bergstrom DW, Beauchamp EG. Intensity and duration of denitrification following application of manure and fertilizer to soil. *J Environ Qual* 1997;**26**(3):706–13.

125. Hammes K, Torn MS, Lapenas AG, Schmidt MWI. Centennial black carbon turnover observed in a Russian steppe soil. *Biogeosciences* 2008;**5**(5):1339–50.

126. Kuzyakov Y, Subbotina I, Chen H, Bogomolova I, Xu X. Black carbon decomposition and incorporation into soil microbial biomass estimated by 14C labeling. *Soil Biol Biochem* 2009;**41**(2):210–9.

127. Wu H, Che X, Ding Z, Hu X, Creamer AE, Chen H, et al. Release of soluble elements from biochars derived from various biomass feedstocks. *Environ Sci Pollut Res* 2016;**23**(2):1905–15.

128. Adekiya AO, Agbede TM, Ejue WS, Aboyeji CM, Dunsin O, Aremu CO, et al. Biochar, poultry manure and NPK fertilizer: sole and combine application effects on soil properties and ginger (*Zingiber officinale* Roscoe) performance in a tropical Alfisol. *Open Agric* 2020;**5**(1):30–9.

129. Sarkhot DV, Ghezzehei TA, Berhe AA. Effectiveness of biochar for sorption of ammonium and phosphate from dairy effluent. *J Environ Qual* 2013;**42**(5):1545–54.

130. Ippolito JA, Stromberger ME, Lentz RD, Dungan RS. Hardwood biochar and manure co-application to a calcareous soil. *Chemosphere* 2016;**142**:84–91.

131. Kammann CI, Schmidt HP, Messerschmidt N, Linsel S, Steffens D, Müller C, et al. Plant growth improvement mediated by nitrate capture in co-composted biochar. *Sci Rep* 2015;**5**, 11080.

132. Hagemann N, Kammann CI, Schmidt HP, Kappler A, Behrens S. Nitrate capture and slow release in biochar amended compost and soil. *PLoS One* 2017;**12**(2), e0171214.

133. Shin J, Park S, Lee S. Optimum method uploaded nutrient solution for blended biochar pellet with application of nutrient releasing model as slow release fertilizer. *Appl Sci* 2019;**9**(9), 1899.

134. Banik C, Koziel JA, De M, Bonds D, Chen B, Singh A, et al. Biochar-swine manure impact on soil nutrients and carbon under controlled leaching experiment using a midwestern mollisols. *Front Environ Sci* 2021;**9**. https://doi.org/10.3389/fenvs.2021.609621.

135. Kocatürk-Schumacher NP, Zwart K, Bruun S, Stoumann Jensen L, Sørensen H, Brussaard L. Recovery of nutrients from the liquid fraction of digestate: use of enriched zeolite and biochar as nitrogen fertilizers. *J Plant Nutr Soil Sci* 2019;**182**(2):187–95.

136. Plaza C, Giannetta B, Fernández JM, López-de-Sá EG, Polo A, Gascó G, et al. Response of different soil organic matter pools to biochar and organic fertilizers. *Agric Ecosyst Environ* 2016;**225**:150–9.

137. Ye J, Zhang R, Nielsen S, Joseph SD, Huang D, Thomas T. A combination of biochar-mineral complexes and compost improves soil bacterial processes, soil quality, and plant properties. *Front Microbiol* 2016;**7**(APR):372. https://doi.org/10.3389/fmicb.2016.00372.

138. Šimanský V, Horák J, Igaz D, Balashov E, Jonczak J. Biochar and biochar with N fertilizer as a potential tool for improving soil sorption of nutrients. *J Soils Sediments* 2018;**18**(4):1432–40.

139. Joseph SD, Camps-Arbestain M, Lin Y, Munroe P, Chia CH, Hook J, et al. An investigation into the reactions of biochar in soil. *Aust J Soil Res* 2010;**48**(6–7):501–15.

140. Prost K, Borchard N, Siemens J, Kautz T, Séquaris JM, Möller A, et al. Biochar affected by composting with farmyard manure. *J Environ Qual* 2013;**42**(1):164–72.

141. Fierer N, Schimel JP, Holden PA. Variations in microbial community composition through two soil depth profiles. *Soil Biol Biochem* 2003;**35**(1):167–76.

142. Kramer C, Gleixner G. Variable use of plant- and soil-derived carbon by microorganisms in agricultural soils. *Soil Biol Biochem* 2006;**38**(11):3267–78.

143. Williams RE, Edwards DR. Effects of biochar treatment of municipal biosolids and horse manure on quality of runoff from fescue plots. *Trans ASABE* 2017;**60**(2):409–17.

144. Zheng H, Wang Z, Deng X, Herbert S, Xing B. Impacts of adding biochar on nitrogen retention and bioavailability in agricultural soil. *Geoderma* 2013;**206**:32–9.

145. Liu Z, Chen X, Jing Y, Li Q, Zhang J, Huang Q. Effects of biochar amendment on rapeseed and sweet potato yields and water stable aggregate in upland red soil. *Catena* 2014;**123**:45–51.

146. Sarkhot DV, Berhe AA, Ghezzehei TA. Impact of biochar enriched with dairy manure effluent on carbon and nitrogen dynamics. *J Environ Qual* 2012;**41**(4):1107–14.

147. Angst TE, Patterson CJ, Reay DS, Anderson P, Peshkur TA, Sohi SP. Biochar diminishes nitrous oxide and nitrate leaching from diverse nutrient sources. *J Environ Qual* 2013;**42**(3):672–82.

148. Kizito S, Luo H, Lu J, Bah H, Dong R, Wu S. Role of nutrient-enriched biochar as a soil amendment during maize growth: exploring practical alternatives to recycle agricultural residuals and to reduce chemical fertilizer demand. *Sustainability* 2019;**11**(11), 3211.

149. McDonald MR, Bakker C, Motior MR. Evaluation of wood biochar and compost soil amendment on cabbage yield and quality. *Can J Plant Sci* 2019;**99**(5):624–38.

150. Qayyum MF, Liaquat F, Rehman RA, Gul M, Ul Hye MZ, Rizwan M, et al. Effects of co-composting of farm manure and biochar on plant growth and carbon mineralization in an alkaline soil. *Environ Sci Pollut Res* 2017;**24**(33):26060–8.
151. Yuan Y, Chen H, Yuan W, Williams D, Walker JT, Shi W. Is biochar-manure co-compost a better solution for soil health improvement and N2O emissions mitigation? *Soil Biol Biochem* 2017;**113**:14–25.
152. Zhang Z, Dong X, Wang S, Pu X. Benefits of organic manure combined with biochar amendments to cotton root growth and yield under continuous cropping systems in Xinjiang, China. *Sci Rep* 2020;**10**(1), 4718.
153. Fischer D, Glaser B. *Synergisms between compost and biochar for sustainable soil amelioration.* InTech; 2012.
154. Agegnehu G, Srivastava AK, Bird MI. The role of biochar and biochar-compost in improving soil quality and crop performance: a review. *Appl Soil Ecol* 2017;**119**: 156–70.
155. Wu H, Lai C, Zeng G, Liang J, Chen J, Xu J, et al. The interactions of composting and biochar and their implications for soil amendment and pollution remediation: a review. *Crit Rev Biotechnol* 2017;**37**(6):754–64.
156. Cai Y, Akiyama H. Effects of inhibitors and biochar on nitrous oxide emissions, nitrate leaching, and plant nitrogen uptake from urine patches of grazing animals on grasslands: a meta-analysis. *Soil Sci Plant Nutr* 2017;**63**(4):405–14.
157. Schmidt HP, Kammann C, Niggli C, Evangelou MWH, Mackie KA, Abiven S. Biochar and biochar-compost as soil amendments to a vineyard soil: influences on plant growth, nutrient uptake, plant health and grape quality. *Agric Ecosyst Environ* 2014;**191**:117–23.
158. Schulz H, Dunst G, Glaser B. No effect level of co-composted biochar on plant growth and soil properties in a greenhouse experiment. *Agronomy* 2014;**4**(1):34–51.
159. Collins HP, Streubel J, Alva A, Porter L, Chaves B. Phosphorus uptake by potato from biochar amended with anaerobic digested dairy manure effluent. *Agron J* 2013;**105** (4):989–98.
160. McKinnon K, Serikstad GL, Eggen T. In: Rahmann G, Aksoy U, editors. *Contaminants in manure—a problem for organic farming?* Istanbul, Turkey: Thünen Institute; 2014. p. 903–4.
161. Bloem E, Albihn A, Elving J, Hermann L, Lehmann L, Sarvi M, et al. Contamination of organic nutrient sources with potentially toxic elements, antibiotics and pathogen microorganisms in relation to P fertilizer potential and treatment options for the production of sustainable fertilizers: a review. *Sci Total Environ* 2017;**607–608**:225–42.
162. Thiele-Bruhn S. Pharmaceutical antibiotic compounds in soils—a review. *J Plant Nutr Soil Sci* 2003;**166**(2):145–67.
163. Grenni P, Ancona V, Barra CA. Ecological effects of antibiotics on natural ecosystems: a review. *Microchem J* 2018;**136**(Suppl C):25–39.
164. Hejna M, Moscatelli A, Onelli E, Baldi A, Pilu S, Rossi L. Evaluation of concentration of heavy metals in animal rearing system. *Ital J Anim Sci* 2019;**18**(1):1372–84.
165. Irshad M, Gul S, Egrinya Eneji A, Anwar Z, Ashraf M. Extraction of heavy metals from manure and their bioavailability to spinach (Spinacia oleracea l.) after composting. *J Plant Nutr* 2014;**37**(10):1661–75.
166. Kumar RR, Park BJ, Cho JY. Application and environmental risks of livestock manure. *J Korean Soc Appl Biol Chem* 2013;**56**(5):497–503.
167. Meng J, Wang L, Zhong L, Liu X, Brookes PC, Xu J, et al. Contrasting effects of composting and pyrolysis on bioavailability and speciation of Cu and Zn in pig manure. *Chemosphere* 2017;**180**:93–9.
168. Liu W, Huo R, Xu J, Liang S, Li J, Zhao T, et al. Effects of biochar on nitrogen transformation and heavy metals in sludge composting. *Bioresour Technol* 2017;**235**:43–9.

169. Liu WR, Zeng D, She L, Su WX, He DC, Wu GY, et al. Comparisons of pollution characteristics, emission situations, and mass loads for heavy metals in the manures of different livestock and poultry in China. *Sci Total Environ* 2020;**734**, 139023.
170. Olowoyo JO, Mugivhisa LL. Evidence of uptake of different pollutants in plants harvested from soil treated and fertilized with organic materials as source of soil nutrients from developing countries. *Chem Biol Technol Agric* 2019;**6**(1), 28.
171. Kuppusamy S, Kakarla D, Venkateswarlu K, Megharaj M, Yoon Y-E, Lee YB. Veterinary antibiotics (VAs) contamination as a global agro-ecological issue: a critical view. *Agric Ecosyst Environ* 2018;**257**:47–59.
172. Riviere JE. *Comparative pharmacokinetics: principles, techniques and applications.* 2nd ed. Hoboken, NJ: Wiley-Blackwell; 2011.
173. Sarmah AK, Meyer MT, Boxall ABA. A global perspective on the use, sales, exposure pathways, occurrence, fate and effects of veterinary antibiotics (VAs) in the environment. *Chemosphere* 2006;**65**(5):725–59.
174. Du L, Liu W. Occurrence, fate, and ecotoxicity of antibiotics in agro-ecosystems. A review. *Agron Sustain Dev* 2012;**32**(2):309–27.
175. Karci A, Balcioğlu IA. Investigation of the tetracycline, sulfonamide, and fluoroquinolone antimicrobial compounds in animal manure and agricultural soils in Turkey. *Sci Total Environ* 2009;**407**(16):4652–64.
176. Łukaszewicz P, Maszkowska J, Mulkiewicz E, Kumirska J, Stepnowski P, Caban M. Impact of veterinary pharmaceuticals on the agricultural environment: a re-inspection. *Rev Environ Contam Toxicol 243*, 2017;89–148.
177. Ok YS, Kim SC, Kim KR, Lee SS, Moon DH, Lim KJ, et al. Monitoring of selected veterinary antibiotics in environmental compartments near a composting facility in Gangwon Province, Korea. *Environ Monit Assess* 2011;**174**(1-4):693–701.
178. Thiele-Bruhn S. *Environmental risks from mixtures of antibiotic pharmaceuticals in soils—a literature review.* Dessau-Roßlau: Umweltbundesamt Texte 32/2019; 2019. p. 120.
179. Krzebietke S, Mackiewicz-Walec E, Sienkiewicz S, Załuski D. Effect of manure and mineral fertilisers on the content of light and heavy polycyclic aromatic hydrocarbons in soil. *Sci Rep* 2020;**10**(1), 4573.
180. Ferrell J, Dittmar PJ, Sellers B, Devkota P, editors. *Herbicide residues in manure, compost, or hay.* Gainesville, FL: University of Florida; 2017.
181. Sobsey DM, Khatib AL, Hill RV, Alocilja E, Pillai S. *Pathogens in animal wastes and the impacts of waste management practices on their survival, transport and fate.* St. Joseph, MI: ASABE; 2006.
182. Sobur AM, Al Momen SA, Sarker R, Taufiqur Rahman AMM, Lutful Kabir SM, Tanvir RM. Antibiotic-resistant *Escherichia coli* and *Salmonella* spp. associated with dairy cattle and farm environment having public health significance. *Vet World* 2019;**12**(7):984–93.
183. Manyi-Loh CE, Mamphweli SN, Meyer EL, Makaka G, Simon M, Okoh AI. An overview of the control of bacterial pathogens in cattle manure. *Int J Environ Res Public Health* 2016;**13**(9), 8431.
184. Chee-Sanford JC, Mackie RI, Koike S, Krapac IG, Lin YF, Yannarell AC, et al. Fate and transport of antibiotic residues and antibiotic resistance genes following land application of manure waste. *J Environ Qual* 2009;**38**(3):1086–108.
185. Jechalke S, Heuer H, Siemens J, Amelung W, Smalla K. Fate and effects of veterinary antibiotics in soil. *Trends Microbiol* 2014;**22**(9):536–45.
186. Pérez-Valera E, Kyselková M, Ahmed E, Sladecek FXJ, Goberna M, Elhottová D. Native soil microorganisms hinder the soil enrichment with antibiotic resistance genes following manure applications. *Sci Rep* 2019;**9**(1), 6760.
187. He Y, Yuan Q, Mathieu J, Stadler L, Senehi N, Sun R, et al. Antibiotic resistance genes from livestock waste: occurrence, dissemination, and treatment. *npj Clean Water* 2020;**3**(1), 4.

188. Forsberg KJ, Reyes A, Wang B, Selleck EM, Sommer MOA, Dantas G. The shared antibiotic resistome of soil bacteria and human pathogens. *Science* 2012;**337**(6098):1107–11.
189. Ngigi AN, Ok YS, Thiele-Bruhn S. Biochar affects the dissipation of antibiotics and abundance of antibiotic resistance genes in pig manure. *Bioresour Technol* 2020;**315**.
190. McKinney CW, Dungan RS, Moore A, Leytem AB. Occurrence and abundance of antibiotic resistance genes in agricultural soil receiving dairy manure. *FEMS Microbiol Ecol* 2018;**94**(3), fiy010.
191. Zhao Q, Wang Y, Wang S, Wang Z, Du XD, Jiang H, et al. Prevalence and abundance of florfenicol and linezolid resistance genes in soils adjacent to swine feedlots. *Sci Rep* 2016;**6**, 32192.
192. Zhu Y-G, Zhao Y, Zhu D, Gillings M, Penuelas J, Ok YS, et al. Soil biota, antimicrobial resistance and planetary health. *Environ Int* 2019;**131**, 105059.
193. Oliver JP, Gooch CA, Lansing S, Schueler J, Hurst JJ, Sassoubre L, et al. Invited review: fate of antibiotic residues, antibiotic-resistant bacteria, and antibiotic resistance genes in US dairy manure management systems. *J Dairy Sci* 2020;**103**(2):1051–71.
194. Beesley L, Moreno-Jiménez E, Gomez-Eyles JL, Harris E, Robinson B, Sizmur T. A review of biochars' potential role in the remediation, revegetation and restoration of contaminated soils. *Environ Pollut* 2011;**159**(12):3269–82.
195. Rajapaksha AU, Chen SS, Tsang DCW, Zhang M, Vithanage M, Mandal S, et al. Engineered/designer biochar for contaminant removal/immobilization from soil and water: potential and implication of biochar modification. *Chemosphere* 2016;**148**: 276–91.
196. Guo M, Song W, Tian J. Biochar-facilitated soil remediation: mechanisms and efficacy variations. *Front Environ Sci* 2020;**8**, 521512.
197. Chen YX, Huang XD, Han ZY, Huang X, Hu B, Shi DZ, et al. Effects of bamboo charcoal and bamboo vinegar on nitrogen conservation and heavy metals immobility during pig manure composting. *Chemosphere* 2010;**78**(9):1177–81.
198. Cui E, Wu Y, Zuo Y, Chen H. Effect of different biochars on antibiotic resistance genes and bacterial community during chicken manure composting. *Bioresour Technol* 2016;**203**:11–7.
199. Wang J, Sui B, Shen Y, Meng H, Zhao L, Zhou H, et al. Effects of different biochars on antibiotic resistance genes during swine manure thermophilic composting. *Int J Agric Biol Eng* 2018;**11**(6):166–71.
200. Li H, Duan M, Gu J, Zhang Y, Qian X, Ma J, et al. Effects of bamboo charcoal on antibiotic resistance genes during chicken manure composting. *Ecotoxicol Environ Saf* 2017;**140**:1–6.
201. Hua L, Wu W, Liu Y, McBride MB, Chen Y. Reduction of nitrogen loss and Cu and Zn mobility during sludge composting with bamboo charcoal amendment. *Environ Sci Pollut Res* 2009;**16**(1):1–9.
202. Oleszczuk P, Hale SE, Lehmann J, Cornelissen G. Activated carbon and biochar amendments decrease pore-water concentrations of polycyclic aromatic hydrocarbons (PAHs) in sewage sludge. *Bioresour Technol* 2012;**111**:84–91.
203. Antonangelo JA, Sun X, Zhang H. The roles of co-composted biochar (COMBI) in improving soil quality, crop productivity, and toxic metal amelioration. *J Environ Manag* 2021;**277**, 111443.
204. Oleszczuk P, Kołtowski M. Changes of total and freely dissolved polycyclic aromatic hydrocarbons and toxicity of biochars treated with various aging processes. *Environ Pollut* 2018;**237**:65–73.
205. Godlewska P, Schmidt HP, Ok YS, Oleszczuk P. Biochar for composting improvement and contaminants reduction. A review. *Bioresour Technol* 2017;**246**:193–202.
206. Du J, Zhang Y, Qu M, Yin Y, Fan K, Hu B, et al. Effects of biochar on the microbial activity and community structure during sewage sludge composting. *Bioresour Technol* 2019;**272**:171–9.

207. Wei L, Shutao W, Jin Z, Tong X. Biochar influences the microbial community structure during tomato stalk composting with chicken manure. *Bioresour Technol* 2014;**154**:148–54.
208. Riaz L, Wang Q, Yang Q, Li X, Yuan W. Potential of industrial composting and anaerobic digestion for the removal of antibiotics, antibiotic resistance genes and heavy metals from chicken manure. *Sci Total Environ* 2020;**718**, 137414.
209. Palansooriya KN, Shaheen SM, Chen SS, Tsang DCW, Hashimoto Y, Hou D, et al. Soil amendments for immobilization of potentially toxic elements in contaminated soils: a critical review. *Environ Int* 2020;**134**, 105046.
210. Li R, Wang Q, Zhang Z, Zhang G, Li Z, Wang L, et al. Nutrient transformation during aerobic composting of pig manure with biochar prepared at different temperatures. *Environ Technol* 2015;**36**(7):815–26.
211. López-Cano I, Roig A, Cayuela ML, Alburquerque JA, Sánchez-Monedero MA. Biochar improves N cycling during composting of olive mill wastes and sheep manure. *Waste Manag* 2016;**49**:553–9.

CHAPTER SEVEN

Biochar role in improving pathogens removal capacity of stormwater biofilters

Renan Valenca, Annesh Borthakur, Huong Le, and Sanjay K. Mohanty*

Department of Civil and Environmental Engineering, University of California Los Angeles, California, United States
*Corresponding author: e-mail address: mohanty@ucla.edu

Contents

1.	Introduction	176
2.	Testing methods to evaluate biochar capacity to remove pathogens in stormwater	179
3.	Pathogen removal processes in biochar-amended filter media	184
	3.1 Attachment and straining	184
	3.2 Die-off and inactivation	185
	3.3 Low remobilization during intermittent flow	188
4.	Challenges	189
	4.1 Not all biochars are made equal	189
	4.2 Chemical weathering could affect removal capacity	190
	4.3 Biological weathering or biofilm development could affect biochar performance	190
	4.4 Pathogen removal depends on how biochar is applied in biofilters	192
	4.5 Limited removal capacity for virus	193
5.	Opportunities	193
	5.1 Selection of best biochar based on biochar properties	193
	5.2 Modifying biochar surface properties	193
	5.3 Lack of field studies	194
	5.4 Change in biofilter microbiome after biochar addition	194
6.	Summary	195
	References	195

Abstract

Stormwater treatment systems such as biofilters have been used to treat and reuse stormwater in water-stressed urban areas. However, the pathogen removal capacity of these systems is low and unreliable. Pathogens are difficult to remove because of many reasons: conventional biofilter amendments have low removal capacity, and previously removed pathogens can grow in biofilters or be remobilized during intermittent infiltration of stormwater. Variable climate affects removal and increases uncertainty to

Advances in Chemical Pollution, Environmental Management and Protection, Volume 7
ISSN 2468-9289
https://doi.org/10.1016/bs.apmp.2021.08.007

Copyright © 2021 Elsevier Inc.
All rights reserved.

175

biofilter performance. Adding biochar to biofilter media can help overcome some of these challenges. Biochar removes pathogens because of hydrophobic interaction and straining, limits remobilization of previously attached bacteria during intermittent flow by increasing residual moisture content, and provides conditions for native microbial communities to strive and out-compete pathogens for nutrients. However, all biochars are not made equal. Thus, bacterial removal capacity varies with biochar properties: removal increases with surface area and fixed carbon content and decreases with volatile matter and ash content. Additionally, the removal efficiency also depends on biochar size and how they are applied such as the presence of compost and compaction conditions. Collectively, these results indicate that biochar with specific properties and application methods can effectively increase the pathogen removal capacity of biofilters in variable climate conditions.

Keywords: Biochar, Fecal indicator bacteria, Microbe, Water scarcity, Resilience, Runoff, Green infrastructure

1. Introduction

Groundwater and surface waters provide most of the water needs in public, industrial, and agricultural sectors.[1,2] However, rapid urbanization and climate change have depleted these water resources and exacerbated water scarcity issues. To alleviate the water deficit, the use of nontraditional water resources such as stormwater has been explored. In most places, gray infrastructures such as concrete canals and pipes have been used to convey stormwater rapidly to minimize flooding. In contrast, green infrastructures are designed to increase infiltration and minimize flooding.[3,4] Among the different types of stormwater treatment systems, infiltration–based systems such as biofilters are popular because of their low footprint and better pollutant removal performance than other GIs.[5] Biofilter consists of a planted top, filter layers, and a drainage layer; all layers serve different functions for pollutant removal (Fig. 1). Biofilters are good at removing suspended sediments but have limited capacity to remove dissolved pollutants including nutrients, some heavy metals, trace organics, and pathogens.[6,7] Among all the pollutants, pathogens or bacterial pathogens are the most difficult to remove because of their small size, persistence, and proliferation inside the stormwater treatment systems.[8,9] Bacterial pathogens can grow in biofilter media due to the presence of nutrients and detach from filter media during intermittent infiltration of stormwater, particularly during the first flush.[10] However, the addition of amendments to filter media can increase removal by adsorption, inactivation, and straining.[11,12]

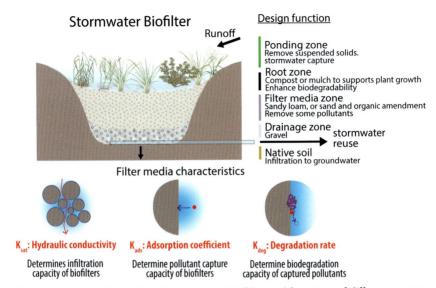

Fig. 1 Schematic of a traditional stormwater biofilter and functions of different components of biofilters and the filter media. *Reprinted and adapted from Mohanty SK, Valenca R, Berger AW, Yu IKM, Xiong X, Saunders TM, et al. Plenty of room for carbon on the ground: Potential applications of biochar for stormwater treatment. Sci Total Environ, 625:1644–1658, Copyright (2018), with permission from Elsevier.*

Amendments for biofilters can be chosen based on three properties, which indicate three unique functions: hydraulic conductivity (K_{sat}) to increase infiltration, adsorption capacity (K_{ads}) to increase pollutant removal from infiltrating water, and biodegradation capacity (K_{deg}) to biologically destroy adsorbed pollutants and recharge the adsorption capacity of the amendment. Normally, the particle size distribution of amendment affects hydraulic conductivity.[13] Compaction could also lower hydraulic conductivity.[14,15] Thus, bulking agents such as coarse sand is used as an amendment to increase infiltration and alleviate compaction with time. An increase in hydraulic conductivity of filter media increases the volume of stormwater infiltrated but it can also minimize the contact time of pollutants with the amendment, thereby reducing treatment of pollutants that exhibit slow removal kinetics.[16] The extent to which filter media could remove pollutants depends on the hydrophobic interaction,[17] cation exchange,[18] or electrostatic attraction.[19] Thus, surface area,[20] surface charge,[21] cation exchange capacity,[22] and organic carbon fractions[23] are used to predict the adsorption capacity of amendments. Natural microorganisms could degrade the pollutants and recharge the surface properties of the amendment.[24] Thus, amendments such as compost or mulch that provide an adequate environment for microbial growth are used to enhance biodegradation,[24] although they may export nutrients to effluent.[25]

One amendment may not fit all the criteria, so it is typical to mix amendments in biofilters to serve different functions.[11] Among different types of engineered-geomedia, biochar has been used in stormwater treatment systems because they can be produced from raw or waste biomass at any place[26] and remove a wide range of pollutants.[27,28] By comparing 6 peer-reviewed studies that investigated pathogens removal using biochar-augmented biofilters[29–34] against typical biofilters constructed without biochar that were reported on the BMP Database,[35] we show that the addition of biochar to biofilters significantly ($P < 0.05$) increase the removal of pathogens (Fig. 2). However, not all biochar is made equal. The removal can vary widely based on biochar properties,[36] design,[37] and conditions at the site.[38] Thus, it is critical to understand why and how biochar improves pathogen removal from stormwater. This chapter describes recent advances in understanding how biochar improves pathogen removal in stormwater treatment systems.

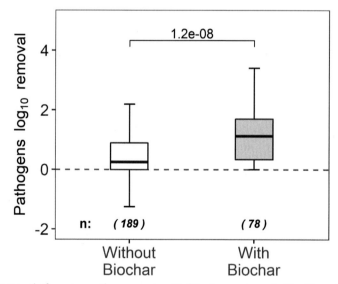

Fig. 2 Removal of varying pathogens using biofiltration systems built with and without the addition of biochar. The horizontal dashed red line indicates no removal of pathogens. Negative values of log_{10} removal values indicate that biofilters are a source of pathogens, while positive values indicate net removal of pathogens. The number of observations of each boxplot (n-value) is represented between parenthesis below the boxplot. Statistical analysis was conducted through the Wilcoxon rank-sum test and is indicated above the boxplot.

2. Testing methods to evaluate biochar capacity to remove pathogens in stormwater

The capacity of biochar to remove pathogens from stormwater is typically assessed by conducting bench-top column experiments under controlled conditions (Table 1). These flow-through column studies are designed to simulate stormwater infiltration in nature,[33] whereas batch experiments are useful to understand how changes in surface or water chemistry can affect maximum sorption capacity or inactivation of pathogens.[49] The capacity of biochar to remove pollutants has been tested without mixing it with other amendments[41,42,45] or with other media in layered and mixture configurations.[37,42,46] The typical application rate for biochar varies between 2% to 33% by volume with soil or sand. The mixture is packed in columns with a diameter ranging from 1.0 cm[32] to 15 cm.[46] The height of the geomedia may vary from 4.5 cm[29] to 180 cm.[42] The depth of the amended layer in the field is typically 45 cm. Thus, the laboratory setup can simulate the designed depth of biofilters. In field studies, rectangular plots are used.[30,37] The mixture is gently packed in biofilters to prevent breaking. However, in some cases such as roadside biofilters, compaction may be necessary or required for soil stability.[14,15] The columns packed with biochar in laboratory are typically subjected to intermittent infiltration of synthetic or natural stormwater pre-contaminated with pathogens or fecal indicator bacteria with concentrations varying from 10^2 to 10^8 colony forming units (CFU) per milliliter. High influent concentration in the laboratory is necessary to determine maximum adsorption capacity. Synthetic stormwater provides greater control on conditions, whereas natural stormwater is useful to form biofilm in the biofilters.[44] Stormwater can be applied on the top of filter media to ensure downward flow by gravity or injected from the bottom with upward flow through the column to simulate saturated flow. While downward flow mimics the flow of stormwater runoff in real-world conditions, upward flow is often adequate to determine maximum removal capacity.[10] Unsaturated flow, which occurs when stormwater is applied from the top, causes underutilization of geomedia[50] as the presence of air and air-water interface may affect bacterial adsorption on biochar.[39] While laboratory column experiments provide an estimate of the biochar's performance in removing pathogens, they do not simulate many real-world field conditions such as variation in stormwater chemistry, influent pathogen concentration, and changing weather patterns such as dry-wet cycles, intense rainfall, and hot climate. These field conditions either reduce[40] or improve the removal of bacteria by biochar.[37]

Table 1 Summary of flow-through column studies that examined the potential of biochar in removing pathogens from stormwater.

Biochar characteristics			Influent water chemistry			Column/Biofilter Setup				
Feedstock	Pyrolysis (°C)	Addition (%)	Type	Pathogen	CFU/mL	Layered or mixed	Internal diameter (cm)	Media Height (cm)	Removal range (%)	References
Poultry litter	350	2 (w/w)	Synthetic stormwater	*E. coli*	1.0×10^7	mix	2.5	10	0–99.9	31
	700									
Pine chips	350									
	700									
Hardwood	500	5–25	Natural wastewater	*Total coliform*	2.6×10^6	mix	50 × 50 rectangular	65	88.7–99.8	30
Wood dust	300	2 (w/w)	Natural stormwater	*E. coli*	10^5	mix	1	10	25.6–87.1	32
	500									
	700									
Softwood	900	30 (v/v)	Synthetic stormwater	*E. coli*	10^3–10^7	mix	2.5	15	61.8–95.2	34
Macadamia Shell	450	10 (w/w)	Synthetic stormwater	*E. coli*	10^7	mix	1.9	5	5.3–67.8	34
Oil Mallee	450									
Phragmites Reed	460									
Rice Husk	650									
Wheat Chaff	550									

Pine wood	350 / 600	1–20 (w/w)	Synthetic stormwater	E. coli	1.2×10^8	mix	NA	4.5	10–80	29
Pine bark	350 / 600									
Softwood	900	5 (w/w)	Synthetic stormwater	E. coli	10^5	mix	2.5	17	80–99.9	39
Sonoma Biochar	350 / 700	5 (w/w)	Synthetic stormwater	E. coli	$1.2–1.7 \times 10^6$	mix	2.5	15	83–99.9	38
Wood chips	350 / 700									
Mix of Monterey pine, Eucalyptus, Bay Laurel, Hardwood and Softwood	395	30 (v/v)	Synthetic stormwater	E. coli, Staph, Salmonella, Bacteriophage MS2	10^5	mix	2.5	15	94.9–99.9	17
Mix of Monterey pine, Eucalyptus, Bay Laurel, Hardwood and Softwood	395	30 (v/v)	Synthetic stormwater	E. coli	$1.5–5.3 \times 10^5$	mix	2.5	10	71.8–97.6	40

Continued

Table 1 Summary of flow-through column studies that examined the potential of biochar in removing pathogens from stormwater.—cont'd

Biochar characteristics			Influent water chemistry			Column/Biofilter Setup				
Feedstock	Pyrolysis (°C)	Addition (%)	Type	Pathogen	CFU/mL	Layered or mixed	Internal diameter (cm)	Media Height (cm)	Removal range (%)	References
Hardwood	NA	100	Natural wastewater	E. coli, Enterococci spp, Bacteriophage MS2, Bacteriophage φX174, Saccharomyces cerevisiae	10^2–10^3	NA	7.5	60	20–99.9	41
Softwood	700	100	Natural wastewater	E. coli, Enterococci	1.9×10^4	layer	5	180	79–95.2	42
Acacia confuse and Celtis sinensis and chemically modified biochars	700	5 (w/w)	Synthetic stormwater	E. coli	0.3–3.2×106	mix	2.5	15	87.9–99.8	43
Softwood	900	5 (w/w)	Natural stormwater	E. coli	10^6	mix	5.1	30.5	93–99.9	14
Softwood	900	15 (v/v)	Natural stormwater	E. coli	10^5	mix	5.1	30.4	60–99.9	15
Oak hardwood	540	30 (v/v)	Synthetic stormwater	E. coli	10^5	mix	2.54	30	87–99.9	36
Wood-based	550									
Yellow pine	990									
Softwood	900									

Mix of Monterey pine, Eucalyptus, Bay Laurel, Hardwood and Softwood	395	30 (v/v)	Natural stormwater	*E. coli, Enterococci*	$1.5–5.5 \times 10^4$	mix	2.5	15	74–94.9	44
Carbon Terra GmbH	NA	100	Natural wastewater	*E. coli, Total coliform*	8.5×10^7	NA	14	60	99–99.5	45
Mix of Monterey pine, Eucalyptus, Bay Laurel, Hardwood and Softwood	395	33.3 (v/v)	Natural stormwater	*E. coli, Enterococci, F + coliphage*	7.8×10^4	layer	50×40 rectangular	30	52–99	37
Pinewood	NA	33 (v/v)	Natural stormwater	*E. coli, Total coliform*	$0.1–5.1 \times 10^4$	layer	15.2	50	43–93.7	46
Poultry litter	350 700	2–10 (w/w)	Synthetic stormwater	*E. coli*	1.3×10^7	mix	2.5	10	9.7–49	47
Wood derived	900 550	30 (v/v)	Synthetic stormwater	*E. coli*	2.2×10^7	mix	7.2	30	99–100	22
Waste wood pellets	520	100	Synthetic stormwater	*E. coli*	$0.1–4.7 \times 10^5$	layer	7.0	23	20–25	48

3. Pathogen removal processes in biochar-amended filter media

Biochars are porous materials with high surface area, but their surface properties can vary widely. The removal of pathogens by biochars depends on the biochar's characteristics as depicted in Fig. 3. The detail of how biochar enables each process is described in the sections below.

3.1 Attachment and straining

Bacteria can be removed by filter media initially by a reversible step governed by weak forces such as van der Waals, electrostatic, and hydrophobic interactions, followed by an irreversible second step that involves direct attachment of bacteria wall or flagella to the surfaces.[51] The surfaces of the

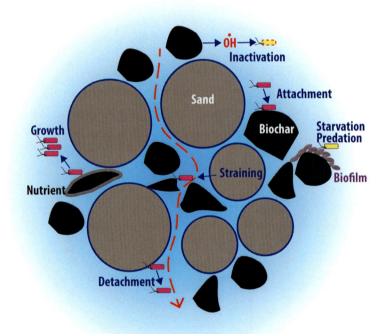

Fig. 3 Mechanisms related to the fate of pathogens in biochar-amended filter media: straining, attachment, detachment, growth, inactivation, predation, and biofilm development.

bacterial cell and the filter media must have opposite charges for electrostatic attraction to be relevant. The biochar surface has hydroxyl and carboxylic acid groups, along with phenolic, quinones, and condensed aromatics groups that make the biochar's surface a net-negative surface.[27] This results in net electrostatic repulsion. Dissolved ions from salts can mask the surface charge and lower the repulsion. Biochar can also increase removal by straining. Grain-to-grain interaction and surface roughness can affect bacterial removal by straining.[34] The extent to which straining is relevant depends on biochar's grain size, roughness, and media porosity along with the size and concentration of bacteria.[52] Biochar's grain size can vary widely based on feedstock size.[53,54] Smaller grain size is helpful to improve bacterial removal by straining. For instance, Mohanty and Boehm[33] examined the removal of *E. coli* in three different types of biofilters: (a) sand-only, (b) sand-biochar, and (c) sand-biochar without fine ($<125\,\mu$m) particles (Fig. 4). They concluded that the addition of biochar to sand increased the removal of *E. coli* significantly when compared to sand-only columns. However, the removal of fine biochar particles ($<125\,\mu$m) increased the transport of *E. coli* in biofilters potentially due to a decrease in straining and surface area available for sorption.[33]

3.2 Die-off and inactivation

Biochar not only removes pathogens from infiltrating stormwater but also affects the fate of removed pathogens. For instance, biochar could prevent the growth of bacteria inside stormwater biofilters.[38] Biochar's feedstock type and pyrolysis temperature can affect the extent to which biochar can prevent pathogen growth. Depending on biochar's pyrolysis temperature, biochar may disrupt the communication between growing bacterial cells by inhibiting the signal of acyl-homoserine lactone which regulates gene expression and alter the extent to which biofilm can form.[55] Biochar could support diverse microbial communities,[56] which can inactivate or kill pathogens via starvation or predation.[57] On the other hand, biochar may induce the growth of bacteria by providing nutrients such as phosphate,[58] although the extent of the growth depends on biochar's feedstock type and biochar properties.[59] A previous study showed that ash in biochar could suppress the growth of previously removed *E. coli* between rainfall events.[36] In between rainfall events, biochar could adsorb more *E. coli* due to an increase in residence time[38] or help inactivate *E. coli*.[60] Biochar could also adsorb metabolites produced by *E. coli*,[56] thereby limiting bacterial growth.

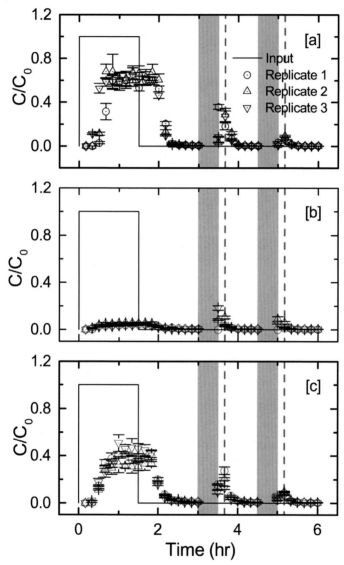

Fig. 4 Transport and mobilization of *E. coli* through columns packed with (a) sand, (b) mixture of sand and biochar, and (c) mixture of sand and biochar where biochar particles smaller than 125 μm were removed. The gray area indicates the 0.5 h pause during which the column was drained, and the dashed lines indicate the timing of the first samples after the pause. The error bar indicates one standard deviation of measurements. *Reprinted (adapted) with permission from Mohanty SK, Boehm AB.* Escherichia coli *removal in biochar-augmented biofilter: effect of infiltration rate, initial bacterial concentration, biochar particle size, and presence of compost. Environ Sci Technol. 2014;48(19): 11535–11542. Copyright (2021) American Chemical Society.*

However, it is expected that a longer duration between rainfall events would allow the bacteria to grow on carbon adsorbent utilizing nutrients in the infiltrating water.[61–63] Excess of bacterial growth[38] and mobilization of bacteria during intermittent infiltration events[10,39] can result in negative removal or net export of indicator bacteria from biofilters.[28] Because biochar can reduce the availability of growth metabolites[56] and remove bacteria by inactivation[64] and adsorption,[38] an addition of biochar to stormwater biofilters would decrease the growth or kill pathogens between rainfall events. But these processes can vary with biochar types. To examine how different biochars may inactivate pathogens during intermittent rainfall, Valenca, Borthakur[36] tested 4 types of biochar and found that *E. coli* did not grow inside the biofilters despite the presence of nutrients. The result indicates that biochar may continue removing pathogens through inactivation, starvation, or predation in between rainfall events (Fig. 5).

Biochar can also remove pathogens by inactivation. Biochar surface can produce hydroxyl radical through the reduction of oxygen and the oxidization of phenolic hydroxyl groups on biochar.[65] These radicals can kill pathogens by compromising the cell wall.[60] Bacterial cell wall properties can affect the inactivation rate. While gram-positive bacteria cell wall is composed of a thick but simple peptidoglycan layer, the cell wall of gram-negative bacteria is composed of a multi-layer of lipid, membrane, and peptidoglycan.

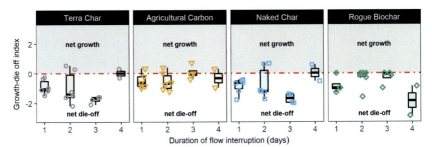

Fig. 5 Growth-die off index (GDI) of filter media as a function of drying duration between infiltration events. GDI was calculated as $-\log_{10}(C_b/C_a)$, where C_b and C_a represent the concentration of *E. coli* in the effluent before and after flow interruption, respectively. Positive GDI values (gray shaded area) represent net-growth of bacteria during flow interruption, while negative GDI values represent net die-off (or decay) or bacteria. *Republished with permission from Valenca, R. et al., Biochar selection for Escherichia coli removal in stormwater biofilters, Am Soc Civil Eng, 2021:147(2); permission conveyed through Copyright Clearance Center, Inc.*

3.3 Low remobilization during intermittent flow

Biochar removal capacity can decrease if some of the attached bacteria are remobilized during the infiltration of stormwater. In fact, intermittent infiltration of stormwater is shown to increase mobilization of attached bacteria from conventional filter media because of an enhanced detachment of bacteria by moving air-water interfaces.[10] Previous studies show that biochar lowers the remobilization of previously attached bacteria by keeping the biofilter moist and increasing the strength of bacterial binding to filter media.[38] The authors analyzed the effect of flow interruption (0.5 h and 21 h) on the remobilization of *E. coli* using sand-only and sand-biochar columns (Fig. 6) and showed that, while sand-only columns remobilized between 10% and 20% of attached *E. coli*, sand-biochar columns remobilized less than 0.1% of attached *E. coli*. However, the presence of natural organic matter (NOM) increased the remobilization of *E. coli* in both columns. The mobilization of bacteria may be enhanced if biochar particles are broken or mobilized and if particulate organic matters are released carrying bacteria.[66] In addition, NOM may compete for the attachment sites and provides a physical barrier for bacteria to access the sites on biochar.[40] The mobilization is sensitive to antecedent weather conditions.[67–69] Weathering processes

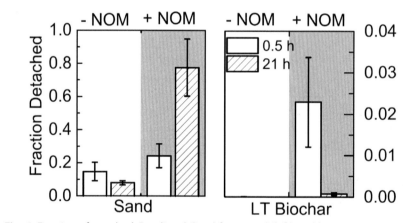

Fig. 6 Fraction of attached *E. coli* mobilized from sand (left) and low-temperature (LT) biochar columns (right) in stormwater with and without NOM during two intermittent flows. The gray background represents results from experiment with NOM. The error bar indicates one standard deviation of results obtained from four replicate column experiments. Note that the scale of y-axis is magnified for LT biochar. *Reprinted from Mohanty SK, Cantrell KB, Nelson KL, Boehm AB, Efficacy of biochar to remove* Escherichia coli *from stormwater under steady and intermittent flow, Water Research, 61:288–296, Copyright (2014), with permission from Elsevier.*

could affect moisture conditions in biofilter and affect biochar state in filter media, both of which could affect bacterial removal.[39] For example, dry-wet and freeze-thaw cycles have been shown to increase bacterial removal by biochar,[70] potentially due to the breaking of biochar by expanding ice or other change in surface properties.[39] Moreover, breakage of large biochar particles may expose newly available active sites for bacterial attachment.[14]

4. Challenges
4.1 Not all biochars are made equal

Biochar's capacity to remove pathogens or pathogen indicators varies by orders of magnitude,[28] which makes it difficult for biofilter designers to select a biochar available in the market. The variability has been attributed to a variation in biochar properties and stormwater chemistry.[17,29,38,71] Unlike activated carbon, biochar properties can vary widely based on preparation conditions and feedstock types.[72] Generally, it is recommended to use wood-based biochar prepared at high pyrolysis temperature[47,73] without removing fine particle size.[33,34,71] Despite constraining these conditions, a previous study[36] showed that bacterial removal could vary (Fig. 7). They showed that the *E. coli* removal capacity of biochar is positively correlated with surface area and carbon content and negatively correlated with ash and organic matter. High removal capacity of biochar has been attributed to an increase in surface hydrophobicity[38,40,43] and surface area[40,43] of biochar, whereas a low removal capacity has been attributed to an increase

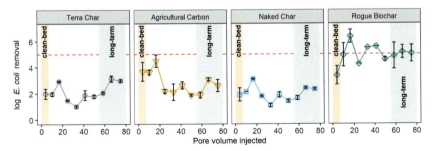

Fig. 7 *E. coli* removal capacity varies with biochar from different vendors. Removal capacity of biochar-augmented filters was investigated during 10 infiltration events. Yellow and gray shaded areas represent clean-bed removal ($n=12$) and long-term removal ($n=12$), respectively. Red dashed line represents detection limit of 1 colony per plate (20 CFU/mL^{-1}). Republished with permission of the Valenca, R. et al., Biochar selection for Escherichia coli removal in stormwater biofilters, Am Soc Civil Eng, 2021:147(2); permission conveyed through Copyright Clearance Center, Inc.

in oxidation of biochar[29] that increase net negative surface charge, and volatile carbon content.[38] These surface properties are influenced by bulk chemical properties of biochar including carbon content, ash content, volatile carbon content, and physical property such as surface area.[74] Thus, these attributes can be used by field managers to select a reliable biochar to remove pathogens.[28]

4.2 Chemical weathering could affect removal capacity

Biochar in biofilters is naturally exposed to dry-wet or freeze-thaw cycles, which can affect bacterial removal by altering surface properties of biochar.[39,75] Under these conditions, the biochar's surface is gradually oxidized, increasing the aliphatic carbon, especially carboxylic acids, and decreasing the aromatic carbon content.[75,76] Additionally, aged biochar has less total carbon and electrical conductivity. Aged biochar particles also have less potassium but more O, Si, N, Na, Al, Ca, Mn, and Fe on their surface due to their interactions with the soil.[77] Finally, biochar may lose fine particles due to weathering cycles.[39,78] The release of particles from biochar can also release the pollutants sorbed onto the particles, making the biochar a secondary source of pollutants in the long term. For instance, Mohanty and Boehm[39] exposed biochar-amended biofilters to varying weather conditions including freeze-thaw cycles, dry-wet cycles in cold (4 °C), and warm (37 °C) conditions (Fig. 8), and found that weathering conditions improved *E. coli* removal. Other studies showed that the weathering of biochar could increase the ability of biochar to adsorb trace metals[79] and nitrogenous compounds.[76]

4.3 Biological weathering or biofilm development could affect biochar performance

Biological weathering occurs via the development of biofilm on biochar's surface,[80] which can alter the performance of biochar in removing contaminants.[39] Biofilm is defined as a broad community of microorganisms–single or multiple species of gram-positive and/or gram-negative bacteria–that grows irreversibly attached to a surface depending on environmental conditions such as nutrient availability.[81] Biofilm development depends on the bacteria's hydrophobicity, surface charge, and outer membrane protein, along with the filter material's charges, chemistry, hydrophobicity, roughness, topography, and stiffness.[82] Compared to smooth surfaces, rough surfaces are more likely to adhere to bacteria and form biofilm because the

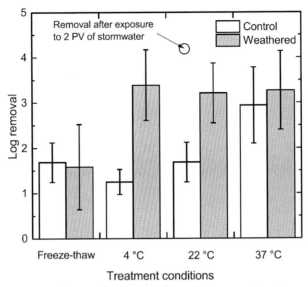

Fig. 8 Removal of *E. coli* by sand-biochar biofilters when exposed to freeze-thaw cycles and dry-wet cycles at 4 °C, 22 °C and 37 °C. *Reprinted from Mohanty SK, Boehm AB., Effect of weathering on mobilization of biochar particles and bacterial removal in a stormwater biofilter, Water Research, 85:208–215, Copyright (2015), with permission from Elsevier.*

enhanced atomic and molecular-scale of a rough surface increases the surface reactivity.[83] Biochar provides excellent support for biofilm development because, besides having high surface roughness and porosity, biochar is less likely to adsorbs metabolites including L-asparagine, L-glutamine, and L-arginine.[56] Afrooz and Boehm[40] investigated the removal of *E. coli* in sand-biochar columns with and without biofilm (Fig. 9) and found that the formation of biofilm may reduce the *E. coli* removal properties of biochar possibly due to the alteration of the hydraulic properties of the biofilter, reduction of the number of attachment sites, and increment of biofilm detachment. However, the laboratory study is typically carried out by a model microorganism that lacks the diverse community to simulate processes such as predation and starvation of pathogen in field conditions. Biofilm in field conditions could increase bacterial removal by offering a polymer-mediated adhesion to bacteria and producing a nutrient-limited environment which could indirectly kill or inactivate the surrounding bacteria. Thus, future studies should evaluate community abundance and diversity on biochar and link them with predation or other removal mechanism.

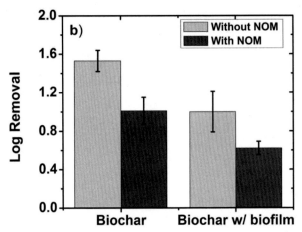

Fig. 9 Log removal of *E. coli* in laboratorial sand-biochar (70% sand, 30% biochar: by volume) columns in presence or absence of biofilm and natural organic matter (NOM). *Reprinted from Afrooz ARMN, Boehm AB,* Escherichia coli *removal in biochar-modified biofilters: effects of biofilm, PLoS One 11(12), open-access, © 2016 Afrooz, Boehm.*

4.4 Pathogen removal depends on how biochar is applied in biofilters

Biochar's capacity to remove pathogens can also depend on how they are packed in biofilters. The packing conditions including compaction energy, presence of other amendments, moisture content during packing, biochar particle size selected, and its application rates. Previous studies showed that varying biochar application rates may alter *E. coli* removal from 60% to 98% depending on the application rate.[38,43,47] The removal varied widely based on the amount of biochar added.[22,33,34,36,40,47] Even biofilters with 100% biochar show a wide removal range: 27%[48] to 99.5%.[45] Low removal occurs when biochar size is larger than 1 mm. Thus, the removal of fine biochar particles can reduce removal capacity significantly.[33,34,41,48] The presence of other amendments in the media filter, such as compost, could also decrease the bacteria removal capacity of the biofilter.[33] However, such effect may be reversed if the biofilter is compacted during packing.[15] Compaction can improve bacteria removal by straining. It appears that compacting biochar with moisture in it improves bacterial removal by minimizing the occurrence of preferential flow.[15] Irrespective of compaction, smaller biochar removes more bacteria.[14] Thus, the particle size distribution of biochar should be standardized to enhance pathogen removal in biochar-amended biofilters.

4.5 Limited removal capacity for virus

Studies investigating virus removal by biochar are scarce.[17] Adsorption mechanism is the main pathway of virus removal in natural or engineered systems. The low size of viruses makes it difficult to remove them by straining.[41] A previous study showed that biochar-augmented biofilters removed more bacteriophage MS2 virus than sand biofilters probably due to the electrostatic and hydrophobic interactions.[17] But another study showed opposite results: sand removed more virus than biochar.[34] It should be noted that the removal of virus from pore water does not eliminate the risk of virus infection. Thus, inactivation of the virus is necessary.[84] A recent study showed that the treatment of greywater with biochar filters reduces the risks of virus infection by 90%.[85] However, the dominant mechanism by which biochar can remove viruses is less clear.

5. Opportunities

5.1 Selection of best biochar based on biochar properties

Choosing biochar with high capacity is essential in meeting the design goal of the removal of bacterial pollutants from contaminated waters. Understanding which properties are related to removal is critical in the selection of appropriate biochar. Generally, an increase in carbon content and surface area, and a decrease in ash content and volatile carbon can increase biochar performance.[36] Thus, there is an opportunity for the vendors to optimize the production condition to produce biochar with these qualities. For instance, washing of biochar[86] and modifying temperature during pyrolysis[87] could lower ash content.

5.2 Modifying biochar surface properties

Modification of biochar's surface could increase their pollutant removal capacity.[88] Biochar is oxidized using strong acids such as phosphoric acids to increase the acidity of the surfaces and to modify their porous structure.[89] Similarly, biochar is also treated with NaOH to increase the oxygen content and basicity[90] and remove ash and condensed organic matter.[89,91,92] Biochar can also be chemically treated to change the functional groups to suit environmental applications. Treating biochar with HNO_3 can form amine groups on the biochar for this purpose.[93] Coating biochar with metal oxides can enhance the sorption capacity of the biochar.[94] Treating biochar with steam increases the surface area of the biochar[95] and also increases the sorption capacity.[96] Moreover, treating biochar with a mixture of CO_2 and ammonia gas increases biochar surface area and pore volume.[97] However, any modification of the biochar's surface could increase the cost of biochar.

5.3 Lack of field studies

Limited studies examined biochar capacity to remove pathogens in the field. A recent study showed that biochar-amended biofilters in the field underperformed compared to the biofilters in the laboratory setting.[37] Another study by Minnesota Pollution Control Agency investigated the feasibility of using biochar in stormwater treatment systems such as stormwater pond filter bench, filter box, and catch basin and they found *E. coli* removal efficiencies of 72% to 93%.[98] The result is similar to the removal reported in lab-scale studies.[22,33,36,40,43] Thus, more field studies should be conducted.

5.4 Change in biofilter microbiome after biochar addition

Biofilter microbiome can affect pathogen removal by predation and competition for resources. Biochar addition can shift in the microbiome in soils by altering the pore water chemistry such as pH,[99] organic carbon fraction and quality,[100] and nutrient availability.[101] Biochar can increase the retention of nitrogen-based contaminants in biofilters. Biochar addition can increase pH and carbon-nitrogen ratio, which could affect the abundance, richness, and diversity of the fungal community and shape fungal association with plants in biofilters. Biochar can also protect the soil microbiome by retaining moisture during dry seasons. Biochar can also adsorb pollutants such as polyaromatic hydrocarbons, trace organics, and heavy metals from stormwater,[102,103] where it can pose toxicity to pathogens. However, the bioavailability of adsorbed heavy metals[104] and sulfamethoxazole[103] could be lower. A slow release of these pollutants could affect pathogen concentration on biochar's surface. A study has used this concept to improve the bacterial removal capacity of biochar by impregnating copper into biochar.[105] Accumulation of toxic material on biochar surface can affect the microbiome responsible for the biodegradation of organic pollutants.[99,101,106–109] Future studies should investigate the role of accumulated toxic pollutants on the biochar microbiome and its impact on pathogen removal.

6. Summary

Biochar is a promising amendment for biofilters to improve pathogen removal from stormwater runoff. Biochar can remove pathogens by multiple processes: straining, attachment, inactivation, growth suppression, and reduced remobilization during intermittent flow. However, the relevance of each process on pathogen removal by biochar can vary based on biochar properties and their weathering under natural conditions. The particle size of biochar plays a critical role in pathogen removal as fine particles significantly increase the overall surface area and increase pathogen removal by straining. Other properties that increase removal include high surface area, high carbon content, whereas volatile carbon content and ash diminish biochar capacity to remove pathogens from stormwater. The removal capacity can further vary with how biochar is applied in biofilters: biochar size distribution, the fraction of biochar in the media mixture, layered or mixing configuration, compaction level, and moisture content during compaction. Although biochar is a recommended amendment to increase bacterial pathogen removal, biochar capacity in removing viruses is limited. In this case, modifying the biochar's surface to inactivate the virus can improve removal. Most results are based on laboratory column studies where some of the field-relevant conditions are not tested. Future studies should validate biochar capacity in field studies.

References

1. Siebert S, Burke J, Faures JM, Frenken K, Hoogeveen J, Doll P, et al. Groundwater use for irrigation - a global inventory. *Hydrol Earth Syst Sci* 2010;**14**(10):1863–80.
2. Lipczynska-Kochany E. Effect of climate change on humic substances and associated impacts on the quality of surface water and groundwater: a review. *Sci Total Environ* 2018;**640**:1548–65.
3. Song YL, Du XQ, Ye XY. Analysis of potential risks associated with urban Stormwater quality for managed aquifer recharge. *Int J Environ Res Public Health* 2019;**16**(17):19.
4. Cramer M, Rinas M, Kotzbauer U, Tranckner J. Surface contamination of impervious areas on biogas plants and conclusions for an improved stormwater management. *J Clean Prod* 2019;**217**:1–11.
5. US EPA. *Low impact development (LID): a literature review.* Washington. DC: US Environ Prot Agency Off Water Low Impact Dev Cent; 2000.
6. Grebel J, Mohanty S, Torkelson A, Boehm A, Higgins C, Maxwell R, et al. Engineered infiltration systems for urban stormwater reclamation. *Environ Eng Sci* 2013;**30**:437–54.
7. LeFevre GH, Paus KH, Natarajan P, Gulliver JS, Novak PJ, Hozalski RM. Review of dissolved pollutants in urban storm water and their removal and fate in bioretention cells. *J Environ Eng* 2015;**141**(1):04014050.

8. Wolfand JM, Bell CD, Boehm AB, Hogue TS, Luthy RG. Multiple pathways to bacterial load reduction by Stormwater best management practices: trade-offs in performance, volume, and treated area. *Environ Sci Technol* 2018;**52**(11):6370–9.
9. Clary J, Jones J, Urbonas B, Quigley M, Strecker E, Wagner T. Can stormwater BMPs remove bacteria? New findings from the international stormwater BMP database. *Stormwater Magazine May* 2008;**5**:1–14.
10. Mohanty SK, Torkelson AA, Dodd H, Nelson KL, Boehm AB. Engineering solutions to improve the removal of Fecal Indicator Bacteria by bioinfiltration systems during intermittent flow of Stormwater. *Environ Sci Technol* 2013;**47**(19):10791–8.
11. Tirpak RA, Afrooz ARMN, Winston RJ, Valenca R, Schiff K, Mohanty SK. Conventional and amended bioretention soil media for targeted pollutant treatment: a critical review to guide the state of the practice. *Water Res* 2021;**189**:116648.
12. Rippy MA. Meeting the criteria: linking biofilter design to fecal indicator bacteria removal. *WIREs Water* 2015;**2**(5):577–92.
13. Trifunovic B, Gonzales HB, Ravi S, Sharratt BS, Mohanty SK. Dynamic effects of biochar concentration and particle size on hydraulic properties of sand. *Land Degrad Dev* 2018;**29**(4):884–93.
14. Le H, Valenca R, Ravi S, Stenstrom MK, Mohanty SK. Size-dependent biochar breaking under compaction: implications on clogging and pathogen removal in biofilters. *Environ Pollut* 2020;**266**:115195.
15. Ghavanloughajar M, Valenca R, Le H, Rahman M, Borthakur A, Ravi S, et al. Compaction conditions affect the capacity of biochar-amended sand filters to treat road runoff. *Sci Total Environ* 2020;**735**:139180.
16. Berger AW, Valenca R, Miao Y, Ravi S, Mahendra S, Mohanty SK. Biochar increases nitrate removal capacity of woodchip biofilters during high-intensity rainfall. *Water Res* 2019;**165**:8.
17. Afrooz A, Pitol AK, Kitt D, Boehm AB. Role of microbial cell properties on bacterial pathogen and coliphage removal in biochar-modified stormwater biofilters. *Environ Sci-Wat Res* 2018;**4**(12):2160–9.
18. Samatya S, Kabay N, Yüksel Ü, Arda M, Yüksel M. Removal of nitrate from aqueous solution by nitrate selective ion exchange resins. *React Funct Polym* 2006;**66**(11):1206–14.
19. Hu QL, Liu HY, Zhang ZY, Xie YH. Nitrate removal from aqueous solution using polyaniline modified activated carbon: Optimization and characterization. *J Mol Liq* 2020;**309**:11.
20. Ahmad M, Rajapaksha AU, Lim JE, Zhang M, Bolan N, Mohan D, et al. Biochar as a sorbent for contaminant management in soil and water: a review. *Chemosphere* 2014;**99**:19–33.
21. Long L, Xue Y, Hu X, Zhu Y. Study on the influence of surface potential on the nitrate adsorption capacity of metal modified biochar. *Environ Sci Pollut Res* 2019;**26**(3):3065–74.
22. Rahman MYA, Nachabe MH, Ergas SJ. Biochar amendment of stormwater bioretention systems for nitrogen and Escherichia coli removal: effect of hydraulic loading rates and antecedent dry periods. *Bioresour Technol* 2020;**310**:123428.
23. Chen DY, Chen XJ, Sun J, Zheng ZC, Fu KX. Pyrolysis polygeneration of pine nut shell: quality of pyrolysis products and study on the preparation of activated carbon from biochar. *Bioresour Technol* 2016;**216**:629–36.
24. Ulrich BA, Vignola M, Edgehouse K, Werner D, Higgins CP. Organic carbon amendments for enhanced biological attenuation of trace organic contaminants in biochar-amended Stormwater biofilters. *Environ Sci Technol* 2017;**51**(16):9184–93.
25. Hurley S, Shrestha P, Cording A. Nutrient leaching from compost: implications for bioretention and other green Stormwater infrastructure. *J Sustain Water Built Environ* 2017;**3**(3):04017006.

26. Alhashimi HA, Aktas CB. Life cycle environmental and economic performance of biochar compared with activated carbon: a meta-analysis. *Resour Conserv Recycl* 2017;**118**:13–26.
27. Mohanty SK, Valenca R, Berger AW, Yu IKM, Xiong XN, Saunders TM, et al. Plenty of room for carbon on the ground: potential applications of biochar for stormwater treatment. *Sci Total Environ* 2018;**625**:1644–58.
28. Boehm AB, Bell CD, Fitzgerald NJM, Gallo E, Higgins CP, Hogue TS, et al. Biochar-augmented biofilters to improve pollutant removal from stormwater – can they improve receiving water quality? *Environ Sci: Water Res Technol* 2020. https://doi.org/10.1039/d0ew00027b.
29. Suliman W, Harsh JB, Fortuna A-M, Garcia-Pérez M, Abu-Lail NI. Quantitative effects of biochar oxidation and pyrolysis temperature on the transport of pathogenic and nonpathogenic Escherichia coli in biochar-amended sand columns. *Environ Sci Technol* 2017;**51**(9):5071–81.
30. de Rozari P, Greenway M, El Hanandeh A. An investigation into the effectiveness of sand media amended with biochar to remove BOD5, suspended solids and coliforms using wetland mesocosms. *Water Sci Technol* 2015;**71**(10):1536–44.
31. Abit SM, Bolster CH, Cantrell KB, Flores JQ, Walker SL. Transport of Escherichia coli, Salmonella typhimurium, and microspheres in biochar-amended soils with different textures. *J Environ Qual* 2014;**43**(1):371–88.
32. Lu L, Chen BL. Enhanced bisphenol a removal from stormwater in biochar-amended biofilters: combined with batch sorption and fixed-bed column studies. *Environ Pollut* 2018;**243**:1539–49.
33. Mohanty SK, Boehm AB. Escherichia coli removal in biochar-augmented biofilter: effect of infiltration rate, initial bacterial concentration, biochar particle size, and presence of compost. *Environ Sci Technol* 2014;**48**(19):11535–42.
34. Sasidharan S, Torkzaban S, Bradford SA, Kookana R, Page D, Cook PG. Transport and retention of bacteria and viruses in biochar-amended sand. *Sci Total Environ* 2016;**548**:100–9.
35. Clary J, Leisenring M, Poresky A, Earles A, Jones J. BMP performance analysis results for the international stormwater BMP database. In: *World Environmental and Water Resources Congress*; 2011. p. 441–9.
36. Valenca R, Borthakur A, Zu Y, Matthiesen EA, Stenstrom MK, Mohanty SK. Biochar selection for Escherichia coli removal in Stormwater biofilters. *J Environ Eng* 2021;**147**(2):06020005.
37. Kranner BP, Afrooz ARMN, Fitzgerald NJM, Boehm AB. Fecal indicator bacteria and virus removal in stormwater biofilters: effects of biochar, media saturation, and field conditioning. *PLoS One* 2019;**14**(9):e0222719.
38. Mohanty SK, Cantrell KB, Nelson KL, Boehm AB. Efficacy of biochar to remove Escherichia coli from stormwater under steady and intermittent flow. *Water Res* 2014;**61**:288–96.
39. Mohanty SK, Boehm AB. Effect of weathering on mobilization of biochar particles and bacterial removal in a stormwater biofilter. *Water Res* 2015;**85**:208–15.
40. Afrooz A, Boehm AB. *Escherichia coli* removal in biochar-modified biofilters: Effects of biofilm. *Plos One* 2016;**11**(12). https://doi.org/10.1371/journal.pone.0167489.
41. Perez-Mercado LF, Lalander C, Joel A, Ottoson J, Dalahmeh S, Vinnerås B. Biochar filters as an on-farm treatment to reduce pathogens when irrigating with wastewater-polluted sources. *J Environ Manage* 2019;**248**:109295.
42. Kaetzl K, Lubken M, Gehring T, Wichern M. Efficient low-cost anaerobic treatment of wastewater using biochar and woodchip filters. *Water* 2018;**10**(7):17.
43. Lau AYT, Tsang DCW, Graham NJD, Ok YS, Yang X, Li XD. Surface-modified biochar in a bioretention system for Escherichia coli removal from stormwater. *Chemosphere* 2017;**169**:89–98.

44. Nabiul Afrooz ARM, Boehm AB. Effects of submerged zone, media aging, and antecedent dry period on the performance of biochar-amended biofilters in removing fecal indicators and nutrients from natural stormwater. *Ecol Eng* 2017;**102**:320–30.
45. Moges ME, Eregno FE, Heistad A. Performance of biochar and filtralite as polishing step for on-site greywater treatment plant. *Manag Environ Qual* 2015;**26**(4):607–25.
46. Ulrich BA, Loehnert M, Higgins CP. Improved contaminant removal in vegetated stormwater biofilters amended with biochar. *Environ Sci: Water Res Technol* 2017; **3**(4):726–34.
47. Bolster CH, Abit SM. Biochar Pyrolyzed at two temperatures affects Escherichia coli transport through a Sandy soil. *J Environ Qual* 2012;**41**(1):124–33.
48. Reddy KR, Xie T, Dastgheibi S. Evaluation of biochar as a potential filter media for the removal of mixed contaminants from urban storm water runoff. *J Environ Eng* 2014;**140**(12):04014043.
49. Valenca R, Ramnath K, Dittrich TM, Taylor RE, Mohanty SK. Microbial quality of surface water and subsurface soil after wildfire. *Water Res* 2020. https://doi.org/10.1016/j.watres.2020.115672.
50. Blecken G-T, Zinger Y, Deletić A, Fletcher TD, Viklander M. Influence of intermittent wetting and drying conditions on heavy metal removal by stormwater biofilters. *Water Res* 2009;**43**(18):4590–8.
51. Goulter RM, Gentle IR, Dykes GA. Issues in determining factors influencing bacterial attachment: a review using the attachment of Escherichia coli to abiotic surfaces as an example. *Lett Appl Microbiol* 2009;**49**(1):1–7.
52. Díaz J, Rendueles M, Díaz M. Straining phenomena in bacteria transport through natural porous media. *Environ Sci Pollut Res* 2010;**17**(2):400–9.
53. Kajina W, Rousset P. Coupled effect of feedstock and pyrolysis temperature on biochar as soil amendment. In: *IC-STAR 40: "Interdisciplinary research enhancement for industrial revolution 40"; 2018-08-29 / 2018-08-30; Belintung, Indonésie. public: s.n*; 2018. p. 5.
54. El-Gamal EH, Saleh M, Elsokkary I, Rashad M, El-Latif M. Comparison between properties of biochar produced by traditional and controlled pyrolysis. *Alex Sci Exch J* 2017;**38**:412–25.
55. Masiello CA, Chen Y, Gao X, Liu S, Cheng H-Y, Bennett MR, et al. Biochar and microbial Signaling: production conditions determine effects on microbial communication. *Environ Sci Technol* 2013;**47**(20):11496–503.
56. Hill RA, Hunt J, Sanders E, Tran M, Burk GA, Mlsna TE, et al. Effect of biochar on microbial growth: a metabolomics and bacteriological investigation in E. coli. *Environ Sci Technol* 2019;**53**(5):2635–46.
57. Matz C, McDougald D, Moreno AM, Yung PY, Yildiz FH, Kjelleberg S. Biofilm formation and phenotypic variation enhance predation-driven persistence of vibrio cholerae. *Proc Natl Acad Sci U S A* 2005;**102**(46):16819–24.
58. Brantley KE, Savin MC, Brye KR, Longer DE. Nutrient availability and corn growth in a poultry litter biochar-amended loam soil in a greenhouse experiment. *Soil Use Manage* 2016;**32**(3):279–88.
59. Yang F, Zhou Y, Liu W, Tang W, Meng J, Chen W, et al. Strain-specific effects of biochar and its water-soluble compounds on bacterial growth. *Appl Sci* 2019; **9**(16):3209.
60. Sun MM, Ye M, Zhang ZY, Zhang ST, Zhao YC, Deng SP, et al. Biochar combined with polyvalent phage therapy to mitigate antibiotic resistance pathogenic bacteria vertical transfer risk in an undisturbed soil column system. *J Hazard Mater* 2019;**365**:1–8.
61. Huggins TM, Haeger A, Biffinger JC, Ren ZJ. Granular biochar compared with activated carbon for wastewater treatment and resource recovery. *Water Res* 2016;**94**:225–32.

62. Velten S, Boller M, Köster O, Helbing J, Weilenmann H-U, Hammes F. Development of biomass in a drinking water granular active carbon (GAC) filter. *Water Res* 2011;**45** (19):6347–54.
63. Wilcox DP, Chang E, Dickson KL, Johansson KR. Microbial growth associated with granular activated carbon in a pilot water treatment facility. *Appl Environ Microbiol* 1983;**46**(2):406–16.
64. Gurtler JB, Boateng AA, Han YX, Douds DD. Inactivation of E. coli O157:H7 in cultivable soil by fast and slow pyrolysis-generated biochar. *Foodborne Pathog Dis* 2014;**11** (3):215–23.
65. Zhang K, Sun P, Faye MCAS, Zhang Y. Characterization of biochar derived from rice husks and its potential in chlorobenzene degradation. *Carbon* 2018;**130**:730–40.
66. Cybulak M, Sokołowska Z, Boguta P, Tomczyk A. Influence of pH and grain size on physicochemical properties of biochar and released humic substances. *Fuel* 2019; **240**:334–8.
67. Rusciano GM, Obropta CC. Bioretention column study: Fecal coliform and total suspended solids reductions. *Trans ASABE* 2007;**50**(4):1261–9.
68. Zhang L, Seagren EA, Davis AP, Karns JS. The capture and destruction of Escherichia coli from simulated urban runoff using conventional bioretention media and Iron oxide-coated sand. *Water Environ Res* 2010;**82**(8):701–14.
69. Chandrasena GI, Deletic A, Ellerton J, McCarthy DT. Evaluating Escherichia coli removal performance in stormwater biofilters: a laboratory-scale study. *Water Sci Technol* 2012;**66**(5):1132–8.
70. Liu Z, Dugan B, Masiello CA, Wahab LM, Gonnermann HM, Nittrouer JA. Effect of freeze-thaw cycling on grain size of biochar. *Plos One* 2018;**13**(1):e0191246.
71. Guan P, Prasher SO, Afzal MT, George S, Ronholm J, Dhiman J, et al. Removal of *Escherichia coli* from lake water in a biochar-amended biosand filtering system. *Ecol Eng* 2020;150. https://doi.org/10.1016/j.ecoleng.2020.105819.
72. Xiao X, Chen B, Chen Z, Zhu L, Schnoor JL. Insight into multiple and multilevel structures of biochars and their potential environmental applications: a critical review. *Environ Sci Technol* 2018;**52**(9):5027–47.
73. Abit SM, Bolster CH, Cai P, Walker SL. Influence of feedstock and pyrolysis temperature of biochar amendments on transport of Escherichia coli in saturated and unsaturated soil. *Environ Sci Technol* 2012;**46**(15):8097–105.
74. Manyà JJ. Pyrolysis for biochar purposes: a review to establish current knowledge gaps and research needs. *Environ Sci Technol* 2012;**46**(15):7939–54.
75. Feng M, Zhang W, Wu X, Jia Y, Jiang C, Wei H, et al. Continuous leaching modifies the surface properties and metal(loid) sorption of sludge-derived biochar. *Sci Total Environ* 2018;**625**:731–7.
76. de la Rosa JM, Rosado M, Paneque M, Miller AZ, Knicker H. Effects of aging under field conditions on biochar structure and composition: Implications for biochar stability in soils. *Sci Total Environ* 2018;613–4. 969–76.
77. Sorrenti G, Masiello CA, Dugan B, Toselli M. Biochar physico-chemical properties as affected by environmental exposure. *Sci Total Environ* 2016;563–4. 237–46.
78. Spokas KA, Novak JM, Masiello CA, Johnson MG, Colosky EC, Ippolito JA, et al. Physical disintegration of biochar: an overlooked process. *Environ Sci Technol Lett* 2014;**1**(8):326–32.
79. Zhong Y, Igalavithana AD, Zhang M, Li X, Rinklebe J, Hou D, et al. Effects of aging and weathering on immobilization of trace metals/metalloids in soils amended with biochar. *Environ Sci: Processes Impacts* 2020;**22**(9):1790–808.
80. Luo Y, Durenkamp M, De Nobili M, Lin Q, Devonshire BJ, Brookes PC. Microbial biomass growth, following incorporation of biochars produced at 350 °C or 700 °C, in a silty-clay loam soil of high and low pH. *Soil Biol Biochem* 2013;**57**:513–23.

81. O'Toole G, Kaplan HB, Kolter R. Biofilm formation as microbial development. *Annu Rev Microbiol* 2000;**54**(1):49–79.
82. Song F, Koo H, Ren D. Effects of material properties on bacterial adhesion and biofilm formation. *J Dent Res* 2015;**94**(8):1027–34.
83. Arnold JW, Bailey GW. Surface finishes on stainless steel reduce bacterial attachment and early biofilm formation: scanning electron and atomic force microscopy study. *Poult Sci* 2000;**79**(12):1839–45.
84. Schijven JF, Hassanizadeh SM. Removal of viruses by soil passage: overview of modeling, processes, and parameters. *Crit Rev Environ Sci Technol* 2000;**30**(1):49–127.
85. Dalahmeh SS, Lalander C, Pell M, Vinnerås B, Jönsson H. Quality of greywater treated in biochar filter and risk assessment of gastroenteritis due to household exposure during maintenance and irrigation. *J Appl Microbiol* 2016;**121**(5):1427–43.
86. Sun K, Kang M, Zhang Z, Jin J, Wang Z, Pan Z, et al. Impact of deashing treatment on biochar structural properties and potential sorption mechanisms of Phenanthrene. *Environ Sci Technol* 2013;**47**(20):11473–81.
87. Ahmed MB, Zhou JL, Ngo HH, Guo WS. Insight into biochar properties and its cost analysis. *Biomass Bioenergy* 2016;**84**:76–86.
88. Rajapaksha AU, Chen SS, Tsang DCW, Zhang M, Vithanage M, Mandal S, et al. Engineered/designer biochar for contaminant removal/immobilization from soil and water: potential and implication of biochar modification. *Chemosphere* 2016;**148**:276–91.
89. Lin Y, Munroe P, Joseph S, Henderson R, Ziolkowski A. Water extractable organic carbon in untreated and chemical treated biochars. *Chemosphere* 2012;**87**(2):151–7.
90. Fan Y, Wang B, Yuan S, Wu X, Chen J, Wang L. Adsorptive removal of chloramphenicol from wastewater by NaOH modified bamboo charcoal. *Bioresour Technol* 2010;**101**(19):7661–4.
91. Liou T-H, Wu S-J. Characteristics of microporous/mesoporous carbons prepared from rice husk under base- and acid-treated conditions. *J Hazard Mater* 2009;**171**(1):693–703.
92. Liu P, Liu W-J, Jiang H, Chen J-J, Li W-W, Yu H-Q. Modification of bio-char derived from fast pyrolysis of biomass and its application in removal of tetracycline from aqueous solution. *Bioresour Technol* 2012;**121**:235–40.
93. Chingombe P, Saha B, Wakeman RJ. Surface modification and characterisation of a coal-based activated carbon. *Carbon* 2005;**43**(15):3132–43.
94. Samsuri AW, Sadegh-Zadeh F, Seh-Bardan BJ. Adsorption of as(III) and as(V) by Fe coated biochars and biochars produced from empty fruit bunch and rice husk. *J Environ Chem Eng* 2013;**1**(4):981–8.
95. Rajapaksha AU, Vithanage M, Zhang M, Ahmad M, Mohan D, Chang SX, et al. Pyrolysis condition affected sulfamethazine sorption by tea waste biochars. *Bioresour Technol* 2014;**166**:303–8.
96. Rajapaksha AU, Vithanage M, Ahmad M, Seo D-C, Cho J-S, Lee S-E, et al. Enhanced sulfamethazine removal by steam-activated invasive plant-derived biochar. *J Hazard Mater* 2015;**290**:43–50.
97. Xiong Z, Shihong Z, Haiping Y, Tao S, Yingquan C, Hanping C. Influence of NH3/CO2 modification on the characteristic of biochar and the CO2 capture. *Bioenergy Res* 2013;**6**(4):1147–53.
98. Matthiesen EA, Nalven S, Spector D, Zhang L. Innovative filter design application targeting *E. coli* and phosphorus removal. *Stormwater* 2019. 01/2019:6.
99. Cui E, Fan X, Li Z, Liu Y, Neal AL, Hu C, et al. Variations in soil and plant-microbiome composition with different quality irrigation waters and biochar supplementation. *Appl Soil Ecol* 2019;**142**:99–109.
100. Singh BP, Cowie AL. Long-term influence of biochar on native organic carbon mineralisation in a low-carbon clayey soil. *Sci Rep* 2014;**4**(1):3687.

101. Jenkins JR, Viger M, Arnold EC, Harris ZM, Ventura M, Miglietta F, et al. Biochar alters the soil microbiome and soil function: results of next-generation amplicon sequencing across Europe. *GCB Bioenergy* 2017;**9**(3):591–612.
102. Bruins MR, Kapil S, Oehme FW. Microbial resistance to metals in the environment. *Ecotoxicol Environ Saf* 2000;**45**(3):198–207.
103. Yao Y, Gao B, Chen H, Jiang L, Inyang M, Zimmerman AR, et al. Adsorption of sulfamethoxazole on biochar and its impact on reclaimed water irrigation. *J Hazard Mater* 2012;209–10. 408–13.
104. Xu Y, Seshadri B, Sarkar B, Wang H, Rumpel C, Sparks D, et al. Biochar modulates heavy metal toxicity and improves microbial carbon use efficiency in soil. *Sci Total Environ* 2018;**621**:148–59.
105. Li YL, Deletic A, DT MC. Removal of E coli from urban stormwater using antimicrobial-modified filter media. *J Hazard Mater* 2014;**271**:73–81.
106. Sun T, Miao J, Saleem M, Zhang H, Yang Y, Zhang Q. Bacterial compatibility and immobilization with biochar improved tebuconazole degradation, soil microbiome composition and functioning. *J Hazard Mater* 2020;**398**:122941.
107. Li X, Song Y, Bian Y, Gu C, Yang X, Wang F, et al. Insights into the mechanisms underlying efficient Rhizodegradation of PAHs in biochar-amended soil: from microbial communities to soil metabolomics. *Environ Int* 2020;**144**:105995.
108. Li X, Yao S, Bian Y, Jiang X, Song Y. The combination of biochar and plant roots improves soil bacterial adaptation to PAH stress: insights from soil enzymes, microbiome, and metabolome. *J Hazard Mater* 2020;**400**:123227.
109. Sarma H, Sonowal S, Prasad MNV. Plant-microbiome assisted and biochar-amended remediation of heavy metals and polyaromatic compounds — a microcosmic study. *Ecotoxicol Environ Saf* 2019;**176**:288–99.

CHAPTER EIGHT

A relationship paradigm between biochar amendment and greenhouse gas emissions

Mohd Ahsaan[a,b], Pratibha Tripathi[a], Anupama[a,b], and Puja Khare[a,b,*]

[a]Crop Production and Protection Division, CSIR-Central Institute of Medicinal and Aromatic Plants, Lucknow, India
[b]Academy of Scientific and Innovative Research (AcSIR), Ghaziabad, India
[*]Corresponding author: e-mail address: kharepuja@rediffmail.com

Contents

1. Introduction	204
2. Strategies adopted for mitigation of GHG emission	206
3. What is biochar?	207
4. Role of biochar in altering the factor responsible for GHGs emissions	209
5. Effect of biochar addition on major soil GHGs emission	212
5.1 Soil CO_2 emission	212
5.2 Soil CH_4 emission	214
5.3 Soil N_2O emission	214
6. Limitations	216
Acknowledgments	217
References	217

Abstract

Agriculture is among one of the potent economic sectors that affects the climate change contributing directly or indirectly approximately 20% of total greenhouse gas (GHG) emissions globally. To reduce GHG emission, mitigation strategies used till date are not feasible or ineffective due to the negative effects on crop productivity, and being costly for operation. Hence a promising and cost-effective mitigation strategy/solution to reduce GHG emission is prerequisite while not compromising the farm productivity in terms of quality or quantity of farm produce. Nowadays, use of biochar for reducing GHG emission has been emerged as a feasible and potential agro-technology from farm waste which can potentially mitigate GHG emissions from agricultural soils as well as improves soil health by enhancing soil C contents. The role of recalcitrant biochar C, impact of biochar on the indigenous soil organic carbon, methanotrophy and underlying responsible mechanisms for soil C sequestration and mitigation of GHGs emission has been discussed in the present chapter.

Keywords: Biochar, Mitigation strategy, Greenhouse gas, Carbon sequestration, Carbon dioxide, Methane, Nitrous oxide (N_2O), Denitrification

1. Introduction

Global warming has become one of the emerging global environmental issues and poses a threat to food security and human health.[1] Various anthropogenic sources such as burning of fossil fuels and industrial activities contributed approximately 78% release of green house gases (GHGs) in the atmosphere.[1,2] Besides, agricultural activities are among potent source contributing to about 11% of total GHGs emission worldwide.[1,3] Mostly, GHGs emission from agriculture occurs at the initial stage of plant growth, soil disturbance, and irrigation. The utilization and transportation of farm products, use of farm inputs (fertilizers, seeds pesticides, etc.) and agro-wastes management (like burning of crop residues and excrement) could also be responsible for the GHGs emission.[4,5] Hence, elevation of global GHGs emissions from this agricultural sector has been also categorized in climate change mitigation technique by minimizing release of GHGs to the atmosphere, through soil and plant.[2] The decomposition of organic wastes significantly contributes to GHG emissions and global warming during agricultural activities.[6,7] The use of bulking agents such as agro-wastes and saw dust[8,9] and amendment of several types of mineral additives like lime, zeolite, bentonite and medical stone during organic waste composting were used to reduce GHG emission[10,11]. However, the effect of these methods for reducing methane (CH_4) and nitrous oxide (N_2O) emission are uncertain. The emission of GHGs from agricultural sector is associated with global C and N cycle due to production of carbon dioxide (CO_2), methane (CH_4) and nitrous oxide (N_2O) as an end or intermediate product.[12] In this process soil aerobic and anaerobic condition are responsible for production of CO_2 and CH_4 respiration while nitrification and denitrification activity of mineral-N led to N_2O emission.[13]

Agriculture practices (irrigation, cultivation and fertilization) have strong influence on GHGs emissions.[2] The nitrogen fertilizers and pesticides together add 6%–1.5% to agriculture's global GHG) production. Implementation of more advanced and conservative fertilizer practices could cut 50% of N_2O emission from agricultural soils. Both CO_2 and N_2O are released during production of synthetic N based fertilizers.[14] Biogeochemical cycling of carbon and nitrogen and land use change play important role in GHG emission from soil.[14] Leguminous roots residue and their nutrient content may promote soil C sequestration through the carbon and nitrogen ratio. Therefore, concentration of atmospheric CO_2 may be decreased by legume crops, as C inputs in soils is related to humus, which represents 40%–60% of soil organic matter

(SOM). Potential annual inputs of GHGs emissions from agricultural soil was higher for CO_2 (5%–20%), CH_4 (15%–30%) and N_2O (80%–90%) into the environment. However, this condition could make global warming potential (GWP) per unit mass of CH_4 and N_2O as more worst as compared to CO_2 are 25 and 198 times higher than CO_2, respectively.[1] Hence, though produced in large amount, the contribution of CO_2 is minor as compared to CH_4 and N_2O.[1,10] Green house gases emissions from agricultural land depend upon in various factors including livestock production, eutrophication of surface waters, nutrient imbalance and nutrient depletion in soil.[15] Therefore, changes in various physio-chemical and biological factors such as soil humidity, soil temperature, soil pH-range, availability of mineral nutrients (labile C, mineral N and O_2), and land-cover properties can control the GHGs emission from agriculture land.[8,9,14,15]

Optimization of low-cost crop production with minimum emission of GHGs without compromising the food security and energy use efficiency are primary objectives of present agriculture. Recently, conversion of agricultural waste into biochar followed by its soil amendment for decreasing soil GHGs emission has been appreciated.[5,16,17] As application of biochar is reported to be beneficial in energy flow by improving soil fertility by retaining soil C content, which decreases emission of GHGs (CH_4 and N_2O).[7,18,19] Moreover, soil amendment of biochar leads to adsorption and retention of NO_3-N and NH_4-H followed by decrease release of soil N[20]. In this chapter we have summarized the effects of biochar on GHG emissions from agricultural soils based on available literature and in our opinion.

Major GHGs related to agriculture which primarily contributes to global warming include CO_2, CH_4 and N_2O. Atmospheric CO_2, biogenic CO_2 (generated through root respiration and decomposition of soil organic matter), and the break-up of carbonate rock are the three main sources of dissolved inorganic carbon.[20] The net atmospheric CO_2 derived through carbonate rocks weathering (karstifiction) accounts 36 million $t\,Ca^{-1}$. About 18% of CO_2 in the air comes from respiration. Soil is a huge C reservoir in a terrestrial ecosystem. Total annual CO_2 flux on average (approximately 31%) was released from arable lands which are very large amount for CO_2 emission.[20]

Methane (CH_4) is one of the important GHGs contributing around 20% to global warming.[1] Generally, the lifetime of CH_4 in the atmosphere is about 12 years. Overall worldwide emission of CH_4 (about 15%–30%) originated from agriculture sources among which major part of CH_4 (12%) and N_2O (7%–11%) emits from soils of paddy fields.[4,14] Methane is a stable C compound produced through the anaerobic decay of organic matter through methanogenic bacteria.[21]

Nitrous oxide (NO_2) emission occurs from soil consisting of agriculture inputs (38%), manure (7%), ruminant animals (32%), and from biomass burning (12%). Emission of N_2O from farm soils into the atmosphere consists of about 84% of all worldwide anthropogenic N_2O emissions.[22] Unbalanced use of N fertilizers and manure resulted low fertilizer use efficiency in plants and elevated emission of GHGs[12].

2. Strategies adopted for mitigation of GHG emission

Physical emission reduction technology chiefly includes modulation in physical agro climatic environment and measures for decreasing GHGs emission from agricultural inputs mainly by field management. Field-management strategies includes water management, fertilizer (types, dosage, and time of application), and tillage. Appropriate water management strategies such as midseason drainage, intermittent irrigation, alternate wetting and drying can substantially decrease GHG emissions.[18] The GHGs emissions from paddy field are major contributors due to high water demand. Hence advanced water management practices during the rice-growing season are the major reported strategy for paddy field to lower CH_4 emissions.[14,21] Significant CH_4 emissions may occur under continuously flooded soils, whereas N_2O emissions result if soils are alternately wet and dry.[23,24] Thus estimated GHG budget exhibits large spatial–temporal variations. Amendment of dairy manure in agricultural soils has been found to result enhanced crop yield through better nutrient uptake but its practical application is not advisable due to emission of GHGs in considerable amount into the environment.[6] A recent field study showed that combined application of biochar and dairy manure has reduced the GHGs emission rate as compared to alone dairy manure in corn silage under boreal climate.[3]

Chemical emission reduction technology is mainly used to control the number of methanogens, nitrifying bacteria, and denitrifying bacteria in the soil. The N_2O emission from soil is related to the soil nitrifying microbes (i.e., autotrophic and heterotrophic bacteria) that dominates the process. The main nitrifying (oxidation of ammonia to nitrite) bacteria groups belongs to *Nitrosomonas, Nitrosococcus*, etc. Nitrification process cannot only reduce the loss of nitrogen fertilizer, but they also increase the utilization rate of nitrogen fertilizer and the yield.[25] Organic fertilizers can be decomposed into single methanogen precursors such as hexanol, hexanoic acid, propionic acid, butyric acid, or petunic acid under the action of saccharide-hydrolyzation and fermenting bacteria. Under anaerobic conditions, methanogens exert influence on these methane precursors and produce CH_4.[26] They can also reduce the nitrite content in crops, improve the quality of crops and reduce the pollution of soil, ground water, and the environment resulting from high fertilization rates. The effect of the chemical method is the most obvious, but the implementation cost is relatively high, and is some cases, the effect of the nitrification inhibitor on the yield increase of crops is not stable.[27] The reduction of N_2O released from the soil through the application of chemical compounds, such as nitrification inhibitors together with nitrogen fertilizer was reported.[9,19] Although a number of mitigation strategies is currently available, an efficient and feasible strategy that not only mitigates CO_2 production but also reduces CH_4 and N_2O emission from soil is still needed.

3. What is biochar?

Biochar is a solid C enriched solid material produced through thermal conversion of organic biomass (wastes) in the absence of or under limited oxygen conditions through pyrolysis (temperature ranges 400–800 °C).[5,7] Conversion of agricultural or animal wastes into biochar has great potential for waste management and is also helpful in related environmental issues mitigation such as GHGs emission.[5] Biochar can be attributed to enhanced sorption capacity for various contaminants, effect on soil pH, water holding capacity (WHC), bulk density and nutrient availability.[28] Biochar has been used as a soil amendment which affects native soil organic matter (SOM) decomposition. Efficiency of biochar to microbial or chemical oxidation rely on various climatic conditions such as feedstock type, pyrolysis temperature, soil pH, biochar pH, soil CEC and soil texture.[29,30] Biochar prepared at low temperature posses less C:N ratio and has been observed to be mineralized early and help ameliorate soil microbial structure. Due to its recalcitrant nature unlike several fertilizers, application of biochar is not required every year revealing its long-lasting benefits in agriculture soil. The biochar aging is one of the most important parameters in reducing GHGs emissions into the atmosphere either though abiotic or biotic processes.[7] Hence, the biochar characteristics decide its effectiveness for mitigating emission of GHGs from the environment.[31] Specific characteristics of biochar such as highly porous structure, acid buffering capacity, water holding and nutrient retention capacities could be connected and undoubtedly could be utilized as a favorable strategy for GHG emission mitigation in agriculture soil.[31] Effect of different biochar types in mitigation of GHG emission is summarized in Table 1.

Table 1 Effect of different biochar types on GHGs emission.

Biochar feedstock type	Pyrolysis temperature (°C)	Biochar application rate	Experiment type/crop	Effect on GHGs	References
Rice husk	Not given	2%–4%	Paddy	Reduced CH_4 emission during rice cultivation	32
Wheat straw	350–550	20 and 40 $t\,ha^{-1}$	Wheat and Paddy	Limit N_2O emission not beyond 2 years	27
Poultry manure and green waste	250	0.5%–2%	Wheat	Significant reduced emission of NH_3	33
Wheat straw	300–550 °C	$1\,g\,L^{-1}$	–	Limited CH_4 production at low acetate concentration	34
Eucalyptus wood	400	29%–51%	Maize	Biochar-N based fertilizers promote C sequestration and reduce GHG emissions	19
Yellow pine wood (Pinus spp.)	500	20 $t\,ha^{-1}$	Maize	Limits emission of major GHGs (CH_4, CO_2 and N_2O)	3
Rice husk and straw	350–400	2.5%	Paddy	Reduced CH_4 emission by decreasing methanogenic but promoting methanotrophic activity	21
Maize straw	500	0.5%–2%	–	Combined application with $450\,mg\,N\,kg^{-1}$ limits CO_2 and N_2O emission	35
Rice straw	Not given	0.5%–2%	Paddy	Reduced CH_4 emission over two growing seasons	36

4. Role of biochar in altering the factor responsible for GHGs emissions

Variable effect of biochar application on major GHGs emission from agriculture field soils has been depicted in Table 2. Temperature and moisture are most important factors that could be potentially best utilized to mitigate GHG emissions from soil while balancing agricultural practices.[24] A positive correlation between soil temperature and CO_2 emission has been recently observed.[38] Soil temperature and soil moisture explained about 74%–86% of the different variations for NO and N_2O emissions, respectively.[24] Temperature and moisture levels of land maintains the substrates availability from plants to soil biota and controls responsible microbial activity for CO_2, N_2O and CH_4 production and oxidation in soils.[13,23,31] Elevated soil temperature favors high emissions due to increased soil respiration rates and leading to decreasing O_2 levels in the soil.[25,38] The change in soil physical, chemical, and biological properties after biochar incorporation has significant effects on the soil's GHG emission. Results of[24] Yang et al. revealed that in the first year, reduction in N_2O emission was observed only at $40\,t\,ha^{-1}$ of biochar amendment whereas at $20\,t\,ha^{-1}$ N_2O emissions was promoted. However in the second year; suppression effect was observed at both the doses of biochar Moreover, under control irrigation conditions emission of N_2O emissions enhanced than flooded soil conditions, but under both conditions biochar led decrease in CH_4 emissions. These reports revealed that potential of biochar in mitigation of GHG emission depends on temperature and water conditions to an extent.

Type of soils also plays a significant role in the emission of greenhouse gases (GHG) to the atmosphere.[19,28,39] When there is a decrease in water levels, drainage has been found to lead to increased emissions of carbon dioxide (CO_2) and methane (CH_4).[31] Basically, drained land surfaces have been observed to reduce CH_4 emissions and high nitrous oxide (N_2O) emissions are particularly associated with nutrient rich organic soils.[25] Mangrove soils are sources of many atmospheric green house gases, such as CO_2, CH_4 and N_2O, etc. which proved that these gas emissions can be further enhanced by anthropogenic nutrients. Semi-arid climate can have excellent properties that enhance soil organic matter benefits for agricultural soils potentially improving the coarse-textured cropping soils to drying climates by increasing water infiltration, nutrient retention capacity and water-holding capacity.[39] In calcareous soils the acidification has been found due to dissolution of carbonate mineral and in acidic soils dissolution of applied limed also

Table 2 Impact of biochar application in mitigating emission of major GHGs under field conditions.

Biochar type	Pyrolysis Temperature (°C)	Crop & soil type	Duration	Biochar application rate (t ha^{-1})	CO_2 emission (kg ha^{-1})	CH_4 emission (kg ha^{-1})	N_2O emission (kg ha^{-1})	References
Rice husks	450	Rice & Orthic Anthrosols	One season	0	1850.9	16.7	0.05	37
				25	1540.8	21.7	0.01	
				50	1458.8	22.8	0.02	
		Wheat & Orthic Anthrosols	One season	0	436.8	−0.16	0.09	
				25	397.7	1.27	0.04	
				50	491.5	0.05	0.03	
Rice straw	600	Paddy & Hydragric Anthrosol	2016 controlled irrigation	0	nd	75.6	5.24	24
				20	nd	53.1	8.67	
				40	nd	70.8	4.63	
			Flooded condition	40	nd	605	2.31	
			2017 controlled irrigation	0	nd	154	4.43	
				20	nd	108	1.86	
				40	nd	130	2.52	
			Flooded condition	40	nd	246	1.67	

Corn residue	400–500	Corn & sandy loam	2015	0	5360	−0.195	0.336	38
				15	4038	−0.702	0.097	
				30	4419	−0.551	−0.028	
				45	4174	0.039	−0.035	
			2016	0	4924	−0.402	0.302	
				15	3975	−0.416	0.170	
				30	3651	−0.661	0.161	
				45	2924	−0.191	0.185	

nd: not determined.

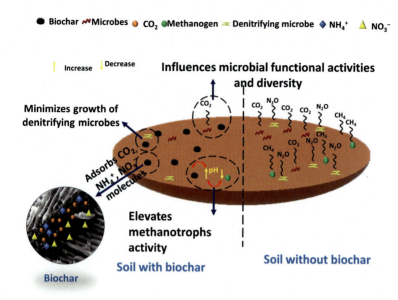

Fig. 1 Mitigation mechanisms of GHGs emissions from agricultural soil by biochar amendments.

contributed to the total CO_2 released from soils. Soil acidity has been frequently targeted as the main contributor to CO_2 emission.[39]

The liming effect or base saturation through pH shift by biochar has been revealed as major mechanism in acidic soil conditions and would be beneficial for mitigation of CO_2[40] and N_2O emission.[41] Under field conditions biochar amendment revealed slight promotion in GHG (N_2O and CO_2) emission followed by its suppression.[38] Yang et al. deciphered two reasons for enhanced N mineralization during first year of biochar application as firstly readily release of biochar intact N content in soil and second biochar facilitated growth promotion of N mineralizing microbes. Biochar addition with pig manure led remarkable decrease in N_2O emission from soil than alone organic amendment, i.e., pig manure.[42] Possible mechanisms employed by biochar for mitigation of GHGs emission from agriculture soil has been shown in Fig. 1.

5. Effect of biochar addition on major soil GHGs emission

5.1 Soil CO_2 emission

Biochar has stable C structure and due to their recalcitrant nature for microbial decomposition, biochar has been recognized as stable C pools and known to sequester carbon via several mechanisms.[43,44] Biochar can modulate C cycle dynamics which includes participation of native soil microorganism and mediates to decrease or increase organic carbon decomposition due to respective enhanced or suppressed priming effect on organic matter, depending on C sources.[43] Moreover, impact of biochar on the microbial decomposition and physical protection of native SOC determine the destiny of native SOC and CO_2 emission.[31,44] Presence of VOC on biochar surface (decided through ratio of O:C and atomic H:C) accounts for its stability against microbial degradation while C:N ratio of biochar reflects its availability as a C source for native microbes.[12,31] As biochar posses high C:N ratio as compared to their feedstocks, due to less available C content and absence of N supply they are not easily decomposed by soil microbes thus an enhanced stable organic C pools is formed with improved soil carbon sequestration. However soil CO_2 emissions are largely dependent on biochar production conditions.[16] Effect of pyrolysis temperature on CO_2 emission has been observed and it was found that low pyrolysis process mitigates more CO_2 (0.6–1.4 Mg CO_2 eq Mg^{-1} feedstock) than generated from fast pyrolysis[16,38] (0.11 Mg CO_2 eq Mg^{-1} feedstock). The possible key factors and mechanistic conditions responsible for biochar mediated altered CO_2 emission include: firstly, labile organic C content of biochar enhances the native soil organic C pool hence biochar addition showed elevated CO_2 emissions from agriculture soils. Second, owing to porous structure biochar posses good adsorption potential and adsorbs native CO_2 molecules of soil and ameliorates CO_2 emission from soil surface. Third, application of biochar modulates various soil characteristics such as soil water content, porosity, pH, CEC, aggregation, soil native C and N cycle, and crop yield which mediate to modulated emission of CO_2. Fourth, addition of biochar in soil influences the microbial functional activities and diversity of native soil microflora responsible for generation of CO_2.

Besides, few studies reporting negative effect of biochar on CO_2 emission from soils has also been observed. A combined meta-analysis of 46 reports showing remarkable increase of 28% in CO_2 emission from biochar added soils revealed strong correlation among biochar and native soil organic carbon content.[44] The authors deciphered that ratio of biochar C and native soil organic carbon below 2 showed a non considerable effect of biochar on CO_2 emission whereas elevated level of CO_2 emission was noticed when ratio is more than 2. A plausible explanation is that reduction in native soil organic carbon by biochar application led to negatively affect C sequestration capacity of biochar suggesting that biochar mediated CO_2 emission from soil is dependent on biochar C and native SOC ratio.

5.2 Soil CH$_4$ emission

The enhanced plant growth through NH_4^+ or NO_3^- based fertilizer can elevate generation of CH_4 by improving availability of organic C which act as methanogenic subtract for fermentation of microbes.[21,25] The addition of biochar has significant effect in decreasing CH_4 emissions from soil.[13,24] Application of biochar with 2% and 4% has shown 45% and 59.9% decreased emission of CH_4 from paddy soils.[32] The responsible factors behind reduced CH_4 emission werean increase in soil pH facilitated by biochar application which elevated methanotrophs activity or decreased the methanogens activity thus minimizing CH_4 emission.[13,36] A balanced activity of methanogens and methanotrophs is thus crucial for reducing CH_4 emission from paddy soils.[21]

Generally, emission of CH_4 caused by the deoxidization released CO_2: CO and volatile organic acids by methanogens under low oxygen conditions.[13,27] Xiao et al.[34] reported that in terrestrial methanogenic conditions application of biochar either elevate or suppress emission of CH_4 it is highly depends on the concentration of acetate (a substrate of methanogenesis and a decomposition product of organic matter). Biochar application generally increases soil pH which favors the growth of methanotrophy cycle (CH_4 oxidation).[31] Addition of biochar was reported to decrease CH_4 emissions from paddy field under waterlogged conditions whereas enhances CH_4 emission under aerobic conditions.[14] Biochar application in forest soils resulted in decreased CH_4 emissions.

Biochar application has been also found to decrease soil bulk density and to increase soil porosity, which favor methanotrophy and CH_4 uptake activity by soil bacteria and could be other possible mechanism for reduced CH_4 emission.[36] Increased soil aeration and porosity induced by biochar application and the consequent promotion of oxic conditions generally decrease CH_4 production, because CH_4 oxidation is an aerobic metabolic process, that is dependent on oxygen availability. Use of biochar combined with limited irrigation technology has been found to enhance crop yield, improved soil fertility without elevating CH_4 and N_2O emissions however biochar promoted CH_4 emission under flooded soil conditions.[24]

5.3 Soil N$_2$O emission

Nitrous oxide in the atmosphere is produced by the conversion of N fertilizer and animal manure into N_2O through nitrification and denitrification process of soil inhabiting microbes. Also, addition of N source can elevate the process besides, limited use of N and fertilizer, dumping of ruminant-urine also increases the N_2O because of the presence of N content in the excreted urine.[25] To minimize N_2O emission from agriculture inputs the knowledge of particular source (soil N or fertilizer N) and microbial process that elevates N_2O and N gas (N_2) is must. Published studies have shown enhanced N mineralization by the application of either farmyard manure or vermicompost moreover; these organic amendments could provide substrate to microbes leading increased soil microbial activities as evident from improved plant nitrogen used efficiency.[7] Inversely, biochar amended soil showed reduced N immobilization and N_2O release followed by improved N mineralization than non treated control.[7] Hence, before practical application of any mitigation technique for reducing N_2O emission from various source understanding of generation, consumption pathways, and responsible parameters is crucial. Various physiochemical properties of soil affect the soil redox status selecting for microbial pathways processing inorganic-N. As enhanced soil pH facilitates more adsorption of NH_4-H or NH_3-H, this leads to reduced release of NH_3. Moreover, binding of N_2O onto functional surface sites of biochar also facilitate reduce N_2O emission.[41] Metal ions (Fe or Cu), are often accumulated on biochar surface during pyrolysis, and bind the N_2O and probably facilitate in the reduction of N_2O emission.[7] Reduced biological nitrification by minimizing ammonia oxidation has been considered as a potential strategy but due to its affect on microbes it should be evaluated carefully.[19] The elevated adsorption capacity of biochar due to the presence of surface functional groups has been found to playa vital role in retaining the soil nitrogen during soil nitrification process.[33] Biochar application as found to enhance C inputs in soil and aggravated enhanced nitrogen retention or reduced nitrogen leaching could be correlated to the reduction of dissimilatory NO_3- into NH_4+.[41] The biochar amendment in dry land soil has decreased N_2O emissions. For example, the application of biochar ($30\,t\,ha^{-1}$) to alpine forest soil significantly decreased (25.5%) cumulative N_2O emissions. The possible reasons for biochar mediated reduced N_2O emission includes: improved soil oxygen level and aeration (oxic condition) after biochar addition minimizes the growth of denitrifyning microbes; adsorbing nature of biochar adsorbs the NH_4+ and NO_3- to its surface which promotes plant production and either inhibits volatilization of ammonia or immobilizes N compounds. Owing to high C:N ratio, biochar supports N immobilization in the soil.[5] All these activities reduce the availability of inorganic N pool to N_2O producing nitrifying or denitrifying microbes. The application of 10% biochar in a forest soil significantly increased N_2O emissions, while no significant effect was observed after the application of 1% biochar. The application of biochar ($5\,t\,ha^{-1}$) in a temperate hardwood forest did not change soil N_2O emission. During the process of denitrification and N_2O generation, NO_3^--N plays crucial role of electron acceptor.[45] Recent studies have also revealed that reduced NO_3--N and enhanced soil NH_4+-N and organic carbon content resulted limited N_2O emission.[24,46] In soil N_2O emission and NO_3^--N contents are positively correlated to each other as limited NO_3-N in soil resulted in decreased denitrification process and followed by reduced production and emission of N_2O.[24] Application of biochar accelerated the anaerobic microsites development which inhibited the conversion of NH_4^+-N to NO_3^--N and results minimized N_2O emissions.[46] Production of N_2O is chiefly occurs via nitrification and denitrification process. Biochars prepared at lower temperatures were reported for reducing ammonia emissions from soil.[33]

Application of biochar in agricultural soils modifies the soil properties which could have an indirect effect on these two processes. In a study, enhanced soil pH after biochar addition was found to minimize CO_2 and N_2O emissions[35] whereas in other report liming effect of biochar has been found to alter native bacterial community structure along with increases N_2O and CO_2 emissions in sandy soils having low pH[29]. Hence due to these contradictory results impact of soil biochar application on N_2O emission processes seems to be very complicated and warrant further investigations to understand underlying mechanism. Overall, Fig. 1 summarizes key parameters and mechanisms of biochar which facilitates reduction in GHGs emission. It can be observed from Fig. 1 that potential of biochar in mitigation of GHG emission is highly dependent on biochar production conditions (feedstocks, pyrolysis temperature, aging duration, etc.) and soil characteristics such as pH, EC, etc.

6. Limitations

Addition of biochar to soils has been investigated by several workers to study greenhouse gas emission mitigation, however at present there are still several conflicting views such as does biochar actually promote or suppress greenhouse gas emissions? Biochar has also been proposed as a negative way to reduce greenhouse gas (GHG) emissions from waste management, and mitigate GHG emissions from agricultural systems, as addition of biochar increases soil C contents revealing that biochar amendment can also enhance the GHG emissions. Biochar amendment has significantly promoted CH_4 emissions during paddy cultivation in upland soils besides it possess no impact on CH_4 emissions on wheat season.[37] The addition of biochar in agricultural soil increases CO_2 emissions through abiotic release of inorganic carbon, the decomposition of labile components of biochar, and the decomposition of organic matters or humus by biochar. Moreover, additions of biochar can affect soil emissions of CO_2 and CH_4 as well as N_2O, as a result of its impact on soil C and N cycles.[18,28,33] However, these results are contradictory with some studies reporting stimulating effects[18,28,33] while others reported inhibiting or no significant effects on soil N_2O and CH_4 fluxes.[34,43] The variable response of soil GHG emissions to biochar amendments has been attributed to differences in biochar properties as well as in environmental conditions in the individual studies.[30] Few recent studies contribute that a combined application of biochar under controlled irrigation in paddy field could be a realistic and sustainable mitigation approach toward GHGs emissions.[24] Adverse impact of biochar on native soil biota has been revealed though growth promotion of pathogenic microbes and suppression of beneficial soil microorganisms.[47] Negative effect of biochar on growth of plant beneficial mycorrhiza fungi has also been observed over a range of soil types.[48] Study of Bolster and Abit[49] deciphered transport of *Escherichia coli* through sandy soils after application of biochar.

Moreover, most of the studies deciphering biochar mediated reduced GHG mitigation were of under laboratory or short duration field experiments.[3,28,30,33,34,38] A long-term study of Liu et al.[27] revealed that single application of biochar though has improved soil quality but neither retain yield benefits nor affect N_2O emission after 2 year. Hence more long-term field evaluations are required for validation of biochar regarding its long-lasting effects.

Acknowledgments

The authors are grateful to Director, CSIR–CIMAP, Lucknow for his encouragement and support. The authors also acknowledge the financial support by the Department of Biotechnology (DBT), New Delhi (BT/PR24706/NER/95/822/2017) under the twinning program. PT is grateful to ICMR, New Delhi for RA fellowship (File No.45/8/2019/MP/BMS).

References

1. IPCC. *Climate change and land: an IPCC special report on climate change, desertification, land degradation, sustainable land management, food security, and greenhouse gas fluxes in terrestrial ecosystems.* Intergovernmental Panel on Climate Change; 2019.
2. Peter C, Helming K, Nendel C. Do greenhouse gas emission calculations from energy crop cultivation reflect actual agricultural management practices?–a review of carbon footprint calculators. *Renew Sustain Energy Rev* 2017;**67**:461–76.

3. Ashiq W, Nadeem M, Ali W, Zaeem M, Wu J, Galagedara L, et al. Biochar amendment mitigates greenhouse gases emission and global warming potential in dairy manure based silage corn in boreal climate. *Environ Pollut* 2020;**265**:114869.
4. Liu S, Zheng Y, Ma R, Yu K, Han Z, Xiao S, et al. Increased soil release of greenhouse gases shrinks terrestrial carbon uptake enhancement under warming. *Glob Chang Biol* 2020;**26**:4601–13.
5. Khare P, Deshmukh Y, Yadav V, Pandey V, Singh A, Verma K. Biochar production: a sustainable solution for crop residue burning and related environmental issues. *Environ Prog Sustain Energy* 2021;**40**:e13529.
6. Matsi TH, Lithourgidis AS, Barbayiannis N. Effect of liquid cattle manure on soil chemical properties and corn growth in northern Greece. *Exp Agric* 2015;**51**:435.
7. Yadav V, Karak T, Singh S, Singh AK, Khare P. Benefits of biochar over other organic amendments: responses for plant productivity (Pelargonium graveolens L.) and nitrogen and phosphorus losses. *Ind Crop Prod* 2019;**131**:96–105.
8. Zhang D, Luo W, Yuan J, Li G, Luo Y. Effects of woody peat and superphosphate on compost maturity and gaseous emissions during pig manure composting. *Waste Manag* 2017;**68**:56–63.
9. Oliveira LV, Higarashi MM, Nicoloso RS, Coldebella A. Use of dicyandiamide to reduce nitrogen loss and nitrous oxide emission during mechanically turned co-composting of swine slurry with sawdust. *Waste Biomass Valoriz* 2020;**11**:2567–79.
10. Awasthi MK, Wang Q, Ren X, Zhao J, Huang H, Awasthi SK, et al. Role of biochar amendment in mitigation of nitrogen loss and greenhouse gas emission during sewage sludge composting. *Bioresour Technol* 2016;**219**:270–80.
11. Wang Q, Wang Z, Awasthi MK, Jiang Y, Li R, Ren X, et al. Evaluation of medical stone amendment for the reduction of nitrogen loss and bioavailability of heavy metals during pig manure composting. *Bioresour Technol* 2016;**220**:297–304.
12. Zhang X, Sun Z, Liu J, Ouyang Z, Wu L. Simulating greenhouse gas emissions and stocks of carbon and nitrogen in soil from a long-term no-till system in the North China plain. *Soil Tillage Res* 2018;**178**:32–40.
13. Liu X, Mao P, Li L, Ma J. Impact of biochar application on yield-scaled greenhouse gas intensity: a meta-analysis. *Sci Total Environ* 2019;**656**:969–76.
14. Li J, Wan Y, Wang B, Waqas MA, Cai W, Guo C, et al. Combination of modified nitrogen fertilizers and water saving irrigation can reduce greenhouse gas emissions and increase rice yield. *Geoderma* 2018;**315**:1–10.
15. Sakadevan K, Nguyen ML. Livestock production and its impact on nutrient pollution and greenhouse gas emissions. *Adv Agron* 2017;**141**:147–84.
16. Field JL, Keske CM, Birch GL, DeFoort MW, Cotrufo MF. Distributed biochar and bioenergy coproduction: a regionally specific case study of environmental benefits and economic impacts. *GCB Bioenergy* 2013;**5**:177–91.
17. Nigam N, Shanker K, Khare P. Valorisation of residue of Mentha arvensis by pyrolysis: evaluation of agronomic and environmental benefits. *Waste Biomass Valoriz* 2018;**9**:1909–19.
18. Feng Y, Xu Y, Yu Y, Xie Z, Lin X. Mechanisms of biochar decreasing methane emission from Chinese paddy soils. *Soil Biol Biochem* 2012;**46**:80–8.
19. Puga AP, Grutzmacher P, Cerri CEP, Ribeirinho VS, de Andrade CA. Biochar-based nitrogen fertilizers: greenhouse gas emissions, use efficiency, and maize yield in tropical soils. *Sci Total Environ* 2020;**704**:135375.
20. Buragiene S, Sarauskis E, Romaneckas K, Adamaviciene A, Kriauciuniene Z, Avizienyte, et al. Relationship between CO2 emissions and soil properties of differently tilled soils. *Sci Total Environ* 2019;**662**:786–95.
21. Nguyen BT, Trinh NN, Bach QV. Methane emissions and associated microbial activities from paddy salt-affected soil as influenced by biochar and cow manure addition. *Appl Soil Ecol* 2020;**152**:103531.

22. Smith K, Crutzen P, Mosier A, Winiwarter W. The global nitrous oxide budget: a reassessment. In: *Nitrous Oxide and climate change.* London: Routledge Taylor and Francis; 2010. p. 63–84.
23. Sriphirom P, Amnat C, Sirintornthep T. Effect of alternate wetting and drying water management on rice cultivation with low emissions and low water used during wet and dry season. *J Clean Prod* 2019;**223**:980–8.
24. Yang S, Xiao Y, Sun X, Ding J, Jiang Z, Xu J. Biochar improved rice yield and mitigated CH_4 and N_2O emissions from paddy field under controlled irrigation in the Taihu Lake region of China. *Atmos Environ* 2019;**200**:69–77.
25. Butterbach-Bahl K, Baggs EM, Dannenmann M, Kiese R, Zechmeister-Boltenstern S. Nitrous oxide emissions from soils: how well do we understand the processes and their controls? *Philos Trans R Soc, B* 2013;**368**:20130122.
26. Liu Y, Tang H, Muhammad A, Huang G. Emission mechanism and reduction counter measures of agricultural greenhouse gases–a review. *Greenhouse Gases Sci Technol* 2019;**9**:160–74.
27. Liu X, Zhou J, Chi Z, Zheng J, Li L, Zhang X, et al. Biochar provided limited benefits for rice yield and greenhouse gas mitigation six years following an amendment in a fertile rice paddy. *Catena* 2019;**179**:20–8.
28. Senbayram M, Saygan EP, Chen R, Aydemir S, Kaya C, Wu D, et al. Effect of biochar origin and soil type on the greenhouse gas emission and the bacterial community structure in N fertilised acidic sandy and alkaline clay soil. *Sci Total Environ* 2019;**660**:69–79.
29. Yadav V, Jain S, Mishra P, Khare P, Shukla AK, Karak T, et al. Amelioration in nutrient mineralization and microbial activities of sandy loam soil by short term field aged biochar. *Appl Soil Ecol* 2019;**138**:144–55.
30. Ramlow M, Cotrufo MF. Woody biochar's greenhouse gas mitigation potential across fertilized and unfertilized agricultural soils and soil moisture regimes. *Gcb Bioenergy* 2018;**10**:108–22.
31. Zhang C, Zeng G, Huang D, Lai C, Chen M, Cheng M, et al. Biochar for environmental management: mitigating greenhouse gas emissions, contaminant treatment, and potential negative impacts. *Chem Eng J* 2019;**373**:902–22.
32. Pratiwi EPA, Shinogi Y. Rice husk biochar application to paddy soil and its effects on soil physical properties, plant growth, and methane emission. *Paddy Water Environ* 2016;**14**:521–32.
33. Mandal S, Donner E, Smith E, Sarkar B, Lombi E. Biochar with near-neutral pH reduces ammonia volatilization and improves plant growth in a soil-plant system: a closed chamber experiment. *Sci Total Environ* 2019;**697**:134114.
34. Xiao L, Liu F, Xu H, Feng D, Liu J, Han G. Biochar promotes methane production at high acetate concentrations in anaerobic soils. *Environ Chem Lett* 2019;**17**:1347–52.
35. Wang L, Yang K, Gao C, Zhu L. Effect and mechanism of biochar on CO2 and N2O emissions under different nitrogen fertilization gradient from an acidic soil. *Sci Total Environ* 2020;**747**:141265.
36. Qi L, Ma Z, Chang SX, Zhou P, Huang R, Wang Y, et al. Biochar decreases methanogenic archaea abundance and methane emissions in a flooded paddy soil. *Sci Total Environ* 2021;**752**:141958.
37. Wang J, Pan X, Liu Y, Zhang X, Xiong Z. Effects of biochar amendment in two soils on greenhouse gas emissions and crop production. *Plant and Soil* 2012;**360**:287–98.
38. Yang W, Feng G, Miles D, Gao L, Jia Y, Li C, et al. Impact of biochar on greenhouse gas emissions and soil carbon sequestration in corn grown under drip irrigation with mulching. *Sci Total Environ* 2020;**729**:138752.
39. Shakoor A, Shahbaz M, Farooq TH, Sahar NE, Shahzad SM, Altaf MM, et al. A global meta-analysis of greenhouse gases emission and crop yield under no-tillage as compared to conventional tillage. *Sci Total Environ* 2021;**750**:142299.

40. Jeffery S, Verheijen FG, van der Velde M, Bastos AC. A quantitative review of the effects of biochar application to soils on crop productivity using meta-analysis. *Agr Ecosyst Environ* 2011;**144**:175–87.
41. Cayuela ML, Sánchez-Monedero MA, Roig A, Hanley K, Enders A, Lehmann J. Biochar and denitrification in soils: when, how much and why does biochar reduce N_2O emissions? *Sci Rep* 2013;**3**:1–7.
42. Shi Y, Liu X, Zhang Q. Effects of combined biochar and organic fertilizer on nitrous oxide fluxes and the related nitrifier and denitrifier communities in a saline-alkali soil. *Sci Total Environ* 2019;**686**:199–211.
43. Zimmerman AR, Gao B, Ahn MY. Positive and negative carbon mineralization priming effects among a variety of biochar-amended soils. *Soil Biol Biochem* 2011;**43**:1169–79.
44. Sagrilo E, Jeffery S, Hoffland E, Kuyper TW. Emission of CO2 from biochar-amended soils and implications for soil organic carbon. *GCB Bioenergy* 2015;**7**:1294–304.
45. Hawthorne I, Johnson MS, Jassal RS, Black TA, Grant NJ, Smukler SM. Application of biochar and nitrogen influences fluxes of CO_2, CH_4 and N_2O in a forest soil. *J Environ Manage* 2017;**192**:203–14.
46. Fidel RB, Laird DA, Parkin TB. Impact of biochar organic and inorganic carbon on soil CO_2 and N_2O emissions. *J Environ Qual* 2017;**46**:505–13.
47. Lehmann J, Rillig MC, Thies J, Masiello CA, Hockaday WC, Crowley D. Biochar effects on soil biota–a review. *Soil Biol Biochem* 2011;**43**:1812–36.
48. Warnock DD, Lehmann J, Kuyper TW, Rillig MC. Mycorrhizal responses to biochar in soil–concepts and mechanisms. *Plant and Soil* 2007;**300**:9–20.
49. Bolster CH, Abit SM. Biochar pyrolyzed at two temperatures affects *Escherichia coli* transport through a sandy soil. *J Environ Qual* 2012;**41**:124–33.

CHAPTER NINE

Biochar for sustainable agriculture: Prospects and implications

Kumar Raja Vanapalli[a], Biswajit Samal[a], Brajesh Kumar Dubey[a,b,*], and Jayanta Bhattacharya[a,c]

[a]School of Environmental Science and Engineering, Indian Institute of Technology, Kharagpur, West Bengal, India
[b]Department of Civil Engineering, Indian Institute of Technology, Kharagpur, West Bengal, India
[c]Department of Mining Engineering, Indian Institute of Technology, Kharagpur, West Bengal, India
*Corresponding author: e-mail address: bkdubey@civil.iitkgp.ac.in

Contents

1. Introduction	222
2. Agronomic properties of biochar	225
3. Biochar induced soil quality improvement	229
3.1 Impact on soil physicochemical properties	229
3.2 Impact on soil nutrient cycling and water use efficiency	230
3.3 Impact on soil microbial ecology	233
3.4 Stability and durability of biochar within the soil	236
4. Biochar-induced crop growth and production	237
4.1 Impact on crop growth and productivity	237
4.2 Impact on crop physiological parameters and quality	242
4.3 Impact on the bioavailability of contaminants in the soil	244
4.4 Impact on GHG emissions from soil	247
5. Economics of biochar application for agriculture	248
6. Concluding remarks and future research needs	250
References	251

Abstract

Biochar application in soil can play a substantial part in altering the soil nutrients dynamics, pollutants, and microbial ecology. Also, strategic biochar treatments in soil may promote sustainable agriculture while providing numerous agronomic, economic, and environmental benefits. Both physical characteristics of biochar including surface area, particle density, and pore size distribution, and chemical characteristics like pH, electrical conductivity, total and plant-available concentrations of carbon, nitrogen, potassium, and phosphorus, cation exchange capacity, and selected minor nutrient contents influence its agronomic potential in soil. Some of the key outputs of biochar amendment include better nutrient management in soil, improved crop growth and productivity,

---

Advances in Chemical Pollution, Environmental Management and Protection, Volume 7
ISSN 2468-9289
https://doi.org/10.1016/bs.apmp.2021.08.008

Copyright © 2021 Elsevier Inc.
All rights reserved.

221

enhanced crop physiological parameters and quality, reduced bioavailability of contaminants in soil and overall decrease in greenhouse gas emissions from soil. Moreover, the associated benefits of carbon sequestration through its highly recalcitrant nature could aid in the mitigation of climate change. The economic benefits of improved crop yield along with reduced costs of frequent fertilizers application and reduced irrigation can help in achieving economic stability and profitability to the farmers without compromising the preservation of environmental ecosystems. However, since most of these effects of biochar are highly dependent on variable factors including soil type, crop type and agroclimatic conditions, etc., long term field investigations focussing on positive and negative implications and potential limitations are necessary to propose biochar for practical agricultural applications.

Keywords: Biochar, Agriculture, Soil amendment, Carbon sequestration, Nutrient dynamics, Crop productivity

1. Introduction

Decline in agricultural productivity due to soil nutrient depletion and reduction in soil organic matter (SOM) has been an important concern of agricultural soils that can be directly linked with global food insecurity. For instance, soil acidification is a major issue in agricultural production systems affecting around 30% of global land surfaces and 40% of the global arable soils.[1] Soil salinization has been estimated to affect 20% (45 million ha) of global irrigated land,[2] with an estimated annual loss of 10 million hectares of world agricultural land destroyed by salt accumulation.[3] Moreover, land degradation has become one of the urgent issues of the African continent, with an estimated annual US $42 billion loss of income and 6 million hectares loss of productive land.[4] Moreover, global soil nutrient deficits have been estimated to be 20 Teragram, with developing countries contributing up to 75%, followed by developed and underdeveloped countries at 14%, 11%, respectively.[5] Since fertility of most of the tropical soils depend on nutrients recycling from organic matter in the soil, economical viability of agriculture without supplementary fertilization cannot sustain for more than 65 years in a temperate prairie, and 6 years in tropical semi-arid forest lands.

Inorganic fertilizers (e.g., nitrogen (N), potassium (K), and phosphorus (P)) had a substantial role in substantially improving agricultural productivity over the last few decades. The paybacks of inorganic fertilizers which were demonstrated by the 'Green Revolution' may have reached a point of diminishing returns. The infeasibilities associated with the extended fallow periods for restoration of soil fertility and organic matter have in turn led to the unsustainability of agriculture systems. To maintain or enhance the agricultural produce to meet the demands of a growing population, intensive inorganic fertilizer use was extensively promoted despite the reduction in land area per capita and associated soil quality declinations.[6] However, this unsustainable practice has been recognized as one of the leading causes of soil deterioration and other environmental complications like rapid organic matter mineralization and a decrease in soil carbon reserves.[7] Moreover, poor nutrient use efficiencies of the soil with low organic matter is known to increase the cost of farming practices, accelerates soil acidification, which in turn decreases crop productivities.

Agriculture is one of the major contributors to greenhouse gas emissions (GHGs) with approximately one-quarter of the global anthropogenic GHGs of 2014.[8] Moreover, 52% of the total global methane from anthropogenic sources and 84% of the nitrous oxide emissions can be attributed to agriculture-related land-use changes and forestry.[9] Also, microbial degradation of some of the pesticides (e.g., organochlorine pesticides) in the soil through enzymatic degradation[10] has been on the decline due to reduced microbial activity in the soil. The reduced microbial activity in the soil can be associated to several environmental factors and the inability of heavily contaminated and degraded soils to support diverse microbial habitat.[11] So, low-cost methods with the ability to stimulate soil microbial activity and enhance pesticide degradation would improve the overall crop productivity and thereby improve the overall sustainability of agriculture.

To secure continuous food supply to the growing population, while maintaining balanced economic, environmental and social viabilities, adopting sustainable agriculture is very much necessary. Sustainable agriculture is the comprehensive integration of farming practices and systems that are economically viable, socially supportive and ecologically sound which in turn promotes the long-term viability of agriculture as a whole.[12] Low nutrient status of soils and soil organic matter (SOM) rapid mineralization are the major constraints inhibiting our progress towards sustainable agriculture.[13] Despite historically positive benefits to soil fertility with the use of organic fertilizers like mulches, composts, and manure, in addition to the added cost of repeated higher dose applications, their rapid decomposition and mineralization lead to global warming.[14] With only a very small slice of the total organic fertilizers applied gets stabilized in the long term, most of SOM gets returned to the atmosphere as CO_2. So, to maintain appropriate levels of SOM and ensure efficient nutrient biological cycling, use of organic and inorganic supplements supported by the knowledge of adaptability to local conditions, with a broader view of maximized agronomic nutrient use efficiency and thereby crop productivity should be of major focus. As an alternative, the use of more stable carbonized materials or their extracts could help in maintaining the necessary levels of organic matter in the soil, responsible for maintaining high levels of SOM over extended periods, thereby improving the overall nutrient use efficiency of the soil.

Pyrolysis— thermal degradation of organic materials in oxygen-depleted conditions, is often considered as an effective method of carbon sequestration from the biomass. The solid by-product of the process—Biochar is expected to retain approximately half of the total carbon contained within the raw material, depending on the feed and process parameters.[15] Its recalcitrant nature can help in carbon retention over extended periods relative to the living organisms, or landfill of unprocessed organic waste. Moreover, the added advantage of mitigation of GHG emissions from the waste through valorisation could aid in countering the problem of climate change. Other benefits of biochar such as improved soil water quality and carbon sequestration, envisage its use even in the absence of many agronomic benefits. However, extensive research is necessary to validate the benefits in variable conditions before proposing it for adaptation and as an economic solution to achieve sustainable agriculture. Thus, the use of biochar as a product or a co-supplement in a blended product can potentially have a wide variety of applications including, remediation against particular environmental pollution, improved resource use efficiency, as an instrument for soil improvement and mitigation of greenhouse gas emissions.[14]

The overarching aim of this chapter is to explore the agronomic potential of biochar in improving the soil properties and to promote crop growth and productivity. Moreover, the influence of biochar on soil nutrient cycle along with its stability and durability are also discussed in detail. Studies depicting the effects of biochar on bioavailability of the contaminants in the soil and green house gas emissions were also briefly elucidated. Moreover, the economics of integrating biochar application into the current agricultural practices and the future research needs for achieving economic and environmental sustainability of the process were also reviewed. We consider that the information provided in this chapter would help in maneuvering biochar at its productive functionality for promoting sustainable agriculture.

2. Agronomic properties of biochar

Fig. 1. showcases the agronomic properties of biochar in soil. Understanding the agronomic characteristics of biochar is necessary to understand its function in the soil and its potential benefits for sustainable agriculture. Biochar is a heterogeneous carbonaceous material made of polycyclic aromatic hydrocarbons and an array of other surface functional groups. Its aromatic ring structure incorporates several hetero-atoms including N, P, O and S, which make its surface heterogenous and chemically reactive.[16] The heterogeneous biochar surfaces can be both acidic and alkaline, hydrophilic and hydrophobic properties, which determine their reaction ability with other substances in the soil solution.

The characteristics of biochar are highly dependent on the feed and process parameters of pyrolysis. The process parameters influencing the inherent properties of biochar are temperature, residence time, heating rate, particle size, type of reactor, oxygen, pressure, and others.[17] The common temperatures adopted for the recovery of chars from organic waste through thermal degradation are between 300 and 800 °C.[18] Both physical characteristics of

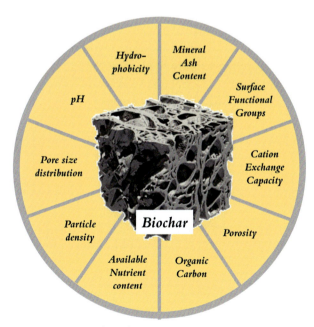

Fig. 1 Agronomic properties of Biochar.

biochar including surface area, particle density, and pore size distribution, and chemical characteristics like pH, electrical conductivity, cation exchange capacity, total and plant available concentrations of Carbon, Nitrogen, and Phosphorus, plant exchangeable cations, and selected minor nutrient and trace element concentrations influence its agronomic potential in soil. Moreover, since significant variations in these properties can be obtained with variable feeds and process parameters, this provides the basis for designing the biochar based on specific applications. For instance, biochars with highly recalcitrant C concentrations and with amorphous structure may be suitable for C sequestration, while porous biochars with high mineral content can be better candidates for improving nutrient use efficiency and agricultural productivity.

The low molar H/C ratios of biochar as compared to the feedstock[19] indicates its polymerization with pyrolysis and so its potential recalcitrance in soil. The highly refractory aryl carbon structure of the biochars can be responsible for the high degree of chemical and microbial stability of its organic carbon.[20] Pyrolysis temperature and the properties of the feedstock are major influencing factors for stability. Stability is determined by the biochar chemical structure, aromaticity and degree of aromatic condensation, the labile carbon content, aliphatic compounds and volatile matter etc., For instance, Cross and Sohi observed an increase in the stability of sugarcane bagasse biochar with pyrolysis temperature and also reported lower stability of chicken manure biochar as compared to the former.[21] The increase in biochar stability with pyrolysis temperature was also observed by other studies.[22] Generally, wood-based biochars are relatively more resistant to biodegradation as compared to those produced from crop residues and animal manures.

The importance of biochar specific surface area is due to its role in organic compounds and metal ions adsorption capacity.[23] The porosity and surface charge characteristics of biochar in relation to its cation and anion exchange capacity are also important factor for nutrient retention.[24] The cation exchange capacity (CEC) of biochar indicates its ability to adsorb and decrease the leaching of cations such as NH_4^+, K^+, and Ca^{2+}, which are essential nutrients for plants.[17] These characteristics are heavily dependent on the temperature of pyrolysis. The CEC of cordgrass biochar increased from 8.1 to 44.5 $Cmol_c$/kg with the rise in pyrolysis temperature from °C 200 to 550 °C, followed by a decrease to 32.4 $Cmol_c$/kg with a further rise in temperature.[25] The aromatization of biochar at higher temperatures and the associated decrease in surface functional groups can be the probable reasons for the decrease in CEC. Therefore, in perspective of improved retention and utilization efficacy of fertilizer cations such as NH_4^+, biochar produced at low temperatures has more potential to be a good soil amender in agriculture. Aging in biochar is known to have a positive effect on the CECs of the biochar[26]; however, the anionic exchange capacities of biochar decrease with aging.

Moreover, pyrolysis results in the volatilization of cellulose and hemicellulose and the formation of macroporous surface at lower temperatures, which with a further rise in temperature transforms into micro and nano pores. The porosity of biochar, which is representative of its pore size distribution encompasses macro- (greater than 50 nm), micro- (less than 2 nm), and nano- (less than 0.9 nm) pores. While the larger pores are key to its functions of aeration, hydrology, and providing a habitat niche for microorganisms, micro pores aid in nutrient and contaminant adsorption and mobility.[13] Its particle size also determines the adsorption ability and selectivity for different nutrients and contaminants. For instance, sorption of trace hydrophobic contaminants in soil was reported to be more efficient in fine dust fraction as compared to biochar particles of large size.[27]

Formation of vascular bundle structure in biochar occurs owing to the release of volatiles, results in the improvement of its specific surface area and pore structure.[28] For instance, with a rise in temperature from 250 °C to 600 °C, biochar's specific surface area produced from sugar cane bagasse increased from 0.56 to $14.1 \, m^2/g$.[29] Similarly, soybean stover biochar's surface area rises from 6 to $420 \, m^2/g$ with the rise in temperature from 300 °C to 700 °C.[30] Significant influences of change in pyrolysis temperature on the yields of biochar's humic and fulvic acids,[31] and other properties like conductivity, pH, P and N concentrations were also reported.[32] For instance, the pH value of corn straw biochar increased from 9.37 to 11.3 and swine manure biochar increased from 7.6 to 11.54, with the rise in temperatures from 300 °C–600 °C[33] to 400–800 °C,[34] respectively. This can be attributed to the free alkali salts released from the feedstock organic matrix at higher temperatures.[30] Moreover, the temperature rise also promotes the formation of unpaired negative charges like hydroxyl and carboxyl functional groups with an ability to attract positive charges as a result of their partial detachment.[35] This tendency improves the cation exchange capacity of biochars, which is a vital property for the adsorption of nutrients or metal contaminants. The biochars produced at high pyrolysis temperatures also tend to have high hydrophobicity, and low O/C and H/C molar ratios promoted by the reduced presence of polar functional groups (–OH and C—O) and a higher abundance of aromatic functional groups.[30]

The effect of choice of feedstock on biochar's specific area was also evident in the study of Lee et al.[36] where surface areas of bagasse and cocopeat produced under similar conditions of pyrolysis were 202 and 13.7 m^2/g. Also, other factors such as recondensation of volatile organic compounds on the char surface, presence of compounds like fulvic acid and humic acid, lipids, that influence the surface binding properties of biochar, its molecular rigidity, and pore access are dependent on the pyrolysis conditions.[37] Higher pyrolysis temperature and time result in higher mineral ash content, indicating the larger concentration of nutrients such as N, P, Mg, K, and Ca. However, there is a significant decrease in the volatile nutrients such as N with the rise in temperature.[38] The nutrient supply and liming potential of the biochars also depend on the feedstock mineral composition.[13] Choice of mineral–rich feedstocks for the production of biochar: like phosphorus in bone biochar[39]; manure based biochars[40] are also important considerations. However, bioavailable concentrations of these minerals in biochar for plant uptake is more relevant to determining its potential improvement in plant growth.

The increasing demand for high-quality biochar has created a need for regulatory control of the product specifications. The European Biochar Foundation has constituted a voluntary biochar quality standards and strategies to ensure acceptable characteristics and a positive environmental footprint. The IBI (International Biochar Initiative) also specifies the tools necessary to validate the biochars characteristics and quality standards. Following such guidelines could ensure minimal contamination from the application of biochars in agriculture. This is essentially important to prevent toxic contamination of agricultural soil from volatile and semi–volatile organic compounds, heavy metals, polycyclic aromatic hydrocarbons, etc., from biochars of contaminated or hazardous waste feedstocks.

The flexibility in the production of biochars with desired properties with change in process parameters makes it an exceptional carbonaceous material with a potential utility in diverse agronomic fields including soil amendment, composting, animal nutrition, odor and GHG emissions mitigation, and the utility of biochar fed animal manure as supplements and organic fertilizers, etc.[41] The significant opportunities presented by the heterogeneity of biochar properties for delivering to specific soil constraints also create hardships to the growth and expansion of the biochar industry. Information regarding the biochar's key characteristics must be vital in its selection for specific applications and so developing our understanding of its agricultural impacts.

3. Biochar induced soil quality improvement

Soil quality indicators can be used to measure its physical, chemical, and biological attributes which can be related to functional soil processes that are sensitive to management practices that yield its overall quality status for plant growth. While bulk density, porosity can be considered as good physical indicators, chemical characteristics such as pH, CEC, nutrient availability can be very dependent on soil and crop management practices. Further, biological properties like earthworm activity, microbial biomass and respiration, ergo-sterol concentrations and soil enzymes have been recommended by previous studies as optimum soil quality indicators.[14]

3.1 Impact on soil physicochemical properties

The increase in soil pH with biochar can be attributed to the inherent alkaline properties of the biochar itself and through enhanced monovalent and divalent cations retention (e.g., Na^+, K^+, Ca^{2+}, and Mg^{2+}) in the soil.[42] Although the soil alkalinization induced by biochar application is expected to promote soil nitrification, but it was observed that soil pH changes do not solely influence its rate of nitrification.[43] However, changes in the P availability can be expected due to the increases in soil pH as it is highly pH–dependent; while insolubility of iron and aluminium phosphate dominate in acidic, insoluble Ca phosphates are predominant in alkaline soils.[13] So, modification of soil pH through biochar amendment to influence the nutrient availability of particularly P, K, and Mg is depicted in the literature.[44] Biochar amendment has reportedly also reduced the mobility of some toxic metals to plant growth, such as copper (Cu) and manganese (Mn), through increasing the soil pH.[45] Apart from the heavy metals, biochar was also responsible for reducing the availability of organic soil contaminants like pesticides and insecticides.[46] Depending on the high C/N ratio, biochar also affects the release of the types of carbon and nitrogen during mineralization,[17] while N immobilization is also expected from the soil inducing N deficiency in plants. However, Yin and Xue[47] also suggested that the recalcitrant nature of the biochar's carbon potentially restricts N immobilization. Further cases of reduced N_2O emissions from the soil was also reported by previous studies.[48] Moreover, biochar's ability to increase the CECs of the soil can be attributed to its charge density per unit surface, which associates to a higher degree of surface oxidation and surface area increase for adsorption of cations.

The mobility of biochar particles within the soil profile is crucial to plant growth and transport through the subsoil and into surface and ground waters. With the reduction in biochar's particle size, an associated movement of biochar into the subsoil was observed with time.[49] A relative rise in the bulk density of soils with depth as observed within some Amazonian dark earth could be correlated to the decrease in organic matter and porosity with depth.[50] However, the incorporation of biochar into the soil to improve mechanical impedance could raise the concerns of waterlogging due to low infiltration and show restricted root growth and plant development. For instance, Hearth et al.[51] reported no significant enhancements in the available water capacity with the application of biochar produced from maize stover at 350, and 550 °C to silt loam soil. This indicated the potential soil micropores blockage by tiny biochar particulates (i.e. mineral fraction or ash). Other studies have also indicated improved soil water permeability and improved root penetration in the shallow root zones of the soil through biochar application.[52] The relatively lower loss of material from biochar in comparison to agricultural residues' black carbon and mineralization of SOM also advocate its benefits to soil. In-depth analysis of soil-biochar interactions influencing the physical structure and nature of the soil in the long term is necessary. The experimental challenges associated with the in-situ application should also be factored in to develop longer-term modeled predictions. Fig. 2. Showcases the biochar soil interactions and their impacts on soil quality parameters.

3.2 Impact on soil nutrient cycling and water use efficiency

Biochar's influence on the availability of nutrients and its potential as a slow-release fertilizer in soil was depicted by previous studies.[44] Heavy metals and nutrient sorption[24] occur mainly through ion exchange from carboxylic and other oxygenated functional groups and cation-π bonding mechanisms.[25] The release of nutrients mainly depends on the biochar's desorption properties, which varies with pyrolysis temperature, properties of feedstock, biochar application rate, and inherent nutrient contents available for adsorption. For instance, divalent and monovalent cation (Na, K, Ca, Mg) nutrient retentivity in the soil is a function of the availability of cation exchangeable sites on the surface of char.[17] Moreover, precipitation of these elements due to the pH change caused by the char also causes reduced nutrient leaching. For instance, K fixation by clay minerals in the soil was reported to reduce the extractable concentrations of K by 20% in the soil.[53] Liard et al.[15]

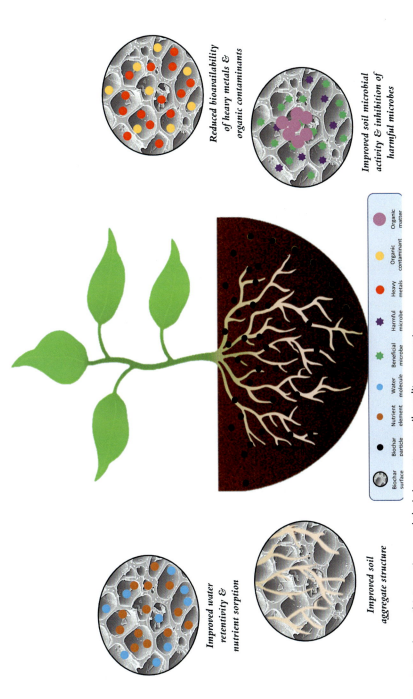

Fig. 2 Biochar soil interactions and their impacts on soil quality parameters.

reported substantial reductions in the recovered masses of K, Ca, and Mg with the addition of biochar at a 20 g/kg rate.

The desorption of PO_4^{3-} exhibited by cacao shell biochar and corncob biochar differed substantially (1484 mg kg^{-1} vs 172 mg kg^{-1}) under similar conditions.[54] Further, the desorption rates of NH_4^+ increased from 18% to 31%, when the pyrolysis temperature of production of hardwood biochar decreased from 600 °C to 400 °C.[55] The higher proportion of hydroxyl and carbonyl functional groups on the surface of the char is reported to promote NH_4^+ adsorption through electrostatic attraction.[56] Moreover, Xu et al. found that with biochar application rate increase (0–10%), the percentage of P desorbed increased from 36% to 41% at low P loads of 20 mg L^{-1}.[57] The study also reported a release of higher than 60% of the adsorbed P from biochar at 100 and 240 mg L^{-1} loading of P. Co-application of biochar with manure also results in high mineralization and high solubilization of organic and inorganic P in soil, respectively,[58] which in turn reduces the bioavailability of P in soil. So, variable types of biochar produced under variable conditions of pyrolysis can be used for the preferential nutrient supply effects in different types of soil with variable nutrient contents.

The ability of biochar to reduce N leaching and increase N use efficiency was also extensively reported by previous studies.[45] The NO_3^- retentivity by the char amendments is a function of its anion exchange capacity. Although functional groups such as carboxylate and phenolate groups in biochar increase its leachability in the soil, inhibition of organic N mineralization results in its low leachability. Moreover, high concentrations of dissolved organic carbon leached from the biochar amendment in the soil also leads to immobilization of NO_3^- N.[15] Biochar also helps in reduced N losses from manure amendment through ammonia volatilization, denitrification, and leaching of nitrate reactions thereby playing the role of a slow-release fertilizer necessary for plant growth. The natural oxidation of biochar is another factor that can impart significant changes in the nutrient release and retention in the soil.[59] Studies conducted by Laird et al.[15] have also shown reduced leaching of total N in the leachate concentrations by 11% with biochar amendment. However, understanding of mechanistic influence (direct/indirect) of biochar on nitrification and N availability for plants would be an absolute necessity to optimize sustainable agricultural practices in the future.

The availability of soil water is an important factor in enhancing agricultural growth and yield, more so in the regions of extreme water stress. Organic matter is often related to the high soil moisture contents due to its inherent water retention ability. Biochar by nature is particularly porous after its initial hydrophobicity at lower temperatures after which it can oxidize, and retain water by absorption.[60] Substantial improvements in soil field capacity have been recorded with biochar amendments, which could be of significant importance to sandy soils, where high water drainage results in poor soil fertility.[13] Although some of the studies have indicated the less likely benefits of biochar applications to soils with high clay content,[61] the potential advantages seen in *Terra Preta* soils from soil or plant water balance improvements is inconsistent with the general notion. For instance, the enhancements in soil water permeability supported by the highly porous biochars are found to be challenging in soils with very high clay content.[52] Major et al. has suggested the substantial influence of physical characteristics of biochar on the pore size distribution of soil, and soil solutions' percolation and flow paths and patterns and the residence time.[62] Especially, the water retention patterns in the root zone of the plants can also influence the nutrient mobility and transport in the soil solution. Other indirect effects of biochar induced changes of soil water content comprise changes in soil oxygen tensions, wetting and drying cycles of soil which influences organic matter decomposition, stability of biochar depending on oxidative breakdown and soil microbial activity. With the impending water crisis due to extensive agriculture and impacts of climate change, conservation of soil moisture content and its availability for plant uptake will be crucial factors in sustainable agriculture. Biochar could be a useful tool to provide opportunities for soil amelioration, particularly in water drought and arid regions with low soil moisture retention is vital in the restoration of plant growth and productivity.

3.3 Impact on soil microbial ecology

The structure and function of microbial communities with different inhabitants such as fungi, bacteria, invertebrates, nematodes, algae etc., in the soil is highly complex. The variability in their presence and abundance with the addition of organic matter (biochar in this case) has a deep effect on soil health, function and productivity.[63] The biochar amendments in soils stimulate soil microbial activity, which in turn affect the soil microbiological properties.[17] Especially, its high surface area provides a highly favorable habitat for microbes to grow, reproduce, and colonize.[11,63]

Rather than being a primary source of substrate metabolism by soil biota, biochar additions were found to alter the substrate availability and enzyme activity on, or around, particles of biochar and also improves the physico-chemical environment in soils, providing a favorable habitat for their growth.[17] However, freshly added biochars can contain suitable substrates to support microbial growth in soil. Moreover, its ability to provide physical protection to microbial communities along with its pH buffering capacity and for maintaining stable and appropriate pH conditions in the microhabitats within biochar particles also influence microbial diversity, abundance, and activity in soil.[64] Other factors including the biochar particle size, type of co-substrate added, C and N availability, soil organic matter content also influence the microbial population growth. Moreover, the reduced toxic effects of Al in soils and improved water holding capacity in biochar amendments also promote microbial activities in the soil.[27]

A significant positive correlation of bacterial-to-fungal ratio with soil C/N ratio was observed with biochar incorporation in sandy loam soils.[63] For instance, A linear increases in the microbial efficiency (CO_2 conversion) was reported by Steiner et al.[65] at different biochar application rates of 50–150 g/kg in soil. In another study, an increase in microbial biomass carbon (MBC) by 29% was reported with willow wood biochar amendment in soil.[66] However, a significant reduction of MBC (28%) was also reported in response to biochar derived from *Eucalyptus marginata* (600 °C) amendment in coarse-textured sandy soil.[67] Some of the potentially harmful and toxic effects of bio-oils and re-condensed organic compounds adsorbed on the surfaces of the biochars could be possible reasons for inhibition in the activity of soil microorganisms. The added risks of pathogen contamination in soils, sediments, and in drinking water through leaching which is promoted by the enhanced transport of viruses with biochar addition is also a matter of concern.

Positive responses of native bacterial communities with biochar amendment in soils are well recognized. For instance, Terra preta soils were about 25% richer with 14 phylogenetic groups in bacterial species relative to 9 in forest soil.[68] Similar positive effects of biochar on diazotrophs (N_2-fixing bacteria) which consequentially leads to enhanced biological nitrogen fixation (BNF) have been observed.[69] Mia et al. has reported a significant increase in the total amount of N fixation (117%) and density of nodule (72%) at 10 t/ha biochar application.[69]

Biochar amendment was also observed to influence the soil fungal activity and community structure.[70] For instance, positive effects of biochar application on the growth of Arbuscular (AM) and ectomycorrhizal (EM) fungi was previously reported.[71] Furthermore, the ability of certain fungal species to produce a range of extracellular enzymes and break down of various compounds in biochar for food was also reported.[72] Further, biochar, mycorrhizal fungi and high N addition promoted mycorrhizal root colonization compared to the mycorrhizae and high N treatment.[73] However, negative impacts of biochars produced at high temperatures on mycorrhizal fungi populations in soil due to their high pH and heavy metal and ash contents were also reported.[74] Significant reductions in arbuscular mycorrhizal fungal (AMF) root abundance with the application of biochars produced from pine, peanut shell, and mango wood was reported with contrasting observations of increasing or decreasing soil P availability.[75]

The biochar's influence on the behavior and growth of earthworms in the soil although significant, is still unclear. One of the factors of influence could be soil pH alteration with the biochar addition for which the mechanisms of interactions responsible are not understood. Although the ability of earthworms to ingesting biochar particles in the soil and excreting their feces is previously reported,[76] their susceptibility to contaminants added from biochar amendment is a matter of concern. For instance, the availability of sorbed PAHs on the surface of biochar for ingestion was hypothesized as a possible reason for the observed reduction in the weight of the earthworms in a PAH contaminated soil.[77] Other hypothesis of organic matter break-down in the gizzard of earthworms facilitated by biochar ingestion was also proposed.[76] The increased abundance of microbes and N processing enzyme activity in the guts of the earthworms were also possible responses of biochar ingestion.[11] The translocation of ingested biochar particles within the soil profile and stabilization of organic particles after passing through the earthworm gut[78] are significant results aiding in the improvement of soil functions irrespective of its limited nutrient value. Moreover, positive impacts of increase in soil enzymes activities, nutrient status and plant growth and fruit yields of proso millet were also reported with co-application of biochar and earthworms.[79] Further, Salem et al.[80] has also reported an increased growth of arbuscular mycorrhizal fungi (AMF) and Plantago lanceolate in the soil during the simultaneous use of biochar and earthworms. The increase in soil pH and heavy metal's content in soil with the biochar addition was reported to be the possible reasons[81] for negative impacts of biochar such as reductions in microbial biomass and abundance, and increase in mortality and genotoxicity of earthworms.[77] However, other studies have proposed these negative effects on earthworm survival were only predominant at very high rates of biochar application.[82] So, this potential toxicity on the earthworm population might be

3.4 Stability and durability of biochar within the soil

The physical and chemical characteristics of biochar, inherent soil properties, and local environmental conditions influence its interactions with soil and so its changes within the soil system.[17] For instance, biochar durability in the soil is negatively affected by the presence of fungi like saprophytes.[83] The C/N ratio of biochar is a principal factor affecting its stability and longevity in soil.[84] So, it can be said that spruce biochar has higher stability in the soil relative to the poultry litter biochar due to its relatively high C/N ratio (250.7 greater than 13.27).[85] Despite its high stability, minute alterations in biochar's physical structure and its chemical composition are observed with prolonged presence in soil. Other additional factors including rain, wind, erosion, interactions with plant roots and fungal hyphae, insects, rodents, and microbial organisms could also result in the biochar's physical fragmentation.[17] Moreover, surface oxidation of the surface functional groups on biochar with time leading to the evolution of carboxylated functional groups could also occur. These functional group transformations could result in the change of the CEC of the char, and which in turn influences its adsorption capacity in the soil. The biochar interactions with mineral ions in the soil leading to the formation of organo-mineral complexes could decrease the biochar's oxidation and degradation rate in the soil and so aid in the preservation of biochar stability in the soil systems.[86]

The durability of biochar is also determined by the inherent properties of the raw material and production conditions. Luo et al. reported higher mineralization rates of biochars produced at 350 °C as compared to the ones produced at 700 °C.[84] Moreover, the study also observed an increase in the mineralization rates to be higher in soils of high pH relative to the acidic soils.[84] Spokas[87] in his study has correlated the stability of the biochars with their corresponding O/C ratios. The study predicted that the biochars with O/C of less than 0.2 as very stable with longer than 1000 years of half-life and biochars with the range of 0.2–0.6 O/C molar ratios are relatively less stable with intermediate half-lives in the range of 100–1000 years. With all due consideration, since biochar exhibits very high stability in the soil, as compared to other organic or inorganic supplements, its long-term soil productivity potential should also be acknowledged.

4. Biochar-induced crop growth and production
4.1 Impact on crop growth and productivity

Table 1 showcases some selected studies depicting the effects of biochar on the crop yield or productivity increases in the soil. The application of biochar to soil was suggested by previous studies to increase agricultural productivity.[17] Factors including soil pH, type, organic/inorganic fertilizer supplementation, the feedstock of biochar, its rate of application, and the crop/plant species could be considered to have the most significant influence on the plant responses in soil. Spokas et al.[102] observed that among forty-four of the published articles analyzed, while about half of the studies reported enhanced crop productivities, others have reported insignificant or a negative effect on crop yield with biochar applications in agriculture soil. These studies assessing biochar's efficiency were based on different soil types and climatic conditions in the study area, biochar types and rate of application. The particle size of biochar over the range of 2–20 mm, however, was not observed to have any significant effect on crop productivities.[103]

For instance, lower than 5% addition of biochar improved the seed germination, yield, and root development of the seashore mallow and sesbania like halophytes.[104] Biochar produced from rice hull has increased the sucrose content and biomass yield in sugarcane plants when applied at a 2% rate in sandy soil.[105] Solaiman et al.[106] also reported increased wheat seed germination (by 93–98%) with the Oil Mallee biochar amendment at 10 t/ha rate; however, decrease in the germination of subterranean clover and mung bean were also observed. Improved root nodulation and soyabean growth were observed in response to bamboo biochar application at less than 10% application rate.[107]

Mc Donald et al.[108] observed variability in the wheat growth response to biochar application in four different soils. The study observed that the highest rate of application in the acidic arenosol has led to low total plant biomass; however, with the rise in the rate of biochar application, increased plant biomass in acidic ferralsol; and insignificant effects in alkaline calcisol and neutral vertisol. This indicates that biochar amendments are not universal in the benefits to all types of soil, and soil type should be an important factor in the design of an agricultural management system. For instance, tropical soils that are acidic with low nutrient retentivity are expected to yield positive benefits relative to different soils of temperate regions.[13] For example, Mohammad et al.[109] reported increased rice

Table 1 Selected studies depicting the effects of biochar on the increase in crop yield or productivity in soil.

S. No	Biochar Feedstock/ Pyrolysis temperature (°C)	Application rate (t/ha)	Crop	Soil Type	Fertilizers added (if any) Kg/ha	Yield/ biomass increase over control (%)	References
1.	Papermill waste/550	10	Wheat	Ferrasol	1.25 g nutricote (N (15.2%), P (4.7%) and K (8.9%))	+250	88
2.	Green waste and poultry litter/550	100	Radish	Alfisol	N (100)	+266 (biomass)	89,90
3.	Not specified	30	Rice	Inceptisol		+294	91
		88		Oxisol		−800	
		88		Oxisol	N (40), K (20), P (20)	−21	
4.	Wood/550	20	Maize	Oxisol	N (156–170), K (84–138), P (30–43)	+(28–143) (1–4 seasons)	92
5.	Forest wood	11	Rice	Ferrasol	N (30), K (50), P (35)	+29 (stover) +73 (grain)	45
6.	Acacia/350–450	10	Barley	Acidic Eutric Nitisol	N (23), P (20)	+45	93
7.	Wheat straw/450	20 40	Soyabean		N (80), P (120), K (100)	+7 +8	94
		20 40	Maize			+6 +7	
		20 40	Peanut			+7 +11	
8.	Eucalyptus	50	Bean	Acidic clayey	Cattle manure (5000 kg), N, P, K (300 kg)	+(53–68) (1–3 seasons)	95

No.	Biochar/pyrolysis temp	Rate (t/ha)	Crop	Soil	Fertilizer	Yield change (%)	Ref
9.	Bamboo/600	7	Maize	Sandy	N (400 kg), K (160 kg), P(500 kg)	Not significant (1–3 seasons)	96
					N (400 kg), K (160 kg), P (500 kg), Vermicompost (20 t/ha)	+ (36–57) (1–3 seasons)	
10.	Willow wood/550	10	Maize	Ferrasol	186 kg Urea, 0.3 L Rutec, 30 L MOP, Zn 7000	+ 29	97
11.	Rice hull/500	13.5	Maize	Silt loam – Saline sodic soil	N (12 kg), P (10 kg), K (10 kg)	Not significant	98
		27				+52	
		67.5				+101	
12.	Spruce chips/550–600	5–30	Wheat	Loamy sand		(Not significant to +23) (1–2 seasons)	99
13.	Commercial charcoal	30	Wheat	Silty loam	22 kg N 50 kg P	+28	100
14.	Pine chips/350	40	Wheat	Fine loamy	N (45 kg), K (80 kg), P(60 kg)	Not significant	85
	Poultry litter/350					+49	
	Pine chips + Poultry litter/350					+(28–76)	
15.	Waste willow wood/550	10	Peanut	Ferrasol	N (26.6 kg), K (66.2 kg) P (31.6 kg), 2.6 kg S, 0.54 kg Zn	+(21 − 23) (1–2 seasons)	101

plants productivity and growth, including their leaf length, height, and grain yield by 32% and 41% in biochar amended (3%) Psammaquent and Plinthudult soils.

Better water use efficiency and soil characteristics by biochar amendments have led to enhanced plant productivities.[52] Moreover, plant nutrients availability optimization,[110] increase in soil microbial activity and biomass[111] and decrease in the exchangeable Al^{3+}[112] with biochar also significantly improves plant growth. For instance, an increase in maize grain by 150 and 98% was observed along with an increase in net water use efficiency (WUE) of 91 and 139% at respective application rates of 20 and 15 t/ha in sandy arid soil.[113] The improved (by 63%) N agronomic efficiency,[114] increased silica uptake by plants,[115] improved P bioavailability[116] are other examples of improved soil nutrient cycles with biochar amendment.

The biochar's ability to alleviate salinity caused reductions in nutrient plant uptakes was also reported to mitigate the adverse effects of soil salinization in salty contaminated soils. For instance, Akhtar et al.[117] observed amelioration of salt stress in salt-affected soils by increasing xylem K^+, adsorbing Na^+ and thus enhancing tuber yields of potatoes.

Usman et al.[118] also reported an increase in tomato yield by 14–43% under saline water irrigation with the application of *Conocarpus* wood waste biochar. Similar beneficial effects of biochar amendment in degraded soils and nutrient-deficient soils were also observed.[52,119] Although improved upland rice grain yields at soil sites with low P availability were observed with biochar soil amendment, probable reduction in grain yield in N-deficient soils were found unless additional N fertilizer is applied.[52] Nitrogen limitation could be the factor for declining yields with biochar since it promotes N immobilization by microbial biomass at high C: N ratios, apart from other growth-limiting factors.

In an overall view, nutrient-poor soils provide a good platform for observing the strong ability of biochar applications in enhancing crop productivities; however, the impacts of biochar on crop productivity in soils that are nutrient-rich is inconsistent. For instance, studies indicating small improvements or even reductions in crop yields in nutrient-rich soils amended with biochar can be found. Other studies have also observed the potential of biochar to mitigate abiotic and biotic plant stresses,[120] alleviate heavy metal uptake under metal stress conditions,[121] improve the antioxidant response of plants to drought and salt accumulation,[122] and decrease in the genotoxicity,[123] thereby promoting plant growth in soil. Finally, the positive effects of biochar on reducing effects of rice blast infestation,[123a] foliar fungal infections[124] in soil should also be acknowledged.

Integrated application of biochar with inorganic or organic fertilizers is found to be an effective technique for improving plant growth and productivity as compared to the lone addition of biochar. Significant increases in yield of maize crop were observed with the co-application of biochar and vermicompost[96] and biochar and lignite fly ash[125] relative to the non-amended soil. Steiner et al.[126] harvested 4–12 times more sorghum and rice yields by co-application of charcoal with compost and/or fertilizer relative to the application of fertilizer alone.

Application of arbuscular mycorrhiza (AM) fungal inoculation along with biochar has led to rise in maize productivity relative to the biochar amendment alone.[127] An associated increase in biomass production by 250% due to biochar addition in fertilizer amended Ferrosol along with an increase in N uptake in wheat can be attributable to improved fertilizer use efficiency. Contrastingly, a fertilizer and biochar amended Calcarosol has reduced radish and wheat biomass although an increase in biomass of soybean was reported.[88] Moreover, the synergistic effect of organic fertilizers and biochar on plant growth and productivity have been reported as mostly insignificant in temperate soils.[128]

The mechanisms involved in the co-application of biochar and fertilizers include direct nutrients supply, moderating soil pH, and improved nutrient use efficiency and thus enhanced plant nutrient uptake for a known fertilizer rate of application through increased cation exchange capacity and water retention capacity of the soil. Especially, improved organic matter stabilization and relatively slow release of essential nutrients from supplements and cations retention could be termed as the long-term advantages of biochar co-application. On the other hand, microbial activity inhibition caused by the release of phytotoxic compounds,[129] and pH-induced micronutrient deficiency.[101] in biochar amended soils could be responsible for negative responses in plant growth.

Since, most of the studies included only short-term field or lab experiments spanning from 1 to 3 years duration, design of experiments running for extended periods are required to test the consistent effects of biochars amendments on crop growth and yields after many subsequent crop cycles. Moreover, with variability in the plant responses with differences in the types of biochars, and soil conditions, special attention is required in biochar selection and its application pattern corresponding to the field conditions for maximum crop productivity.

4.2 Impact on crop physiological parameters and quality

Crop growth is a factor of plant physiological responses to biochar amendments. Studies have observed significant to no significant variability in the physiological parameters of plants including leaf chlorophyll content, leaf osmotic potential, photosynthetic and transpiration rates, sub-stomatal CO_2 concentration, membrane stability index, stomatal pore aperture and density, and lycopene, carotenoids, sugars, anthocyanin, ascorbic acid amino acids, and protein contents etc., with changes in biochar amendments.[130] For instance, Haider et al.[131] also observed an increase in leaf osmotic potential and, reduced stomatal resistance with biochar applications in poor sandy soils under well-watered and drought conditions. On the other hand, reduction in leaf chlorophyll content of upland rice with biochar soil amendment on nutrient-poor soils was also reported by Asai et al.[52] Moreover, other studies also observed no significant influence of biochar application on plant physiological parameters like the mid-day leaf water potential, photosystem II (Fv/Fm) photochemical efficiency, and water potential of roots,[117] tree water use, vapor pressure deficit, plant water status as depicted by leaf water potential, stomatal conductance and photosynthetic capacity.[132]

The positive effects of biochar in plant nutrient supply causing higher P and N availability in the soil has improved dry mass accumulation in the vegetative parts leading to taller plants with larger stem diameters and an increased root mass fraction.[133] Moreover, improved water relations in the plants grown in biochar amendments also contributed to an increase in the biomass as supported by less negative pre-dawn and midday water potentials of plants.[117] The improved soil fertility in biochar amendments in non-water limited situations, vegetative mass like lateral branching is enhanced, but not fruit mass. For instance, Akhtar et al.[130] biochar had positive effects on tomato fruit quality in deficit irrigation; however, under full irrigation treatment fruit quality of processing tomato had not significantly improved. The improved nutrient and water availability caused by biochar treatments with/without fertilizers also caused improved fruit development. Biochar and other organic fertilizers act as activators or precursors of growth substances, phytohormones, and plants secondary compounds, which facilitates the increase in fruit size.

Increased organic acid (citric acid and malic acid) concentrations (titratable acidity) in fruit tissues could be due to the additional C supplied by biochar aiding in the production of these organic acids, that influence the fruit acidity. Akhtar et al.[130] and Al-Harbi et al.[134] observed significant improvements in total acidity (TA) and total sugar contents (TS) values of pepper with co-application of biochar and compost. Petrucelli et al.[135] on the other hand, reported no substantial improvements in the total soluble solids (TSS) and TA, by biochar treatments. These two properties depict the organoleptic fruits quality in terms of relative sweetness and acidity.[135] Vitamin C content, which depends on numerous factors like cultivar, plant nutrition, and plant maturity was also increased with biochar treatment under different irrigation treatments[118]; however, Al-Harbi et al.[134] observed insignificant variation in the vitamin C contents of peppers with biochar addition. These positive influences of biochar on the fruit quality can be correlated to the overall improvements in the plant physiology resulting in better quality fruits.[130]

Strong variations observed in the secondary metabolites like phenolic compounds with the biochar and other organic amendments could be the result of the hormesis effect: low doses of phytotoxic chemicals from biochar.[135] These compounds including polyphenols and flavonoids have a critical part in the development, adaptation, and management of plants growth, and in overcoming stress environments.[136] Alterations of maturity, hormonal regulation, ripening physiological effects could also be possible impacts of biochar addition.[135] However, comprehensive studies to interpret the mechanisms of effects of biochar on the accumulation of the secondary metabolite should be of focus.

The potential improvements in the macronutrient availability induced by biochar application in the soil could have resulted in overall improvements in plant mineral contents. For instance, Vaccari et al.[100] observed significant improvements in the plant nutrient (N, Mg, K, and P) contents in biochar treatments throughout the crop cycle. With respect to improved overall physiological parameters in the plants by biochar promoting plant mineral contents, a higher supply of nutrients supported by organic soil amendment application had greater influence as compared to the improved water use efficiency as observed by Al-Harbi et al.[134] in the case of pepper fruit nutrient contents. Similarly, higher immobilization of nutrients in biochar amendments could also be observed in the leaf nutrient concentrations as observed by other studies[97]; however, Eyles et al.[132] has reported no significant effects of biochar amendment on N concentrations and micronutrients in the leaf. Akhtar et al.[130] have observed reduced leaf N content in the biochar treated plants and proposed a decrease in N availability due to high NH_4^+ sorption on the surface of the biochar as a probable reason for the same.

Moreover, the reduced concentrations of heavy metals in the tomato fruits with biochars as observed in the wastewater and manure amendments, indicates the reduced bioavailability and so enhances the plants' capacity to obtain less concentrations of heavy metals per unit biomass produced. This can be considered as one of the beneficial properties of biochar for its application in heavy metal contaminated soils or co-application with heavy metal loaded supplements.

4.3 Impact on the bioavailability of contaminants in the soil

Table 2 showcases some selected studies depicting the biochar effects on the bioavailability of organic and inorganic pollutants in soil. The interference of inorganic and organic contaminants into the food chain has key socio-economic and environmental implications

So, mitigating pollutant bioavailability and uptake to food crops will be an essential part of promoting sustainable agriculture. The sorption ability of biochar can be used as an effective strategy to reduce the mobility, volatilization, leaching of contaminants in soil, and their uptake by plants. The potential of carbonaceous adsorbents to decrease the bioavailability of organic contaminants such as), polycyclic aromatic hydrocarbons (PAHs), polychlorinated dibenzo-p-dioxins/dibenzofurans (PCDD/Fs, pesticides and hormones is demonstrated by many previous studies.[76] Pesticide sorption from the soil solution and reducing its microbial bioavailability, and thus decreasing the biological and chemical degradation of the pesticides in soil has been recorded as one of the effective functions of biochar amendment in soil.[46]

For instance, a decrease in dieldrin concentration from 0.06 to $0.04\,mg\,kg^{-1}$ in soil was observed by Saito et al.[141] with woodchip biochar amendment. Similar effects of reduced bioavailability of organic contaminants due to biochar amendment on simazine,[142] carbofuran, chlorpyrifos and fipronil,[143] and sulfamethazine (SMT)[144] were also reported. Based on the inherent characteristics of the biochar and the chemical structure of pesticide, the mechanisms of biochar sorption include hydrogen bonds, hydrophobic effects, pore-filling, electrostatic attractions, separation into uncarbonized fractions, and π-π interactions. For example, the adsorption of carbaryl and atrazine by biochar was attributed to hydrophobic effects, π-π electron donor-acceptor interactions, and pore-filling.[145] In contrast, when some of the pesticides exist as weak bases or neutral molecules, the formation of weak hydrogen bonds to carboxyl functional groups or to the surface of clay through their heterocyclic nitrogen (N) atoms could be proposed as a mechanism of sorption.[146]

Table 2 Selected studies depicting the effects of biochar on the bioavailability of organic and inorganic contaminants in soil.

S. No	Biochar Feedstock	Application rate (%)	Organic/inorganic pollutant	Removal efficiency or reduced bioavailability (%)	Soil type	References
1.	Tomato green waste	5	Cd^{2+}	34	Mollisol	137
2.	Sugarcane bagasse	1.5	Cd^{2+}	63	–	27
		1.5	Cr^{2+}	85		
3.	Rice straw	3	Cd^{2+}	73	Acidic soil	138
4.	Rice straw	3	Pb^{2+}	70		139
			Cu^{2+}	37		
	Castor leaves	3	Pb^{2+}	41		
			Cu^{2+}	22		
5.	Miscanthus	10	Cd^{2+}	71		121
			Pb^{2+}	92		
			Zn^{2+}	87		
6.	Paper mill sludge	5	Atrazine	515	Ferrosol	46
	Poultry manure	1	Atrazine	220		
	Paper mill sludge	1	Diuron	448		
7.	Saw dust	2	Fluometuron	340–365	Sandy loam	140
8.	Woodchip/Macadania nutshell/hardwood	10	Bentazone	13–40	Silty loam	23
	Woodchip/Macadania nutshell/hardwood	10	Aminocyclopyrachlor	18–240		

The heavy metals sorption onto biochar, on the other hand, occurs through diverse mechanisms including complexation, ion exchange, physical adsorption, etc., The surface functionality of biochar is known to support the formation of metal complexes with various functional groups. Alternatively, ion exchange of adsorbed cations (Ca^{2+}, K^+, Mg^{2+}, etc.,) with heavy metal contaminants in soil can also occur. For instance, alterations in biochemical and physiological characteristics of the spinach plant through enhanced biomass yield and photosynthesis were reported as probable reasons for the reduction of Cd toxicity reduction using cotton stalk biochar.[147] Similarly, reduced bioavailability, phytoavailability, mobility of different heavy metals in soil using biochars were made by previous studies such as Pb and As with soybean stover and pine needle biochar.[148] Through better soil nutrient management in soil, biochar aids in optimal plant growth and productivity in many upland contaminated soils.[149] However, their efficacy is dependent upon factors such as pyrolysis conditions, biochar rate of application, and feedstock type occupy a prominent role in the distribution of different heavy metals in the soil. Fellet et al.[150] observed better performance of manure pellets biochar in the immobilization of toxic inorganic elements in the mine tailings as compared to biochars from fir tree pellets and orchard pruning residues origins. The relative competitiveness among different metals with biochar also effect their ability of sorption and phytoavailability. Other mechanisms of biochar include heavy metal immobilization through soil pH modification and alteration of other soil properties, changes in the redox state of heavy metals, etc.,.[139] For instance, Rehmann et al.[151] has observed a decrease in the phytoavailable concentrations of Ni using Eucalyptus wood biochar produced at 450 °C through an increase of soil pH. Conversely, biochars' efficiency in decreasing the hazard of heavy metals and other organic contaminants toxicity might vary significantly with time from blockage of pores by insoluble compounds; comprehensive exploration of the longevity of its advantages in the soil is necessary.

The high sorption ability of biochar to organic pesticides in soils was also reported to decrease the herbicide efficiency, causing demands for their increased rates of application.[152] The biochars high labile organic C content has also been marked as a probable reason for blocking the biochar pores resulting in reduced atrazine sorption. This necessitates caution and careful selection of biochars and their application rates to optimize the sorption properties for variable uses. While low adsorption characteristics are needed for agronomical use of pre-emergent herbicides, high adsorption

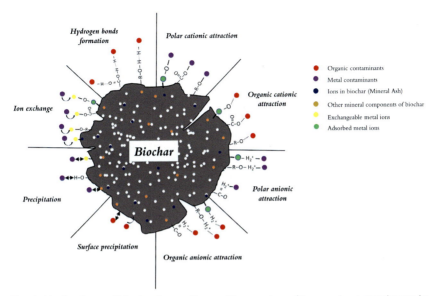

Fig. 3 Mechanisms of biochar interactions with organic and inorganic contaminants in the soil.

characteristics help in decreasing the bio and phytoavailability of toxic organic pollutants.[11] Fig. 3. Depicts the mechanisms of biochar interactions with organic and inorganic contaminants in the soil.

4.4 Impact on GHG emissions from soil

Although photosynthesis is an efficient process of CO_2 sequestration, the instability of the captured C and release into the atmosphere through microbial decay and respiration makes it incompetent for long-term C sequestration. Production of biochar—a relatively more stable form of carbon, considerably increases the efficacy of long-term C sequestration. Production of biochar also leads to sequestrating of about half of the initial C in biomass relative to combustion (3%) and microbial decomposition (less than 10–20% after 5–10 years).[153] Since the net energy of pyrolysis is also highly positive, coupling it with the biochar's application to soil allows the atmospheric CO_2 sequestration, with storage of more C than released. For instance, Liu et al.[154] observed decreased rates of CO_2 emissions from the bamboo and rice straw biochar amended paddy soil. Kimetu et al.[155] reported net negative CO_2 emissions from the soil organic carbon (SOC) of SOC poor soils with biochar amendment; however, the same study reported net positive CO_2 emissions from SOC-rich soil.

Similarly, Major et al.[92] observed an increase in carbon dioxide emissions from unplanted plots amended with biochar. In contrast, no significant effect of biochar amendments or enhanced CO_2 emissions at earlier stages of application were also recorded in a few cases.[142]

Apart from carbon dioxide, biochar amendments were also efficient in decreasing methane (CH_4) and nitrous oxide (N_2O) emissions from soils.[156] For instance, Rondon et al.[157] stated almost complete suppression of CH_4 emissions and reduction in N_2O emissions (by 50, 80%) from biochar amendments in grass stands and soybean croplands. Similarly, reductions in CH_4 emissions from paddy soil is also reported by Liu et al.[154] through a short-term incubation experiment. However, these impacts of biochar addition on CH_4 and N_2O were highly variable depending on the crops and soils, local conditions, and feedstock.[158] For example, Spokas and Reicosky[159] reported an inconsistent trend of greenhouse gas (GHG) emissions from 16 different types of biochar amendments in their 100-day incubation experiment. So, these discrepancies in the results necessitate caution and care to accurately interpret and extrapolate the applicability of biochar for mitigation of GHGs from soil to large field scale applications.

Other beneficial applications of biochar in the field of agriculture include its use as a manure/compost additive and as a feed additive for animals. Biochar is noted for enhancing composting process performance and mitigation of GHG emissions (NH_3, N_2O, CO_2)[160] and other odorous substances (H_2S, VOCs)[161] from composting apart from retaining major nutrients in the compost. The inclusion of biochar as a dietary supplement for farm animals has also reportedly induced improved digestion, weight gain, feed conversion ratio, and also acted as a medicine against intoxication from several viral and bacteriological diseases, apart from the mitigation of GHGs emissions from animals.[162]

5. Economics of biochar application for agriculture

The practical key to sustainable agriculture is to achieve economic stability and profitability to help farmers build a better quality of life without compromising the need for the preservation of environmental ecosystems. The scope of biochar to enhance agricultural productivity is an important driver to integrate sustainable agriculture principles into the current agricultural practices. The differences in life cycle costs and environmental impacts of biochar in particular regulate its economic viability in environment and agriculture uses.[163] The associated costs of feedstock collection, processing and transport, pyrolysis setup, and associated costs of biochar's handling and application must balance the environmental benefits and monetary gains of crop production improvements to make the practice sustainable. Creation of closed-loop systems where the waste from agriculture can be tuned into feedstock for biochar, thereby offering a sustainable process to support the high-value food crops production through reduced capital costs. For instance, Dunlop et al.[164] has created a closed-loop system where tomato crop green waste biochar was used as a substrate for soilless, hydroponic tomato production. The use of other waste materials like food or organic fraction from municipal solid waste could also be considered as an economically beneficial endeavor. Moreover, successful design and running of pyrolysis plant at the agricultural site may reduce the feedstock load by 20–30% of the wet waste mass, reducing associated conveyance costs and other environmental benefits like waste valorisation, which otherwise has to be transported to landfill.

Also, the decrease in the requirement for frequent application of chemical fertilizers on biochar amended soils, could significantly reduce their investment in chemical fertilizers and will encourage the transformation to sustainable soil fertility with enhanced nutrient management in the soil essential for plant growth.[14] The improved health of the soil can also provide farmers with more choices for crop selection. Moreover, the economic benefits of reduced irrigation requirement, reduced bioavailability of heavy metals provide a viable alternative and a cost-effective strategy in arid and contaminated soils.

For instance, Shackley et al. has estimated the price for a ton of biochar production to be around 148–389 British pounds.[165] Lehmann (2009), depicted biochar sequestration to have the very lowest carbon costs, relative to the forest and geological sequestration, and similar costs to the burial of logs.[166] Also, recovery of bio-oil from the pyrolysis along with biochar and incorporating suitable energy applications for the same, would increase the economic sustainability of the pyrolysis process and reduce the overall net cost of biochar produced. Studies depicting qualitative enhancement of the liquid oil recovered from biomass pyrolysis through the addition of plastics in the feed have harnessed interest in recent times.[167] The char obtained from the process has also shown superior agronomic properties with high mineral content and proposed to be used as a good soil amender.[18, 168]

Some of the studies such as Li et al.[169] have reported significant return on farmers' investment in biochar only in long term after consecutive cropping seasons, because of the high cost associated with large biochar application rates. Furthermore, as balanced nutrient capital of biochar is uncertain, and success of improved crop productivity with biochar amendment seems questionable. In that case, the additional costs of complementarity of biochar with additional inputs (fertilizers, composts, etc.) would require comprehensive evaluation. Moreover, to overcome this constraint, the development of artificially nutrient doped biochars and biochar composites with enhanced mineral contents, surface functionality, exchangeable cations and other agronomic properties can be considered as an alternative. Hameeda et al.[170] has also proposed that the use of the manure-biochar mixture as an organic fertilizer in agriculture supported by irrigation with either wastewater or groundwater can fetch relatively very higher agricultural yields at low costs. However, a comprehensive economical assessment between the utilitarian benefits of these materials and traditionally used fertilizers will require a long-term investment in agricultural experimentation. This will prove the economic feasibility of biochar amendments as a viable alternative.

6. Concluding remarks and future research needs

The biochar's ability to enhance soil fertility and improve crop yields were demonstrated in several studies; however, there are clear knowledge gaps in the interpretation of the mechanism of biochar interactions with soil and its biota. Prominently, the longevity of biochar in soil and its implications to soil processes and mechanisms need further investigation as field studies involving realistic conditions are scarce. Future work is anticipated on the long-term impacts of aging in biochar on soil nutrient retention and plant availability due to changing surface properties over time in the soil, especially in field conditions.

Quantitative determination of nutrient dynamics in soils with biochar application through efficient kinetics modeling using both field and laboratory-based studies is necessary to predict the nutrient availability in soil and to interpret the various processes affecting soil nutrient availability and fertility. Comprehensive field studies of biochars as a long-term sink for atmospheric carbon dioxide sequestration, and its long- term environmental fate in the soil should also be assessed. Optimization studies for best pyrolysis conditions and dose of application suitable for specific farm conditions are needed for better commercialization of biochar usage. Similarly, the potential risks associated with long term application including accumulation of residual pesticides in soil following the release of sequestered pesticides from biochar through field trials in future studies are very much essential.

It can be observed that the significant improvements in the crop productivity by biochar additions is specific to a wide range of factors including soil type, crop type and agro-climatic conditions, etc., therefore, the development of clear hypothesis testing mechanisms of action would be a more reliable method of effect evaluation than just outcome-based observations. Determination of potential detrimental effects of biochars in agricultural soil is vital to propose for real practical applications. Understanding the negative implications based on the limitations in biochar applications with respect to various environmental conditions provides us with the necessary information to maneuver its productive functionality for promoting sustainable agriculture.

References

1. Kunhikrishnan A, Thangarajan R, Bolan NS, Xu Y, Mandal S, Gleeson DB, et al. Chapter one—Functional relationships of soil acidification, liming, and greenhouse gas flux. In: Sparks DL, editor. *Advances in agronomy. Advances in agronomy*, vol. 139. Academic Press; 2016. p. 1–71. https://doi.org/10.1016/bs.agron.2016.05.001.
2. Machado RMA, Serralheiro RP. Soil salinity: effect on vegetable crop growth. management practices to prevent and mitigate soil salinization. *Horticulturae* 2017;**3**(2):30. https://doi.org/10.3390/horticulturae3020030.
3. Pimentel D, Berger B, Filiberto D, Newton M, Wolfe B, Karabinakis E, et al. Water resources: agricultural and environmental issues. *Bioscience* 2004;**54**(10):909–18. https://doi.org/10.1641/0006-3568(2004)054[0909:WRAAEI]2.0.CO;2.
4. Bationo A, Lungu O, Naimi M, Okoth P, Smaling E, Lamourdia T. African soils: their productivity and profitability of fertilizer use. In: *Background paper for the African fertilizer summit 9-13th June 2006, Abuja, Nigeria*; 2006.
5. Tan ZX, Lal R, Wiebe KD. Global soil nutrient depletion and yield reduction. *J Sustain Agric* 2005;**26**(1):123–46. https://doi.org/10.1300/J064v26n01_10.
6. Srivastava AK. Integrated nutrient management: concept and application in citrus. In: *Citrus II. Tree and Forestry Science and Biotechnology*. vol. 3. Global Science Books; 2009. p. 32–58 [SPECIAL ISSUE 1].
7. Liu E, Yan C, Mei X, He W, Bing SH, Ding L, et al. Long-term effect of chemical fertilizer, straw, and manure on soil chemical and biological properties in northwest China. *Geoderma* 2010;**158**(3):173–80. https://doi.org/10.1016/j.geoderma.2010.04.029.
8. Smith P, Martino Z, Cai D. Agriculture. In: *Climate change 2007: mitigation*. Cambridge University; 2007.
9. Semida WM, Beheiry HR, Sétamou M, Simpson CR, Abd El-Mageed TA, Rady MM, et al. Biochar implications for sustainable agriculture and environment: a review. *S Afr J Bot* 2019;**127**:333–47. https://doi.org/10.1016/j.sajb.2019.11.015.
10. Gregory SJ, Anderson CWN, Camps Arbestain M, McManus MT. Response of plant and soil microbes to biochar amendment of an arsenic-contaminated soil. *Agric Ecosyst Environ* 2014;**191**:133–41. https://doi.org/10.1016/j.agee.2014.03.035.

11. Shaaban M, Van Zwieten L, Bashir S, Younas A, Núñez-Delgado A, Chhajro MA, et al. A concise review of biochar application to agricultural soils to improve soil conditions and fight pollution. *J Environ Manag* 2018;**228**:429–40. https://doi.org/10.1016/j.jenvman.2018.09.006.

12. Lichtfouse E, Navarrete M, Debaeke P, Souchère V, Alberola C, Ménassieu J. Agronomy for sustainable agriculture. A review. *Agron Sustain Dev* 2009;**29**(1):1–6. https://doi.org/10.1051/agro:2008054.

13. Atkinson CJ, Fitzgerald JD, Hipps NA. Potential mechanisms for achieving agricultural benefits from biochar application to temperate soils: a review. *Plant Soil* 2010;**337** (1):1–18. https://doi.org/10.1007/s11104-010-0464-5.

14. Agegnehu G, Srivastava AK, Bird MI. The role of biochar and biochar-compost in improving soil quality and crop performance: a review. *Appl Soil Ecol* 2017;**119**:156–70. https://doi.org/10.1016/j.apsoil.2017.06.008.

15. Laird D, Fleming P, Wang B, Horton R, Karlen D. Biochar impact on nutrient leaching from a midwestern agricultural soil. *Geoderma* 2010;**158**(3–4):436–42. https://doi.org/10.1016/j.geoderma.2010.05.012.

16. Suliman W, Harsh JB, Abu-Lail NI, Fortuna A-M, Dallmeyer I, Garcia-Perez M. Influence of feedstock source and pyrolysis temperature on biochar bulk and surface properties. *Biomass Bioenergy* 2016;**84**:37–48. https://doi.org/10.1016/j.biombioe.2015.11.010.

17. Lehmann DJ. *Biochar for environmental management: science and technology*. Earthscan; 2012.

18. Vanapalli KR, Bhattacharya J, Samal B, Chandra S, Medha I, Dubey BK. Optimized production of single-use plastic–eucalyptus wood char composite for application in soil. *J Clean Prod* 2021;**278**:123968. https://doi.org/10.1016/j.jclepro.2020.123968.

19. Bakshi S, Banik C, Laird DA. Estimating the organic oxygen content of biochar. *Sci Rep* 2020;**10**(1):13082. https://doi.org/10.1038/s41598-020-69798-y.

20. Solomon D, Lehmann J, Thies J, Schäfer T, Liang B, Kinyangi J, et al. Molecular signature and sources of biochemical recalcitrance of organic C in Amazonian dark earths. *Geochim Cosmochim Acta* 2007;**71**(9):2285–98. https://doi.org/10.1016/j.gca.2007.02.014.

21. Cross A, Sohi SP. A method for screening the relative long-term stability of biochar. *GCB Bioenergy* 2013;**5**(2):215–20. https://doi.org/10.1111/gcbb.12035.

22. Crombie K, Mašek O, Sohi SP, Brownsort P, Cross A. The effect of pyrolysis conditions on biochar stability as determined by three methods. *GCB Bioenergy* 2013;**5**(2):122–31. https://doi.org/10.1111/gcbb.12030.

23. Cabrera A, Cox L, Spokas K, Hermosín MC, Cornejo J, Koskinen WC. Influence of biochar amendments on the sorption–desorption of aminocyclopyrachlor, bentazone and pyraclostrobin pesticides to an agricultural soil. *Sci Total Environ* 2014;**470–471**:438–43. https://doi.org/10.1016/j.scitotenv.2013.09.080.

24. Kookana RS, Sarmah AK, Van Zwieten L, Krull E, Singh B. Chapter three—Biochar application to soil: agronomic and environmental benefits and unintended consequences. In: Sparks DL, editor. *Advances in agronomy*. vol. 112. Academic Press; 2011. p. 103–43. https://doi.org/10.1016/B978-0-12-385538-1.00003-2.

25. Harvey OR, Herbert BE, Rhue RD, Kuo L-J. Metal interactions at the biochar-water interface: energetics and structure-sorption relationships elucidated by flow adsorption microcalorimetry. *Environ Sci Technol* 2011;**45**(13):5550–6. https://doi.org/10.1021/es104401h.

26. Heitkötter J, Marschner B. Interactive effects of biochar ageing in soils related to feedstock, pyrolysis temperature, and historic charcoal production. *Geoderma* 2015;**245–246**:56–64. https://doi.org/10.1016/j.geoderma.2015.01.012.

27. Bashir S, Hussain Q, Akmal M, Riaz M, Hu H, Ijaz SS, et al. Sugarcane bagasse-derived biochar reduces the cadmium and chromium bioavailability to mash bean and enhances the microbial activity in contaminated soil. *J Soils Sediments* 2018;**18**(3):874–86. https://doi.org/10.1007/s11368-017-1796-z.

28. Tomczyk A, Sokołowska Z, Boguta P. Biochar physicochemical properties: pyrolysis temperature and feedstock kind effects. *Rev Environ Sci Biotechnol* 2020;**19**(1):191–215. https://doi.org/10.1007/s11157-020-09523-3.
29. Ding W, Dong X, Ime IM, Gao B, Ma LQ. Pyrolytic temperatures impact lead sorption mechanisms by bagasse biochars. *Chemosphere* 2014;**105**:68–74. https://doi.org/10.1016/j.chemosphere.2013.12.042.
30. Ahmad M, Lee SS, Dou X, Mohan D, Sung J-K, Yang JE, et al. Effects of pyrolysis temperature on soybean stover- and peanut shell-derived biochar properties and TCE adsorption in water. *Bioresour Technol* 2012;**118**:536–44. https://doi.org/10.1016/j.biortech.2012.05.042.
31. Trompowsky PM, de Benites VM, Madari BE, Pimenta AS, Hockaday WC, Hatcher PG. Characterization of humic like substances obtained by chemical oxidation of eucalyptus charcoal. *Org Geochem* 2005;**36**(11):1480–9. https://doi.org/10.1016/j.orggeochem.2005.08.001.
32. Chandra S, Bhattacharya J. Influence of temperature and duration of pyrolysis on the property heterogeneity of rice straw biochar and optimization of pyrolysis conditions for its application in soils. *J Clean Prod* 2019;**215**:1123–39. https://doi.org/10.1016/j.jclepro.2019.01.079.
33. Yuan J-H, Xu R-K, Zhang H. The forms of alkalis in the biochar produced from crop residues at different temperatures. *Bioresour Technol* 2011;**102**(3):3488–97. https://doi.org/10.1016/j.biortech.2010.11.018.
34. Tsai W-T, Liu S-C, Chen H-R, Chang Y-M, Tsai Y-L. Textural and chemical properties of swine-manure-derived biochar pertinent to its potential use as a soil amendment. *Chemosphere* 2012;**89**(2):198–203. https://doi.org/10.1016/j.chemosphere.2012.05.085.
35. Singh B, Singh BP, Cowie AL. Characterisation and evaluation of biochars for their application as a soil amendment. *Aust J Soil Res* 2010;**48**(6–7):516–25. https://doi.org/10.1071/SR10058.
36. Lee Y, Park J, Ryu C, Gang KS, Yang W, Park YK, et al. Comparison of biochar properties from biomass residues produced by slow pyrolysis at 500°C. *Bioresour Technol* 2013;**148**:196–201. https://doi.org/10.1016/j.biortech.2013.08.135.
37. Pignatello JJ, Kwon S, Lu Y. Effect of natural organic substances on the surface and adsorptive properties of environmental black carbon (char): attenuation of surface activity by humic and fulvic acids. *Environ Sci Technol* 2006;**40**(24):7757–63. https://doi.org/10.1021/es061307m.
38. Li A, Liu H, Wang H, Xu HB, Jin LF, Liu JL, et al. Effects of temperature and heating rate on the characteristics of molded bio-char. *BioResources* 2016;**11**:3259–74. https://bioresources.cnr.ncsu.edu/. accessed 2021-04-23.
39. El Refaey A, Mahmoud A, Saleh M. Bone biochar as a renewable and efficient p fertilizer: a comparative study. *Alex J Agric Res* 2015;**30**(3):127–37.
40. Agbede TM, Adekiya AO, Eifediyi EK. Impact of poultry manure and NPK fertilizer on soil physical properties and growth and yield of carrot. *J Hortic Res* 2017;**25**(1):81–8. https://doi.org/10.1515/johr-2017-0009.
41. Kalus K, Koziel JA, Opaliński S. A review of biochar properties and their utilization in crop agriculture and livestock production. *Appl Sci* 2019;**9**(17):3494. https://doi.org/10.3390/app9173494.
42. Novak JM, Lima I, Xing B, Gaskin JW, Steiner C, Das KC, et al. Characterization of designer biochar produced at different temperatures and their effects on a loamy sand. *Ann Environ Sci* 2009;**3**:195–206.
43. Berglund LM, DeLuca TH, Zackrisson O. Activated carbon amendments to soil alters nitrification rates in scots pine forests. *Soil Biol Biochem* 2004;**36**(12):2067–73. https://doi.org/10.1016/j.soilbio.2004.06.005.
44. Ding Y, Liu Y, Liu S, Li Z, Tan X, Huang X, et al. Biochar to improve soil fertility. A review. *Agron Sustain Dev* 2016;**36**(2):36. https://doi.org/10.1007/s13593-016-0372-z.

45. Steiner C, Glaser B, Teixeira WG, Lehmann J, Blum WEH, Zech W. Nitrogen retention and plant uptake on a highly weathered central Amazonian ferralsol amended with compost and charcoal. *J Plant Nutr Soil Sci* 2008;**171**(6):893–9. https://doi.org/10.1002/jpln.200625199.

46. Martin SM, Kookana RS, Van Zwieten L, Krull E. Marked changes in herbicide sorption–desorption upon ageing of biochars in soil. *J Hazard Mater* 2012;**231–232**: 70–8. https://doi.org/10.1016/j.jhazmat.2012.06.040.

47. Yin C, Xu Z. Biochar: nutrient properties and their enhancement. In: *Biochar for environmental management science and technology*. Taylor and Francis Group; 2009.

48. Chang J, Clay DE, Clay SA, Chintala R, Miller JM, Schumacher T. Biochar reduced nitrous oxide and carbon dioxide emissions from soil with different water and temperature cycles. *Agron J* 2016;**108**(6):2214. https://doi.org/10.2134/agronj2016.02.0100.

49. Hammes K, Torn M, Lapenis A, Schmidt M. Centennial black carbon turnover in a russian steppe soil. *Biogeosciences* 2008;**5**:661–83. https://doi.org/10.5194/bgd-5-661-2008.

50. Martins GC. Soil physical characterization. In: Teixeira WG, Lehmann J, Kern DC, Glaser B, Wodos WI, editors. *Amazonian Dark Earths: origin properties management*. Dordrecht: Springer Netherlands; 2003. p. 271–86. https://doi.org/10.1007/1-4020-2597-1_15.

51. Herath HMSK, Camps-Arbestain M, Hedley M. Effect of biochar on soil physical properties in two contrasting soils: an alfisol and an andisol. *Geoderma* 2013;**209–210**:188–97. https://doi.org/10.1016/j.geoderma.2013.06.016.

52. Asai H, Samson BK, Stephan HM, Songyikhangsuthor K, Homma K, Kiyono Y, et al. Biochar amendment techniques for upland rice production in northern Laos: 1. Soil physical properties, leaf SPAD and grain yield. *Field Crop Res* 2009;**111**(1):81–4. https://doi.org/10.1016/j.fcr.2008.10.008.

53. Zhao X, Wang J, Wang S, Xing G. Successive straw biochar application as a strategy to sequester carbon and improve fertility: a pot experiment with two rice/wheat rotations in paddy soil. *Plant Soil* 2014;**378**(1):279–94.

54. Hale SE, Alling V, Martinsen V, Mulder J, Breedveld GD, Cornelissen G. The sorption and desorption of phosphate-P, ammonium-N and nitrate-N in cacao shell and corn cob biochars. *Chemosphere* 2013;**91**(11):1612–9. https://doi.org/10.1016/j.chemosphere.2012.12.057.

55. Zhang H, Voroney RP, Price GW. Effects of temperature and processing conditions on biochar chemical properties and their influence on soil C and N transformations. *Soil Biol Biochem* 2015;**83**:19–28. https://doi.org/10.1016/j.soilbio.2015.01.006.

56. Chandra S, Medha I, Bhattacharya J. Potassium-iron rice straw biochar composite for sorption of nitrate, phosphate, and ammonium ions in soil for timely and controlled release. *Sci Total Environ* 2020;**712**:136337. https://doi.org/10.1016/j.scitotenv.2019.136337.

57. Xu G, Sun J, Shao H, Chang SX. Biochar had effects on phosphorus sorption and desorption in three soils with differing acidity. *Ecol Eng* 2014;**62**:54–60. https://doi.org/10.1016/j.ecoleng.2013.10.027.

58. Gul S, Whalen JK. Biochemical cycling of nitrogen and phosphorus in biochar-amended soils. *Soil Biol Biochem* 2016;**103**:1–15. https://doi.org/10.1016/j.soilbio.2016.08.001.

59. Cheng C-H. Natural oxidation of black carbon in soils: changes in molecular form and surface charge along a climosequence. *Geochim Cosmochim Acta* 2008;**72**:1598–610.

60. Cheng C-H, Lehmann J, Thies JE, Burton SD, Engelhard MH. Oxidation of black carbon by biotic and abiotic processes. *Org Geochem* 2006;**37**(11):1477–88. https://doi.org/10.1016/j.orggeochem.2006.06.022.

61. Woolf D. *Biochar as a soil amendment: a review of the environmental implications.* Swansea University; 2008. p. 31.
62. Major J, Steiner C, Downie A, Lehmann J, Joseph S. Biochar effects on nutrient leaching. In: *Biochar for environmental management: Science and technology.* Routledge; Taylor and Francis Group; 2009. p. 271.
63. Thies JE, Rillig MC. Characteristics of biochar: biological properties- biochar for environmental management. *Sci Technol* 2009;**1**:85–105.
64. Azeem M, Hale L, Montgomery J, Crowley D, Jr MEM. Biochar and compost effects on soil microbial communities and nitrogen induced respiration in turfgrass soils. *PLoS One* 2020;**15**(11), e0242209. https://doi.org/10.1371/journal.pone. 0242209.
65. Steiner C, de Arruda MR, Teixeira WG, Zech W. Soil respiration curves as soil fertility indicators in perennial central amazonian plantations treated with charcoal, and mineral or organic fertilisers. *Trop Sci* 2007;**47**(4):218–30. https://doi.org/10.1002/ts.216.
66. Ameloot N, De Neve S, Jegajeevagan K, Yildiz G, Buchan D, Funkuin YN, et al. Short-term CO2 and N2O emissions and microbial properties of biochar amended sandy loam soils. *Soil Biol Biochem* 2013;**57**:401–10. https://doi.org/10.1016/j.soilbio. 2012.10.025.
67. Dempster DN, Gleeson DB, Solaiman ZM, Jones DL, Murphy DV. Decreased soil microbial biomass and nitrogen mineralisation with eucalyptus biochar addition to a coarse textured soil. *Plant Soil* 2012;**354**(1–2):311–24. https://doi.org/10.1007/ s11104-011-1067-5.
68. Kim J-S, Sparovek G, Longo RM, De Melo WJ, Crowley D. Bacterial diversity of terra preta and pristine forest soil from the Western Amazon. *Soil Biol Biochem* 2007;**39** (2):684–90.
69. Mia S, van Groenigen JW, van de Voorde TFJ, Oram NJ, Bezemer TM, Mommer L, et al. Biochar application rate affects biological nitrogen fixation in red clover conditional on potassium availability. *Agric Ecosyst Environ* 2014;**191**:83–91. https://doi.org/ 10.1016/j.agee.2014.03.011.
70. Wu X, Ren L, Luo L, Zhang J, Zhang L, Huang H. Bacterial and fungal community dynamics and shaping factors during agricultural waste composting with zeolite and biochar addition. *Sustainability* 2020;**12**(17):7082.
71. Holste EK, Kobe RK, Gehring CA. Plant species differ in early seedling growth and tissue nutrient responses to arbuscular and ectomycorrhizal fungi. *Mycorrhiza* 2017;**27**(3):211–23. https://doi.org/10.1007/s00572-016-0744-x.
72. Liu, Liu F, Ravnskov S, Rubæk GH, Sun Z, Andersen MN. Impact of wood biochar and its interactions with mycorrhizal fungi, phosphorus fertilization and irrigation strategies on potato growth. *J Agron Crop Sci* 2017;**203**(2):131–45. https://doi.org/10.1111/ jac.12185.
73. LeCroy C, Masiello CA, Rudgers JA, Hockaday WC, Silberg JJ. Nitrogen, biochar, and mycorrhizae: alteration of the symbiosis and oxidation of the char surface. *Soil Biol Biochem* 2013;**58**:248–54. https://doi.org/10.1016/j.soilbio.2012.11.023.
74. Glaser B. Microbial response to charcoal amendments and fertilization of a highly weathered tropical soil. In: Birk J, Steiner C, Teixiera W, Zech W, Woods WI, Teixeira WG, Rebellato L, editors. *Amazonian Dark Earths: Wim Sombroek's vision.* Dordrecht: Springer Netherlands; 2009. p. 309–24. https://doi.org/10.1007/978-1-4020-9031-8_16.
75. Warnock DD, Mummey DL, McBride B, Major J, Lehmann J, Rillig MC. Influences of non-herbaceous biochar on arbuscular mycorrhizal fungal abundances in roots and soils: results from growth-chamber and field experiments. *Appl Soil Ecol* 2010;**46**(3): 450–6. https://doi.org/10.1016/j.apsoil.2010.09.002.

76. Gu J, Zhou W, Jiang B, Wang L, Ma Y, Guo H, et al. Effects of biochar on the transformation and earthworm bioaccumulation of organic pollutants in soil. *Chemosphere* 2016;**145**:431–7. https://doi.org/10.1016/j.chemosphere.2015.11.106.
77. Gomez-Eyles JL, Sizmur T, Collins CD, Hodson ME. Effects of biochar and the earthworm Eisenia fetida on the bioavailability of polycyclic aromatic hydrocarbons and potentially toxic elements. *Environ Pollut* 2011;**159**(2):616–22. https://doi.org/10.1016/j.envpol.2010.09.037.
78. Shan J, Wang Y, Gu J, Zhou W, Ji R, Yan X. Effects of biochar and the geophagous earthworm metaphire guillelmi on fate of 14C-catechol in an agricultural soil. *Chemosphere* 2014;**107**:109–14. https://doi.org/10.1016/j.chemosphere.2014.03.030.
79. Paz-Ferreiro J, Fu S, Méndez A, Gascó G. Interactive effects of biochar and the earthworm pontoscolex corethrurus on plant productivity and soil enzyme activities. *J Soils Sediments* 2014;**14**(3):483–94. https://doi.org/10.1007/s11368-013-0806-z.
80. Salem M, Kohler J, Wurst S, Rillig MC. Earthworms can modify effects of hydrochar on growth of plantago lanceolata and performance of arbuscular mycorrhizal fungi. *Pedobiologia* 2013;**56**(4):219–24. https://doi.org/10.1016/j.pedobi.2013.08.003.
81. Weyers SL, Spokas KA. Impact of biochar on earthworm populations: a review. *Appl Environ Soil Sci* 2011;**2011**, e541592. https://doi.org/10.1155/2011/541592.
82. Anyanwu A, Jd C, Nworie O, Chamba E. Influence of biochar aged in acidic soil on ecosystem engineers and two tropical agricultural plants. *Ecotoxicol Environ Saf* 2018;**153**:116–26. https://doi.org/10.1016/j.ecoenv.2018.02.005.
83. Verheijen F, JEFFERY SL, BASTOS AC, van der Velde M, DIAFAS I. *Biochar application to soils—a critical scientific review of effects on soil properties, processes and functions.* EU Science Hub—European Commission; 2010. https://doi.org/10.2788/472.
84. Luo Y, Durenkamp M, De Nobili M, Lin Q, Brookes PC. Short term soil priming effects and the mineralisation of biochar following its incorporation to soils of different PH. *Soil Biol Biochem* 2011;**43**(11):2304–14. https://doi.org/10.1016/j.soilbio.2011.07.020.
85. Sigua GC, Novak JM, Watts DW, Johnson MG, Spokas K. Efficacies of designer biochars in improving biomass and nutrient uptake of winter wheat grown in a hard setting subsoil layer. *Chemosphere* 2016;**142**:176–83. https://doi.org/10.1016/j.chemosphere.2015.06.015.
86. Rocha Oliveira F, Patel A, Jaisi D, Adhikari S, Lu H, Khanal S. Environmental application of biochar: current status and perspectives. *Bioresour Technol* 2017;**246**:110–22. https://doi.org/10.1016/j.biortech.2017.08.122.
87. Spokas KA. Review of the stability of biochar in soils: predictability of O:C molar ratios. *Carbon Manag* 2010;**1**(2):289–303. https://doi.org/10.4155/cmt.10.32.
88. Van Zwieten L, Kimber S, Morris S, Chan KY, Downie A, Rust J, et al. Effects of biochar from slow pyrolysis of papermill waste on agronomic performance and soil fertility. *Plant Soil* 2010;**327**(1):235–46.
89. Chan KY, Van Zwieten L, Meszaros I, Downie A, Joseph S. Agronomic values of greenwaste biochar as a soil amendment. *Aust J Soil Res* 2007;**45**(8):629–34. https://doi.org/10.1071/SR07109.
90. Chan KY, Van Zwieten L, Meszaros I, Downie A, Joseph S. Using poultry litter biochars as soil amendments. *Soil Res* 2008;**46**(5):437–44.
91. Noguera D, Rondón M, Laossi K-R, Hoyos V, Lavelle P, Cruz de Carvalho MH, et al. Contrasted effect of biochar and earthworms on rice growth and resource allocation in different soils. *Soil Biol Biochem* 2010;**42**(7):1017–27. https://doi.org/10.1016/j.soilbio.2010.03.001.
92. Major J, Rondon M, Molina D, Riha SJ, Lehmann J. Maize yield and nutrition during 4 years after biochar application to a colombian savanna oxisol. *Plant Soil* 2010;**333**(1–2):117–28. https://doi.org/10.1007/s11104-010-0327-0.

93. Agegnehu G, Nelson PN, Bird MI. The effects of biochar, compost and their mixture and nitrogen fertilizer on yield and nitrogen use efficiency of barley grown on a nitisol in the highlands of Ethiopia. *Sci Total Environ* 2016;**569–570**:869–79. https://doi.org/10.1016/j.scitotenv.2016.05.033.

94. Liu L, Wang Y, Yan X, Li J, Jiao N, Hu S. Biochar amendments increase the yield advantage of legume-based intercropping systems over monoculture. *Agric Ecosyst Environ* 2017;**237**:16–23.

95. Raboin L-M, Razafimahafaly AHD, Rabenjarisoa MB, Rabary B, Dusserre J, Becquer T. Improving the fertility of tropical acid soils: liming versus biochar application? A long term comparison in the highlands of Madagascar. *Field Crop Res* 2016; **199**:99–108. https://doi.org/10.1016/j.fcr.2016.09.005.

96. Doan TT, Henry-des-Tureaux T, Rumpel C, Janeau J-L, Jouquet P. Impact of compost, vermicompost and biochar on soil fertility, maize yield and soil erosion in northern Vietnam: a three year mesocosm experiment. *Sci Total Environ* 2015;**514**:147–54. https://doi.org/10.1016/j.scitotenv.2015.02.005.

97. Agegnehu G, Bass AM, Nelson PN, Bird MI. Benefits of biochar, compost and biochar–compost for soil quality, maize yield and greenhouse gas emissions in a tropical agricultural soil. *Sci Total Environ* 2016;**543**:295–306. https://doi.org/10.1016/j.scitotenv.2015.11.054.

98. Kim H-S, Kim K-R, Yang JE, Ok YS, Owens G, Nehls T, et al. Effect of biochar on reclaimed tidal land soil properties and maize (Zea Mays L.) response. *Chemosphere* 2016;**142**:153–9. https://doi.org/10.1016/j.chemosphere.2015.06.041.

99. Tammeorg P, Simojoki A, Mäkelä P, Stoddard FL, Alakukku L, Helenius J. Short-term effects of biochar on soil properties and wheat yield formation with meat bone meal and inorganic fertiliser on a boreal loamy sand. *Agric Ecosyst Environ* 2014;**191**:108–16. https://doi.org/10.1016/j.agee.2014.01.007.

100. Vaccari FP, Maienza A, Miglietta F, Baronti S, Di Lonardo S, Giagnoni L, et al. Biochar stimulates plant growth but not fruit yield of processing tomato in a fertile soil. *Agric Ecosyst Environ* 2015;**207**:163–70. https://doi.org/10.1016/j.agee.2015.04.015.

101. Agegnehu G, Bass AM, Nelson PN, Muirhead B, Wright G, Bird MI. Biochar and biochar-compost as soil amendments: effects on peanut yield, soil properties and greenhouse gas emissions in tropical North Queensland, Australia. *Agric Ecosyst Environ* 2015;**213**:72–85. https://doi.org/10.1016/j.agee.2015.07.027.

102. Spokas KA, Cantrell KB, Novak JM, Archer DW, Ippolito JA, Collins HP, et al. Biochar: a synthesis of its agronomic impact beyond carbon sequestration. *J Environ Qual* 2012;**41**(4):973–89. https://doi.org/10.2134/jeq2011.0069.

103. Moreira A. In: Lehmann J, Kern D, German L, Mccann J, Martins GC, Lehmann J, Wodos WI, editors. *Soil fertility and production potential. in Amazonian dark earths: origin properties management.* Dordrecht: Springer Netherlands; 2003. p. 105–24. https://doi.org/10.1007/1-4020-2597-1_6.

104. Zheng R, Li C, Sun G, Xie Z, Chen J, Wu J, et al. The influence of particle size and feedstock of biochar on the accumulation of Cd, Zn, Pb, and As by Brassica Chinensis L. *Environ Sci Pollut Res* 2017;**24**(28):22340–52. https://doi.org/10.1007/s11356-017-9854-z.

105. Alvarez-Campos O, Lang TA, Bhadha JH, McCray JM, Glaz B, Daroub SH. Biochar and mill ash improve yields of sugarcane on a sand soil in Florida. *Agric Ecosyst Environ* 2018;**253**:122–30. https://doi.org/10.1016/j.agee.2017.11.006.

106. Solaiman ZM, Murphy DV, Abbott LK. Biochars influence seed germination and early growth of seedlings. *Plant Soil* 2012;**353**(1):273–87. https://doi.org/10.1007/s11104-011-1031-4.

107. Scheifele M, Hobi A, Buegger F, Gattinger A, Schulin R, Boller T, et al. Impact of pyrochar and hydrochar on soybean (Glycine Max L.) root nodulation and biological nitrogen fixation. *J Plant Nutr Soil Sci* 2017;**180**(2):199–211. https://doi.org/10.1002/jpln.201600419.

108. Macdonald LM, Farrell M, Zwieten LV, Krull ES. Plant growth responses to biochar addition: an Australian soils perspective. *Biol Fertil Soils* 2014;**50**(7):1035–45. https://doi.org/10.1007/s00374-014-0921-z.

109. Muhammad N, Aziz R, Brookes PC, Xu J. Impact of wheat straw biochar on yield of rice and some properties of psammaquent and plinthudult. *J Soil Sci Plant Nutr* 2017;**17**(3):808–23. https://doi.org/10.4067/S0718-95162017000300019.

110. Gaskin, Steiner C, Harris K, Das KC, Bibens B. Effect of low-temperature pyrolysis conditions on biochar for agricultural use. *Trans ASABE* 2008;**51**(6):2061–9. https://doi.org/10.13031/2013.25409.

111. Jaiswal AK, Elad Y, Paudel I, Graber ER, Cytryn E, Frenkel O. Linking the below-ground microbial composition, diversity and activity to soilborne disease suppression and growth promotion of tomato amended with biochar. *Sci Rep* 2017;**7**(1):1–17. https://doi.org/10.1038/srep44382.

112. Glaser B, Lehmann J, Zech W. Ameliorating physical and chemical properties of highly weathered soils in the tropics with charcoal—a review. *Biol Fertil Soils* 2002;**35**(4):219–30. https://doi.org/10.1007/s00374-002-0466-4.

113. Uzoma KC, Inoue M, Andry H, Fujimaki H, Zahoor A, Nishihara E. Effect of cow manure biochar on maize productivity under sandy soil condition. *Soil Use Manag* 2011;**27**(2):205–12. https://doi.org/10.1111/j.1475-2743.2011.00340.x.

114. Abbas A, Yaseen M, Khalid M, Naveed M, Aziz MZ, Hamid Y, et al. Effect of biochar-amended urea on nitrogen economy of soil for improving the growth and yield of wheat (Triticum Aestivum L.) under field condition. *J Plant Nutr* 2017;**40**(16):2303–11. https://doi.org/10.1080/01904167.2016.1267746.

115. Wang M, Wang JJ, Wang X. Effect of KOH-enhanced biochar on increasing soil plant-available silicon. *Geoderma* 2018;**321**:22–31. https://doi.org/10.1016/j.geoderma.2018.02.001.

116. Ahmad M, Usman ARA, Al-Faraj AS, Ahmad M, Sallam A, Al-Wabel MI. Phosphorus-loaded biochar changes soil heavy metals availability and uptake potential of maize (Zea Mays L.) plants. *Chemosphere* 2018;**194**:327–39. https://doi.org/10.1016/j.chemosphere.2017.11.156.

117. Akhtar SS, Andersen MN, Liu F. Biochar mitigates salinity stress in potato. *J Agron Crop Sci* 2015;**201**(5):368–78. https://doi.org/10.1111/jac.12132.

118. Usman ARA, Al-wabel MI, Ok YS, Al-harbi A, Wahb-allah M, El-naggar AH, et al. Conocarpus biochar induces changes in soil nutrient availability and tomato growth under saline irrigation. *Pedosphere* 2016;**26**(1):27–38. https://doi.org/10.1016/S1002-0160(15)60019-4.

119. Kimetu JM, Lehmann J, Ngoze SO, Mugendi DN, Kinyangi JM, Riha S, et al. Reversibility of soil productivity decline with organic matter of differing quality along a degradation gradient. *Ecosystems* 2008;**11**(5):726–39. https://doi.org/10.1007/s10021-008-9154-z.

120. Yusif SA, Dare MO, Haruna S, Haruna FD. Evaluation of growth performance of tomato in response to biochar and arbuscular mycorrhizal fungi (Amf) inoculation. *Nig J Basic Appl Sci* 2017;**24**(2):31. https://doi.org/10.4314/njbas.v24i2.5.

121. Houben D, Evrard L, Sonnet P. Beneficial effects of biochar application to contaminated soils on the bioavailability of Cd, Pb and Zn and the biomass production of rapeseed (Brassica Napus L.). *Biomass Bioenergy* 2013;**57**:196–204. https://doi.org/10.1016/j.biombioe.2013.07.019.

122. Ramzani PMA, Shan L, Anjum S, Khan W-D, Ronggui H, Iqbal M, et al. Improved quinoa growth, physiological response, and seed nutritional quality in three soils having different stresses by the application of acidified biochar and compost. *Plant Physiol Biochem* 2017;**116**:127–38. https://doi.org/10.1016/j.plaphy.2017.05.003.

123. Rees F, Dhyèvre A, Morel JL, Cotelle S. Decrease in the genotoxicity of metal-contaminated soils with biochar amendments. *Environ Sci Pollut Res* 2017;**24**(36):27634–41. https://doi.org/10.1007/s11356-017-8386-x.

123a. Carvalho KM, Martin DF. Removal of aqueous selenium by four aquatic plants. *J Aquat Plant Manag* 2001;**39**:33–6.

124. Elad Y, David DR, Harel YM, Borenshtein M, Kalifa HB, Silber A, et al. Induction of systemic resistance in plants by biochar, a soil-applied carbon sequestering agent. *Phytopathology* 2010;**100**(9):913–21. https://doi.org/10.1094/PHYTO-100-9-0913.

125. Masto RE, Ansari MA, George J, Selvi VA, Ram LC. Co-application of biochar and lignite fly ash on soil nutrients and biological parameters at different crop growth stages of zea mays. *Ecol Eng* 2013;**58**:314–22. https://doi.org/10.1016/j.ecoleng.2013.07.011.

126. Steiner C, Teixeira WG, Lehmann J, Nehls T, de Macêdo JLV, Blum WEH, et al. Long term effects of manure, charcoal and mineral fertilization on crop production and fertility on a highly weathered central amazonian upland soil. *Plant Soil* 2007;**291**(1):275–90. https://doi.org/10.1007/s11104-007-9193-9.

127. Mau AE, Utami SR. Effects of biochar amendment and arbuscular mycorrhizal fungi inoculation on availability of soil phosphorus and growth of maize. *J Degrade Min Land Manage* 2014;**1**(2):69–74. https://doi.org/10.15243/jdmlm.2014.012.069.

128. Schmidt H-P, Kammann C, Niggli C, Evangelou MWH, Mackie KA, Abiven S. Biochar and biochar-compost as soil amendments to a vineyard soil: influences on plant growth, nutrient uptake, plant health and grape quality. *Agric Ecosyst Environ* 2014;**191**:117–23. https://doi.org/10.1016/j.agee.2014.04.001.

129. Niu Y, Chen Z, Müller C, Zaman MM, Kim D, Yu H, et al. Yield-scaled N2O emissions were effectively reduced by biochar amendment of sandy loam soil under maize—wheat rotation in the North China Plain. *Atmos Environ* 2017;**170**:58–70. https://doi.org/10.1016/j.atmosenv.2017.09.050.

130. Akhtar SS, Li G, Andersen MN, Liu F. Biochar enhances yield and quality of tomato under reduced irrigation. *Agric Water Manag* 2014;**138**:37–44. https://doi.org/10.1016/j.agwat.2014.02.016.

131. Haider G, Koyro H-W, Azam F, Steffens D, Müller C, Kammann C. Biochar but not humic acid product amendment affected maize yields via improving plant-soil moisture relations. *Plant Soil* 2015;**395**(1):141–57. https://doi.org/10.1007/s11104-014-2294-3.

132. Eyles A, Bound SA, Oliver G, Corkrey R, Hardie M, Green S, et al. Impact of biochar amendment on the growth, physiology and fruit of a young commercial apple orchard. *Trees* 2015;**29**(6):1817–26. https://doi.org/10.1007/s00468-015-1263-7.

133. Poorter H, Niklas KJ, Reich PB, Oleksyn J, Poot P, Mommer L. Biomass allocation to leaves, stems and roots: meta-analyses of interspecific variation and environmental control. *New Phytol* 2012;**193**(1):30–50. https://doi.org/10.1111/j.1469-8137.2011.03952.x.

134. Al-Harbi AR, Obadi A, Al-Omran AM, Abdel-Razzak H. Sweet peppers yield and quality as affected by biochar and compost as soil amendments under partial root irrigation. *J Saudi Soc Agric Sci* 2020;**19**(7):452–60. https://doi.org/10.1016/j.jssas.2020.08.002.

135. Petruccelli R, Bonetti A, Traversi M, Faraloni C, Valagussa M, Pozzi A. Influence of biochar application on nutritional quality of tomato (Lycopersicon Esculentum). *Crop Pasture Sci* 2015;**66**:747–55. https://doi.org/10.1071/CP14247.

136. Lattanzio V, Kroon PA, Quideau S, Treutter D. Plant phenolics—secondary metabolites with diverse functions. In: *Recent advances in polyphenol research*. John Wiley & Sons, Ltd; 2008. p. 1–35. https://doi.org/10.1002/9781444302400.ch1.

137. Yasmin Khan K, Ali B, Cui X, Feng Y, Yang X, Joseph Stoffella P. Impact of different feedstocks derived biochar amendment with cadmium low uptake affinity cultivar of pak choi (Brassica Rapa Ssb. Chinensis L.) on phytoavoidation of cd to reduce potential dietary toxicity. *Ecotoxicol Environ Saf* 2017;**141**:129–38. https://doi.org/10.1016/j.ecoenv.2017.03.020.

138. Bashir S, Zhu J, Fu Q, Hu H. Cadmium mobility, uptake and anti-oxidative response of water spinach (ipomoea aquatic) under rice straw biochar, zeolite and rock phosphate as amendments. *Chemosphere* 2018;**194**:579–87. https://doi.org/10.1016/j.chemosphere.2017.11.162.

139. Rizwan MS, Imtiaz M, Chhajro MA, Huang G, Fu Q, Zhu J, et al. Influence of pyrolytic and non-pyrolytic rice and castor straws on the immobilization of Pb and Cu in contaminated soil. *Environ Technol* 2016;**37**(21):2679–86. https://doi.org/10.1080/09593330.2016.1158870.

140. Cabrera A, Cox L, Spokas KA, Celis R, Hermosín MC, Cornejo J, et al. Comparative sorption and leaching study of the herbicides fluometuron and 4-chloro-2-methylphenoxyacetic acid (MCPA) in a soil amended with biochars and other sorbents. *J Agric Food Chem* 2011;**59**(23):12550–60. https://doi.org/10.1021/jf202713q.

141. Saito T, Otani T, Seike N, Murano H, Okazaki M. Suppressive effect of soil application of carbonaceous adsorbents on dieldrin uptake by cucumber fruits. *Soil Sci Plant Nutr* 2011;**57**(1):157–66. https://doi.org/10.1080/00380768.2010.551281.

142. Jones DL, Murphy DV, Khalid M, Ahmad W, Edwards-Jones G, DeLuca TH. Short-term biochar-induced increase in soil CO2 release is both biotically and abiotically mediated. *Soil Biol Biochem* 2011;**43**(8):1723–31. https://doi.org/10.1016/j.soilbio.2011.04.018.

143. Yang X-B, Ying G-G, Peng P-A, Wang L, Zhao J-L, Zhang L-J, et al. Influence of biochars on plant uptake and dissipation of two pesticides in an agricultural soil. *J Agric Food Chem* 2010;**58**(13):7915–21. https://doi.org/10.1021/jf1011352.

144. Rajapaksha AU, Vithanage M, Zhang M, Ahmad M, Mohan D, Chang SX, et al. Pyrolysis condition affected sulfamethazine sorption by tea waste biochars. *Bioresour Technol* 2014;**166**:303–8. https://doi.org/10.1016/j.biortech.2014.05.029.

145. Zhang P, Sun H, Yu L, Sun T. Adsorption and catalytic hydrolysis of carbaryl and atrazine on pig manure-derived biochars: impact of structural properties of biochars. *J Hazard Mater* 2013;**244–245**:217–24. https://doi.org/10.1016/j.jhazmat.2012.11.046.

146. Inyang M, Gao B, Zimmerman A, Zhang M, Chen H. Synthesis, characterization, and dye sorption ability of carbon nanotube–biochar nanocomposites. *Chem Eng J* 2014;**236**:39–46. https://doi.org/10.1016/j.cej.2013.09.074.

147. Younis U, Malik SA, Rizwan M, Qayyum MF, Ok YS, Shah MHR, et al. Biochar enhances the cadmium tolerance in spinach (Spinacia Oleracea) through modification of Cd uptake and physiological and biochemical attributes. *Environ Sci Pollut Res* 2016;**23**(21):21385–94. https://doi.org/10.1007/s11356-016-7344-3.

148. Ahmad M, Ok YS, Kim B-Y, Ahn J-H, Lee YH, Zhang M, et al. Impact of soybean stover- and pine needle-derived biochars on Pb and as mobility, microbial community, and carbon stability in a contaminated agricultural soil. *J Environ Manag* 2016;**166**:131–9. https://doi.org/10.1016/j.jenvman.2015.10.006.

149. Kim H-S, Kim K-R, Kim H-J, Yoon J-H, Yang JE, Ok YS, et al. Effect of biochar on heavy metal immobilization and uptake by lettuce (Lactuca Sativa L.) in agricultural soil. *Environ Earth Sci* 2015;**74**(2):1249–59. https://doi.org/10.1007/s12665-015-4116-1.

150. Fellet G, Marmiroli M, Marchiol L. Elements uptake by metal accumulator species grown on mine tailings amended with three types of biochar. *Sci Total Environ* 2014;**468–469**:598–608. https://doi.org/10.1016/j.scitotenv.2013.08.072.

151. Rehman MZ, Rizwan M, Ali S, Fatima N, Yousaf B, Naeem A, et al. Contrasting effects of biochar, compost and farm manure on alleviation of nickel toxicity in maize (Zea Mays L.) in relation to plant growth, photosynthesis and metal uptake. *Ecotoxicol Environ Saf* 2016;**133**:218–25. https://doi.org/10.1016/j.ecoenv.2016.07.023.

152. Cheng C-H, Lin Z-P, Huang Y-S, Chen C-P, Chen C-T, Menyailo OV. Reduction of diuron efficacy with biochar amendments. *IJESD* 2016;**7**(7):480–5. https://doi.org/10.18178/ijesd.2016.7.7.824.

153. Lehmann J, Rondon M. Bio-char soil management on highly weathered soils in the humid tropics. In: *Biological approaches to sustainable soil systems*. vol. 113. Taylor and Francis Group; 2006. p. e530 [517].

154. Liu Y, Yang M, Wu Y, Wang H, Chen Y, Wu W. Reducing CH4 and CO2 emissions from waterlogged paddy soil with biochar. *J Soils Sediments* 2011;**11**(6):930–9. https://doi.org/10.1007/s11368-011-0376-x.

155. Kimetu JM, Lehmann J, Kimetu JM, Lehmann J. Stability and stabilisation of biochar and green manure in soil with different organic carbon contents. *Soil Res* 2010;**48**(7):577–85. https://doi.org/10.1071/SR10036.

156. Karhu K, Mattila T, Bergström I, Regina K. Biochar addition to agricultural soil increased CH4 uptake and water holding capacity—results from a short-term pilot field study. *Agric Ecosyst Environ* 2011;**140**(1):309–13. https://doi.org/10.1016/j.agee.2010.12.005.

157. Rondon MA, Lehmann J, Ramírez J, Hurtado M. Biological nitrogen fixation by common beans (Phaseolus Vulgaris L.) increases with bio-char additions. *Biol Fertil Soils* 2007;**43**(6):699–708. https://doi.org/10.1007/s00374-006-0152-z.

158. Zimmerman AR, Gao B, Ahn M-Y. Positive and negative carbon mineralization priming effects among a variety of biochar-amended soils. *Soil Biol Biochem* 2011;**43**(6):1169–79. https://doi.org/10.1016/j.soilbio.2011.02.005.

159. Spokas KA, Reicosky DC. Impacts of sixteen different biochars on soil greenhouse gas production. *Ann Environ Sci* 2009;**3**:179–93.

160. Brennan RB, Healy MG, Fenton O, Lanigan GJ. The effect of chemical amendments used for phosphorus abatement on greenhouse gas and ammonia emissions from dairy cattle slurry: synergies and pollution swapping. *PLoS One* 2015;**10**(6), e0111965. https://doi.org/10.1371/journal.pone.0111965.

161. Dias BO, Silva CA, Higashikawa FS, Roig A, Sánchez-Monedero MA. Use of biochar as bulking agent for the composting of poultry manure: effect on organic matter degradation and humification. *Bioresour Technol* 2010;**101**(4):1239–46. https://doi.org/10.1016/j.biortech.2009.09.024.

162. Kammann CI, Linsel S, Gößling JW, Koyro H-W. Influence of biochar on drought tolerance of chenopodium quinoa willd and on soil–plant relations. *Plant Soil* 2011;**345**(1–2):195–210. https://doi.org/10.1007/s11104-011-0771-5.

163. Palansooriya KN, Ok YS, Awad YM, Lee SS, Sung J-K, Koutsospyros A, et al. Impacts of biochar application on upland agriculture: a review. *J Environ Manag* 2019;**234**:52–64. https://doi.org/10.1016/j.jenvman.2018.12.085.

164. Dunlop SJ, Arbestain MC, Bishop PA, Wargent JJ. Closing the loop: use of biochar produced from tomato crop green waste as a substrate for soilless, hydroponic tomato production. *HortScience* 2015;**50**(10):1572–81. https://doi.org/10.21273/HORTSCI.50.10.1572.

165. Shackley S, Hammond J, Gaunt J, Ibarrola R. The feasibility and costs of biochar deployment in the UK. *Carbon Manag* 2011;**2**(3):335–56. https://doi.org/10.4155/cmt.11.22.

166. Lehmann J, Czimczic C, Laird D, Sohi S. Stability of biochar in soil. In: *Biochar for environmental management: science and technology*. Routledge; Taylor and Francis Group; 2009. p. 183–205.

167. Abnisa F, Wan Daud WMA. A review on co-pyrolysis of biomass: an optional technique to obtain a high-grade pyrolysis oil. *Energy Convers Manag* 2014;**87**:71–85. https://doi.org/10.1016/j.enconman.2014.07.007.
168. Vanapalli KR, Bhattacharya J, Samal B, Chandra S, Medha I, Dubey BK. Single-use LDPE—eucalyptus biomass char composite produced from co-pyrolysis has the properties to improve the soil quality. *Process Saf Environ Prot* 2020;**149**:185–98. https://doi.org/10.1016/j.psep.2020.10.051.
169. Li C, Xiong Y, Qu Z, Xu X, Huang Q, Huang G. Impact of biochar addition on soil properties and water-fertilizer productivity of tomato in semi-arid region of inner Mongolia, China. *Geoderma* 2018;**331**:100–8. https://doi.org/10.1016/j.geoderma.2018.06.014.
170. Hameeda, Gul S, Bano G, Manzoor M, Chandio TA, Awan AA. Biochar and manure influences tomato fruit yield, heavy metal accumulation and concentration of soil nutrients under wastewater irrigation in arid climatic conditions. *Cogent Food Agric* 2019;**5**(1):1576406. https://doi.org/10.1080/23311932.2019.1576406.

> **CHAPTER TEN**

Structure and function of biochar in remediation and as carrier of microbes

Kim Yrjälä[a,c,*] and Eglantina Lopez-Echartea[b]

[a]Zhejiang A & F University, State Key Laboratory of Subtropical Silviculture, Hangzhou, Zhejiang, China
[b]Department of Biochemistry and Microbiology, University of Chemistry and Technology, Prague, Czech Republic
[c]Department of Forest Sciences, University of Helsinki, Helsinki, Finland
*Corresponding author: e-mail address: kim.yrjala@helsinki.fi

Contents

1.	Introduction	264
2.	Structures of biochar	265
	2.1 Chemical elements	265
	2.2 The surfaces of biochars	267
3.	Application of biochar in remediation	274
	3.1 Inorganic contamination	274
	3.2 Organic contamination	276
	3.3 Phytoremediation	279
	3.4 Application of biochar on advanced oxidation process	282
4.	Microbes and biochar	283
	4.1 Microbial colonization of biochar	283
	4.2 Biochar as carrier of microbes	285
References		288

Abstract

Degradation of land and contamination of soils with environmental pollutants is a continuous threat to the wellbeing of man and the environment. Prevention of pollution is still not priority and cost-effective means of cleaning up contaminated land and water are much in demand. Many technological means are available for cleanup, but usually they are very costly and intended remediation tends to be postponed due to lack of funds. New nature-based materials are appearing that should be cost-effective in remediation but need practical development since they require extra knowledge about structure-function relations. In situ or on-site bioremediation can be implemented using plants and associated beneficial microbes. The development of biochars for use in environmental remediation is emerging for both inorganic and organic pollutants and their combination with phytoremediation is a good opportunity. This chapter presents the current knowledge in the area of sustainable remediation and discuss the possibilities of using biochar as a carrier of remedial microbes.

Advances in Chemical Pollution, Environmental Management and Protection, Volume 7
ISSN 2468-9289
https://doi.org/10.1016/bs.apmp.2021.09.002

Copyright © 2021 Elsevier Inc.
All rights reserved.

263

Keywords: Phytoremediation, Bacterial immobilization, Pyrolysis, Biochar structure, Heavy metals, Organic contaminants

1. Introduction

About sixty years after Rachel Carson published her famous book 'Silent spring'[1] and with it the beginning of the environmental movement, the problem with pesticides, herbicides, coloring dyes from the industry and now pharmaceuticals still pose a clear threat to human and environmental health. Carson's book was evidently an inspiration for the emergence of both the sustainability idea and the circular economy concept, where recycling of industrial products is in focus.[2] The organic waste is becoming a resource for designing and making new products for bioenergy and bio-based smart materials[3] Thermal treatment of biomass is receiving increasing attention for producing new carboneous products like biochar. When using pyrolysis for organic waste at 300–650 °C in oxygen-depleted atmosphere, the biomass carbon gets transformed to biochar, oil and gas, where the process conditions and source feedstock determine the ratios of these three products. Different waste streams at local municipality level can be detected, collected and innovatively used for such thermal treatment, e.g., coconut waste, rice husk, other agricultural waste or wood bark. The water content of agroforestry waste requires pretreatment before pyrolysis, which requires energy and resources.[3] Biochars produced from agricultural and forestry waste have interestingly been used in environmental remediation of both organic and inorganic contamination in soils and water. This is due to the adsorptive characteristics of porous biochars that can bind both, organic compounds like pesticides and petroleum compounds, and also heavy metals like Cd, Zn, Cu and Cr.[4] Microbes are involved in all these processes, influencing the state of contaminants and in the case of organics they are even able to degrade recalcitrant aromatic compounds like PAHs, benzenes and toluene.[5] Along with an increasing amount of biochar related publications about its application to remediation, it has become evident that it is possible to design biochars for specific uses. This is possible through careful choice of feedstock, suitable activation of the manufactured carbon and optimization of pyrolysis conditions.[6] The porosity of biochar having large surface areas is important for adsorbing contaminants, but also acts as surfaces for microbes to attach. Very little is known, however, about how the surface structure of biochar favors certain microbes providing a microenvironment for them to multiply and stay in close contact with this stable carboneous compound in soil.

This review provides an in-depth assessment of the carbon structure effects on remediation, seeking suitable designer biochars for use in environmental remediation, considering different waste streams as a source of feedstock, and how microbes can improve decontamination. The large surface area of biochar can provide attachment sites for microbes and engineered carbon could function as a carrier of microbes in environmental clean-up applications. An interesting option is biochar-enhanced phytoremediation of soil contamination, where the biochar may relieve plant stress caused by contaminants and lower their bioavailability.

2. Structures of biochar

Biochar is a solid, carbon-rich product acquired from pyrolysis under an oxygen-limited atmosphere and high temperature.[7] The structure of biochar is fundamental for the endeavor in developing innovative practical environmental use or for industrial use as carbon sequestration material.

The attention for biochar applications in agriculture and forestry, environmental remediation, and materials science[8] is increasing. This includes the development of functional materials taking advantage of relatively cheap biochar and its sustainable properties.[9] Biochars from different sources would enable multipurpose applications, but that requires deeper understanding of biochar development for specific tasks. Biomass from multiple sources are available for biochar production with targeted purposes. The question is, how to choose the right precursors to obtain biochars for specific purposes.[6] During pyrolysis, the structure of biomass transforms and develops in various ways that shapes the properties of biochar. For adequate biochar development, a structure and function knowledge is obligatory and the systematic evaluation of suitable sources and design of biochars is asked for. A great opportunity, with increasing structure function information is to not only to functionalize biochar physically and chemically, but to use it as a carrier for microbes in field applications.[10]

2.1 Chemical elements

Heterogeneous biochar can according to Xiao et al.[6] be defined as an aggregation of multiple molecules, containing both, skeletal and small extracted molecules. The major structures of biochar contain commonly: C, H, O, and N, which actually are part of global geochemical cycles. In specific biochars P, Si, S and Fe occur at different amounts and some of these elements are nutrients for plants.

Text box, Carbon. The elements are generally carbonate and bicarbonate in the inorganic phase and of aliphatic carbon. Aromatic carbon and functionalized carbon is in the organic phase.[11] Cellulose/hemicellulose and lignin biomass becomes first aliphatic carbon and later at high pyrolysis temperatures they become aromatic by dewatering and cracking reactions.[12]

Text box, Silicon. Rice straw, rice husk, and corn straw are good feedstock for biochar production yielding silicon.[13] The silicon becomes dehydrated and polymerized in charring and upon further pyrolysis partially crystallized. Silicon in biochars contribute to the sorption and retention of heavy metals in aqueous systems.[14]

Text box, Hydrogen. Three types of hydrogen are present in biochars: aliphatic hydrogen, aromatic hydrogen, and active functionalized hydrogen. Hydrogen bonds $(D-H\cdots:A)$ form between surface of biochar and ionizable molecules by sharing protons. Hydrogen is an important building block of biochar structures connected to ionizable molecules in sorption.[15]

Text box, Oxygen. Oxygen is contained in metallic oxides, hydroxides and in inorganic constellations such as carbonate, bicarbonate, and sulfate. This oxygen is largely stable resulting in alkalinity of biochars. Since natural oxidation processes conveys oxygen into the organic structures of biochars, the O/C atomic ratio is considered as an indicator of aging/oxidization in biochars.[16]

Text box Nitrogen. Functional groups containing nitrogen are acylamide, amine, imine, nitro, nitroso, pyridine and pyrrole groups. Nitrogen-sorption, nitrification, nitrogen fixation as well as N2O greenhouse gas emissions are essential for soil functioning in connection to biochar amendments. Biochars show a high affinity for ammonium nitrogen (NH_4^+) in cation-exchange, but no clear influence on nitrate-N removal.[17] Biochars can sorb ammonia that becomes bioavailable which makes it potent to maintain nitrogen in soil.[18]

Text box Phosphorus. Phosphorus does not volatilize at pyrolysis temperatures for biochar. Common pyrolysis temperatures tend to enrich phosphorus in biochar. Ngo et al.[19] reported an overwhelming dominance of inorganic phosphorus compared to organic phosphorus in bamboo-derived biochar produced at 500–600 °C.

2.2 The surfaces of biochars

The chemical and biological reactions and interactions occur on the surface of biochar. The surfaces have aliphatic and aromatic functional groups including hydroxyl, epoxy, carboxyl, acyl, carbonyl, ether, ester, amido, sulfonic and azyl groups. The functional groups on biochar provide sorption sites for metal cations[14] and ionizable organic pollutants.[20]

2.2.1 Surface area and porosity

Metal sorption capacity of biochar resides mainly on the surface area and porosity. During pyrolysis of biomass, micropores form due to water loss in dehydration process.[21] The pore varies from nano- (<0.9 nm), micro- (<2 nm), to macro-pores (>50 nm) that is important for metal sorption. The pyrolysis temperature is fundamental for surface structure, but the conformation of biochar feedstock is also important (Table 1).

2.2.2 pH and surface charge

Pyrolysis temperature and feedstock affect pH to similar extent as porosity and surface area that will contribute to usability in remediation (Table 2). The biochars are mostly alkaline with exceptions depending on feedstock.

Biochar's pH increases with increasing pyrolysis temperature (Fig. 1). Biochar's pH correlates positively with pyrolysis temperature in oak wood biochar,[24] biosolids, wheat, corn, and maize residues.[24]

Surface charge is an important property influencing metal sorption to biochar. In metal removal from water, the pH of solution has a strong influence on the surface charge. The zero charge (pHPZC) of biochar in solution is when pH results in net zero charge on the surface. When solution pH is $>$ pHPZC, then the biochar is negatively charged and binds to positively charged metal cations such as Cd(II), Pb(II), and Hg(II). When solution pH is $<$ pHPZC, biochar is positively charged and will then bind metal anions such as $HAsO_4^{2-}$ and $HCrO_4^-$. Using biochars produced from canola, corn, soybean and peanut straw at temperatures of 300, 500 and 700 °C the pH of solution was 3–7 and they all were negatively charged.[11]

2.2.3 Surface functional groups

There are different ways to treat available biomass to produce biochar that further needs activation and modification for different applications (Fig. 2). The abundance of surface functional groups in biochar correlates negatively

Table 1 Physic-chemical properties of biochars produced from various feedstock under different pyrolysis temperatures[22]

Feed stock	Temperature (C)	Surface area (m^2g^{-1})	Porosity (cm^3g^{-1})	pH	Atomic ratio			Content of mineral elements (%)				Reference
					H/C	O/C	N/C	K	Ca	Mg	P	
Wheat straw	300	116			0.55	0.29						23
	400	189										
	500	309			0.36	0.09						
	600	438										
	700	363			0.15	0.05						
Oak wood	350	450		4.84	0.06	0.26	0.001					24
	600	642		4.91	0.02	0.1	0.001					
Corn stover	350	293		5.88	0.07	0.37	0.013					24
	600	527		6.71	0.03	0.21	0.012					
Municipal biosolids	400	5.49		8.46	1.01		0.12					25
	450	7.21		9.74	0.87		0.12					
	500	7.73		9.75	0.68		0.11					
	550	8.45		10.5	0.58		0.11					
	600	5.99		11.7	0.43		0.09					
Maize straw	300	1	0.01	9.84	0.07	0.49	0.026					26
	450	4	0.01	10.5	0.06	0.35	0.023					
	600	70	0.06	11.4	0.03	0.26	0.02					

Material	Temp	C1	C2	C3	C4	C5	C6	C7	C8	C9	C10	Ref
Pine needle	100	0.65			1.44	0.62	0.012					27
	200	6.22			1.91	0.48	0.013					
	250	9.52			1.08	0.4	0.012					
	300	19.9			0.75	0.28	0.014					
	400	112	0.044		0.45	0.17	0.013					
	500	236	0.095		0.33	0.14	0.012					
	600	207	0.076		0.26	0.1	0.01					
	700	491	0.19		0.18	0.1	0.011					
Broiler litter manure	350	59.5	0									28
	700	94.2	0.018									
Municipal biosolids	500	25.4	0.056	8.81	0.48	0.45	0.075	0.85	5.93	1.47	1.82	29
	600	20.3	0.053	9.54	0.22	0.3	0.064	0.85	6.27	1.55	1.88	
	700	32.2	0.068	11.1	0.15	0.3	0.048	0.99	6.44	1.64	2.04	
	800	48.5	0.09	12.2	0.03	0.17	0.026	0.93	6.58	1.66	1.93	
	900	67.6	0.099	12.2	0.09	0.12	0.029	0.87	6.96	1.75	2.02	
Poultry litter manure	400	5.4	0.003	9.5				3.88	2.83	1.73	1.22	30
	600	6.3	0.003	10.4				5.88	3.59	2.4	1.54	
Swine manure	400	5.8	0.008	10				1.62	2.03	1.57	0.97	30
	600	10.6	0.011	10.4				3.53	2.89	2.13	1.55	

Continued

Table 1 Physic-chemical properties of biochars produced from various feedstock under different pyrolysis temperatures[22]—cont'd

Feed stock	Temperature (C)	Surface area (m^2g^{-1})	Porosity (cm^3g^{-1})	pH	Atomic ratio			Content of mineral elements (%)				Reference
					H/C	O/C	N/C	K	Ca	Mg	P	
Conocarpus wastes	200			7.37	0.06	0.41	0.011	0.04	4.34	0.34	0.08	31
	400			9.67	0.04	0.18	0.012	0.05	5.18	0.4	0.09	
	600			12.2	0.02	0.08	0.009	0.09	6.47	0.48	0.11	
	800			12.4	0.01	0.06	0.011	0.12	6.75	0.78	0.13	
Wastewater sludge	300			5.32	1.05	0.24	0.111	0.1	3.47	0.35	2.5	32
	400			4.87	0.76	0.17	0.102	0.11	4.17	0.43	2.8	
	500			7.27	0.52	0.02	0.09	0.18	4.62	0.46	3.3	
	700			12	0.3	0	0.05	0.2	5.35	0.54	3.6	
Oak wood	200			4.6			0.014	0.13	0.39	0.04	0.03	33
	400			6.9			0.021	0.38	1.18	0.15	0.06	
	600			9.5			0.029	0.44	1.39	0.18	0.06	
Wheat straw	200			6.11	1.42	0.58	0.018					34
	400			10.8	0.63	0.2	0.012					
	600			11	0.26	0.13	0.011					

Table 2 Diversity of biochar's feedstock for remediation of organic pollutants.

Feedstock	Pyrolysis temperature and time	Pollutant	Matrix	References
bulrush straw	3 h at 300 °C	PHCs	soil	Wang et al.[105]
ponderosa pine wood	900 °C	PHCs	soil	Mukome et al.[49]
wheat straw	6 h at 300 °C	PAHs		Cao et al.[50]
food waste digestate	100 min. at 800 °C	azo dye	waste water	Huang et al.[51]
food waste	7 h at 300 °C	organic dyes	waste water	Chu et al.[52]
pine	750 °C	PFAS	loamy sand and sandy clay loam	Askeland et al.[53]
waste timber	500–650 °C	PFAS	soil	Silvani et al.[54]
coconut shell	No information	PFAS	soil	Silvani et al.[54]
wood shrub	600 °C	PFAS	soil	Silvani et al.[54]
pine and spruce pellet	700 °C	octadecane and octadecanoic acid	sandy soil	Hallin et al.[106]
rice husk	± 600 °C	pyrene, PCB and p,p'-DDE	soil	Bielská et al.[55]
mixed wood shavings	20 min at 700 °C	pyrene, PCB and p,p'-DDE	soil	Bielská et al.[55]
maize straw	350 °C	PBDE	liquid solution	Du et al.[56]
wheat straw	500 °C	hexachlorobenzene	soil	Song et al.[64]

Continued

Table 2 Diversity of biochar's feedstock for remediation of organic pollutants.—cont'd

Feedstock	Pyrolysis temperature and time	Pollutant	Matrix	References
caragana	4 h at 500 °C	PCP	liquid cultures	Zhang et al.[57]
empty fruit bunches	1, 2 and 3 h at 300, 500 and 700 °C	imidazolinones	soil	Yavari et al.[107]
rice husk	1, 2 and 3 h at 300, 500 and 700 °C	imidazolinones	soil	Yavari et al.[107]
Different bark types	750 and 900 °C	organic micropollutants	wastewater	Hagemann et al.[59]
cotton gin waste	2 h at 350, 500 and 700 °C	pharmaceuticals	liquid solution	Ndoun et al.[108]
guayule bagasse	2 h at 350, 500 and 700 °C	pharmaceuticals	liquid solution	Ndoun et al.[108]
bamboo biomass	2 h at 380 °C	sulfonamide antibiotics	liquid solution	Ahmed et al.[109]
bamboo wood	2 h at 380 °C	chloramphenicol	deionized water, lake water and synthetic wastewater	Ahmed et al.[110]
eucalyptus wood	2 h at 380 °C	chloramphenicol	deionized water, lake water and synthetic wastewater	Ahmed et al.[110]
holm oak	500 °C	tetracyclines and sulfonamides	liquid solution	García-Delgado et al.[111]

Structure and function of biochar in remediation

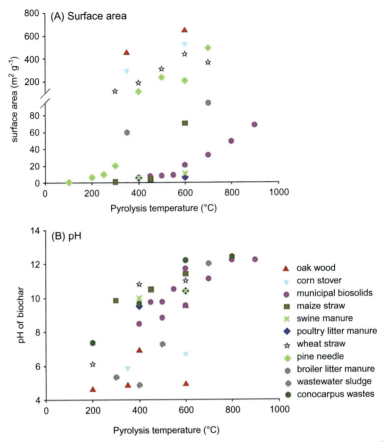

Fig. 1 Surface area and pH of biochars produced from various feedstock at pyrolysis temperature ranging from 100 °C to 900 °C. Data are from Table 1.[22]

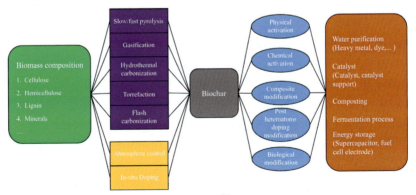

Fig. 2 Production and application of biochar.[36]

with increasing temperature, primarily due to higher degree of carbonization. With increasing temperature, atomic ratios of H/C, O/C, and N/C decrease, suggesting decrease in abundances of hydroxyl, carboxylic, and amino groups. Biochar produced at a relatively high pyrolysis temperature absorbs organic pollutants better due to increased surface areas, microporosity and hydrophobicity, notwithstanding the biochar yield, is reduced.[35] Biochar produced at a low temperature is more suitable for inorganic or polar organic pollutants having oxygen-containing functional groups, promoting electrostatic attraction and precipitation.[37]

3. Application of biochar in remediation
3.1 Inorganic contamination

According to the national survey of soil pollution in China in 2014, the percentage of soil pollution sites (both agricultural and industrial) that exceeded the second level of Soil Environmental Quality Standard for the permitted limits were 16.1%. Among these sites 82.8%, were HM pollution sites.[38] Biochar can effectively decrease HMs uptake by crops[39–41] reducing the mobility/bioavailability of HMs in contaminated soils. The mechanisms of how biochar influences HM mobility/bioavailability are less known just as the potential risks associated with biochar application.[38] The diverse sources of raw materials or feedstock and the diverse pyrolysis temperatures may in turn affect binding between biochar and metals. That include differences in, e.g., pH, organic carbon content, cation exchange capability (CEC), microporous structure, specific surface area (SSA), active functional groups and mineral contents. Biochar may influence both metal mobility and bioavailability in soils by[1]: direct interactions between metals and biochar, and[2] impacting soil characteristics and thus indirectly affecting metal availability, i.e., indirect interactions (Fig. 3).

Potassium containing compounds (e.g., KOH and K_2CO_3) in chemical activation have been propagated as useful method for modification of biochar to increase the total surface area and porosity, but also to add active functional groups to pristine biochar.[25,42] Studies have emphasized catalytic effect during pyrolysis limiting the formation of volatile compounds that improve the biochar yield.[43] The use of biochars of potassium (K)-rich feedstock like banana peels (BB) and cauliflower leaves (CB) affected formation of minerals in biochar.[44] Multi-element Copper (Cu(II)), Cadmium (Cd(II)) and Led (Pb(II)) sorption experiments revealed adsorptive property of BB and CB biochars. Biochar adsorption was quite high (61.4% and

Structure and function of biochar in remediation

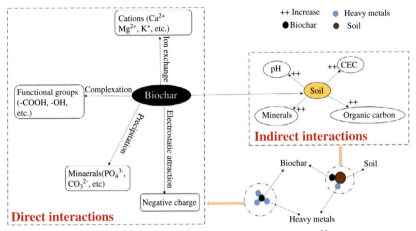

Fig. 3 Possible mechanisms of biochar-metal interactions in soils.[38]

64.6% for BB and CB, respectively) evidently due to the catalytic action of K according to XRD patterns during the pyrolytic conversion of the feedstock. Electrostatic attraction was reported as the predominant mechanism for metal sorption capacity of BB. The formation of K-minerals during pyrolysis can contribute in sorption process through precipitation onto these minerals or exchanging K^+ with other Cu(II), Pb(II) and Cd(II) ions.

As and Pb contamination in soil in soil is challenging. Mechanisms involved in carbon dynamics, soil quality and bioavailability of As and Pb after a 473-day amendment was the aim in a pot study.[45] Sandy soils were treated with three different biochars, rice straw biochar (RB), wood biochar (WB) and grass residue biochar (GB). Each of three biochars (i.e., GB, RB, and WB) was applied to 100 g dry weight of each soil at a rate of 30 t ha − 1 (22.2 g kg − 1 soil). The application of RB, WB and GB significantly decreased the bioavailability of Pb by 36.8%, 82.6%, and 58.3%, respectively, in the sandy soil. Possible suggested mechanisms include ion exchange, physical adsorption, electrostatic attraction, reduction, (co)precipitation, and complexation. Suggested mechanisms for Pb immobilization in the sandy loam soil were: (I) the electrostatic attraction of the Pb cations on the negatively charged biochar surfaces, and (II) Pb-biochar surface complexation, as divalent elements, with affinity to form complexes with O-containing functional groups. Application of the biochars to the sandy soil also enriched the microbial communities where the humification index was higher in the biochar treated soil. The increased cumulative CO_2-C in the sandy soils treated with biochars correlated with the enriched microbial communities. All biochars significantly reduced the Pb bioavailability in the sandy loam soil.

Multimetal contamination is common and the effect of modified coconut shell biochar (MCSB) on the availability of metals and soil biological activity in multi-metal (cadmium (Cd), nickel (Ni) and zinc (Zn)) contaminated soil was evaluated.[46] Hydrochloric acid pickling and ultrasonication significantly improved surface functional groups and pore structure. In 63-day incubation at 25 °C using 0%, 2.5% and 5% addition of MCSB or CSB, the acid soluble Cd, Ni and Zn decreased by 30.1%, 57.2% and 12.7%, respectively, in highest 5% MCSB addition. This showed that modified biochar immobilized metals better than unmodified CSB. The maximum bacterial number in 5% MCSB treatment increased by 149.4% compared to control.

Bamboo has been put forward as an important and suitable plant for use in sustainable development.[47] In a laboratory leaching study the effect of biochars derived from bamboo (BB) and pig (PB) were investigated with dibutylphtalate (DBP), Cd and Pb.[48] Soil columns with low or high organic carbon (LOC; 0.35% C: HOC; 2.24% C) were spiked with DBP, Cd, and Pb. Addition of PB to the LOC soil significantly ($P < 0.05$) reduced the leaching by up to 88% for DBP, 38% for Cd, and 71% for Pb, but influence was insignificant in the HOC soil. The higher efficacy of PB to reduce the leaching of DBP, Cd and Pb in the LOC soil was suggested to be related to PB's higher specific surface area, and the surface alkalinity. They thought pH and minerals in PB contributed to the higher sorption effect. Leaching of DBP, Cd and Pb were significantly ($P < 0.05$) higher in the LOC soil than in the HOC soil. This study revealed that the effectiveness of biochars was dependent on the soil organic carbon content.

3.2 Organic contamination

3.2.1 Biochar remediation of petroleum hydrocarbons

Contamination by petroleum hydrocarbons (PHCs) is of great environmental concern, as these are one of the most widely used, transported and spilled chemicals worldwide. Oil spills affect the surrounding vegetation and they can even leak to the groundwater. A recent study found that different types of biochars in combination with fertilizer were an effective remediation strategy for crude oil contaminated soil.[49] This study highlighted the importance of biochar feedstock as samples amended with ponderosa pine wood biochar achieved the highest degradation rate, but the walnut shell biochar was inhibitory for PCH biodegradation (Fig. 4).[49]

Structure and function of biochar in remediation

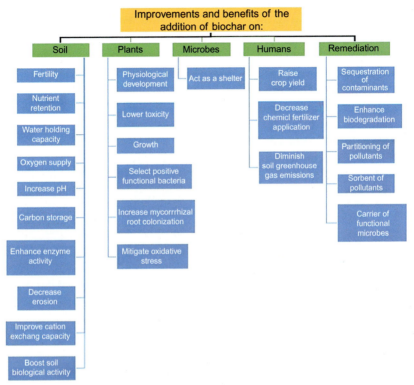

Fig. 4 Improvements and benefits on soil, microbes, plants and on bioremediation upon the addition of biochar.

3.2.2 Biochar remediation of other organic pollutants and emerging contaminants

The removal of other organic pollutants such as polycyclic aromatic hydrocarbons (PAHs), dyes, fungicides, emerging contaminants or micropollutants constitute a big challenge. Based on the properties of biochar and its proven efficiency on removing PHCs it also has successfully been tested on other organic contaminants. Remediation of PAH-contaminated soil with wheat straw biochar revealed that this type of biochar inhibited the removal of three-ring phenanthrene but accelerated four-ring benzo[a] pyrene removal.[50] The addition of the biochar may inhibit enzymes activity in soil like dehydrogenases and stimulate others like polyphenol oxidase.[50] Other type of organics like azo dyes used in the textile industry have effectively been treated with biochar. Huang et al.[51] tested the addition of food waste digestate biochar and peroxymonosulfate to treat wastewater containing azo dye pollutant. They achieved a >99% removal already in

10 min. They attributed the high removal efficiency to the catalytic sites in the biochar, which could activate peroxymonosulfate to produce reactive oxygen species. Food waste magnetic biochar was used as a catalyst for removal of organic dyes with a heterogeneous Fenton-like processes alone or in combination with H_2O_2 and ultrasound Chu et al.[52] A combination of ultrasound, H_2O_2 and the magnetic biochar as catalyst gave best results.

A study on pine derived biochar sorption kinetics of *per-* and poly-fluoroalkyl substances (PFAS) in two types of soil revealed that in all cases, the addition of biochar reduced the leachable fraction of such compounds.[53] Silvani et al.[54] carried out another PFAS study testing different types of biochars (activated and unactivated) remediation with positive results. Activated biochar amendments improved the removal compared to unactivated biochar, and the study identified total organic carbon as a key determinant for the removal.[54] The adsorption of pyrene, polychlorinated biphenyl (PCB) and dichlorodiphenyldichloroethylene (p,p'-DDE) was investigated using two biochars (mixed wood shavings–high technology and rice husk–low technology production processes).[55] Both biochars had similar adsorption for single compounds and the mixture in the presence and absence of soil.[55]

Maize straw biochar improved removal of polybrominated diphenyl ether (PBDE) from soil by adsorption and its efficiency increased with addition of 4-monobromodiphenylether-degrading strain *Sphingomonas* sp. DZ3 immobilized into the biochar.[56] Reductive dechlorination of PCP was studied with several types of biochar (rice husk, bamboo, caragana, and garbage) and only caragana-derived biochar showed stable electron transfer activity for PCP dechlorination.[57] The study concluded that the beneficial effects of caragana biochar on the dechlorination of PCP depended mainly on the electrical conductivity of the biochar.

The efficacy of using a combined biofilm-biochar approach was evaluated to remove organic (naphthenic acids (NAs)) and inorganic (metals) contaminants from process water (OSPW) generated by Canada's oil sands mining operations.[58] A microbial community obtained from an OSPW sample was cultured as biofilms on carbonaceous materials. Biochar samples, from softwood bark (SB) and Aspen wood (N3) were used for NA removal studies performed with and without biofilms and in the presence and absence of contaminating metals. SB and N3 assays showed similar NA removal in 6-day experiment (>30%). Biodegradation by SB-associated biofilms increased NA removal to 87% in the presence of metals.

Biofilm-associated biochar showed metal sorption and up to four times higher immobilization of Fe, Al and As. The results suggested this combined approach is promising for OSPW treatment.

Organic micropollutants (OMPs) bioremediation has not yet been studied much and the impact of biochar upon these pollutants has just recently been addressed. Hagemann et al.[59] compared activated carbon and wood-based activated biochar removal of OMPs in wastewater treatment. The results suggested that activated biochar had similar or even better removal of OMPs and dissolved organic carbon than commercial activated carbon. The study emphasized the possibility of using locally produced activated biochar as a strategy to improve waste water treatment and reduce the negative impact of using activated carbon.[59]

3.3 Phytoremediation

Phytoremediation is a technology that has successfully been applied to treat several types of pollutants. This bioremediation technology has been used with different soil amendments such as biosurfactants, compost, biochar or a specific degrading bacterial consortia to improve the removal of contaminants.[60] These amendments tend to increase pollutants bioavailability, soil nutrients, humidity, and enhance bacterial degradation. Specifically biochar has been used in phytoremediation to improve plants growth and physiological development.[61] Biochar enhanced phytoremediation also tends to select beneficial functional bacteria that could prevent nutrient loss and benefit plant nutrition.[62] Biochar has enhanced biodegradation of petroleum hydrocarbons[63] and it may remediate pollutants by adsorption, partitioning and sequestration. Biochar can serve as a mechanism to decrease the bioavailability of organic contaminants such as PCBs in soil.[61]

3.3.1 Biochar and phytoremediation of organic pollutants

As mentioned earlier the application of biochar to soil has several benefits for plants and their use in phytoremediation has proven to be very effective. For instance the addition of compost, biochar and hydrocarbon degrading bacterial consortia with Italian ryegrass removed up to 85% of total PHCs in soil.[60] Song et al.[64] found that the joint action of ryegrass in biochar-amended soils enhanced the removal of hexachlorobenzene especially in the rhizosphere. The study suggested that the biochar-plant synergy could be an effective strategy for the treatment of soils contaminated with volatile and semi-volatile organic contaminants.[64] Biochar has been reported as reducer of toxicity of PHCs for plants such as *Spartina anglica* when it improved plant growth, increased shoot height, root vitality and total chlorophyll.[65] This study also reported that PHCs degradation efficiency improved with planting of *S. anglica* and with the addition of rice husk-biochar, rhamnolipid and rhamnolipid modified biochar.[65] The addiction of biochar increased the plant growth and the degradation of PCHs in soil was more effective.[66]

3.3.2 Biochar and phytoremediation of heavy metals

Plants can extract or stabilize HM from contaminated sites through different mechanisms: phytoextraction, phytoaccumulation, phytostabilization or phytovolatilization.[67,68] Table 3 presents detailed information about phytoremediation strategies applied to the remediation of heavy metals. In general phytoremediation is cheap, aesthetic, receives public acceptance and is environmental friendly.

Biochar capacity to aid in remediation of HMs is attributed to its high porosity, high binding capacity, nutrient retention, pH–precipitation effect and great water holding capacity.[68,69] Biochar itself cannot decrease the concentration of HMs in soil but it can reduce the bioavailability and phytotoxicity.[37] Consequently, plants would have a much better chance of extracting HMs from soil and therefore lowering their concentrations from soil. The following section provides some examples of studies about use of biochar in phytoremediation.

One of the most abundant and toxic heavy metals in the environment is Cd. Several plant species in combination with biochar have been tested for its removal. Alphalpha in combination with biochar decreased soil available Cd.[70] Tea waste biochar was beneficial to the phytoremediation with ramie of Cd contaminated sediments.[71] This type of biochar also alleviated Cd induced growth inhibition and oxidative damage in ramie seedlings by reducing reactive oxygen species accumulation.[71]

Due to its toxic effects to humans and wildlife, lead is one of the heavy metals with the highest concern and consequently its remediation is urgent. Cd contamination is threatening rice cultivation in certain areas of China. The reduction of Cd and Pb uptake by rice (*Oryza sativa* L) was aimed in a contaminated field.[72] The soil amendments used included lime, DaSan Yuan (DASY), DiKang No.1 (DEK1), biochar, Fe-biochar, Yirang, phosphorus fertilizer, (Green Stabilizing Agent) GSA-1, GSA-2, GSA-3, and GSA-4, applied at 1% rate in a field experiment. GSA-4 treatment showed best effects to reduce Cd and Pb phytoavailability in soil and uptake by early

Table 3 Phytoremediation strategies for remediation of heavy metal contaminated sites[69]

Phytoremediation Strategies	Description and Objective	Possible target HM contaminants	Some of the plant species for phytoremediation	Limitations	Utilization of Phytoremediated biomass
Phytoextraction (Phytoaccumulation)	Accumulation of heavy metal pollutants in harvestable biomass, i.e., shoots. It is a promising technique and is also known as phytomining	Pb, Ni, Cu, Hg, Cr, Cd, and As	*Alyssum lesbiacum, Arabidopsis halleri, Helianthus annuus, Tagetes patula, Celocia cristata,* and *Brassica napus*	Dependence on the growing conditions essential for plants and microorganisms, a long period is required to completely remediate the sites	Biodiesel production, ornament, and oil production.
Phytostabilization	Limiting the mobility and bioavailability of heavy metal pollutants in soil by plant roots. It does not reduce the concentration of heavy metals present in contaminated soil but prohibits their off-site movement	Zn, Cd, Pb, Zn, and Cu	*Silene vulgaris, Agrostis capillaris, Festuca* spp., and *Agrostis spp.*	Phytostabilization needs continuous monitoring to avoid undesired HM mobilization or leaching.	Animal feed and human food
Phytovolatilization	The process of conversion of toxic HM pollutants into a gaseous form and their discharge into the atmosphere.	Hg and Se	*Arabidopsis thaliana, Liriodendron tulipifera, Brassica pekinensis, Brassica juncea, H. annuus,* and *Nicotiana tabacum*	Limited to only volatile HM (Hg and Se) pollutants, it does not remove the pollutant completely as it is transferred from one segment (soil) to another (atmosphere).	Animal feed, human food, model organism (*A. thaliana*), and medicine.
Rhizofiltration	Using the plant root systems to absorb, concentrate, and precipitate HM contaminants.	Pb, Cd, Cu, Ni, Zn and Cr	*Helianthus spp., N. tabacum, Spinacia oleracea, Secale cereale,* and *B. juncea*	Plants need to be grown in a greenhouse first and then transferred to the remediation site.	Animal feed, human food, and medicinal use.

rice Treatment with GSA-4 improved rice growth (56%) and grain yield (42%). The enhancement effects on grain yield may be result from the positive effects of GSA-4 application on increasing photosynthesis (116%) and transpiration rate (152%) as compared to the control. Significant reduction in Cd and Pb uptake in shoot (42% and 44%) and in grains (77% and 88%), was observed. Biochar reduced metal availability by increasing surface sorption on biochar. Possible mechanisms of reduced metal phytoavailability by biochar application include precipitation of metal-organo compounds.. Negative correlation was recorded between DTPA extractable Cd/Pb and soil pH that directly depended on applied amendments. The combined amendment (GSA-4) immobilized heavy metals effectively in contaminated paddy field to secure safe rice production.

Usually heavy metal contamination comes in a mixture of metals and metalloids, especially from mining, and remediation of mixed contamination is of great importance. Mixed feedstock biochar had positive effects on phytoremediation with *Solanum nigrum* towards a mine tailings containing several heavy metals including Cu, Zn, Cd, Hg, Pb and Mn.[73] The study demonstrated that the application of biochar increased metal uptake by plant roots and decreased phytotoxicity.[73] Another phytoremediation field study with biochar and several plant species demonstrated that soybean displayed best results towards the extraction of most of the HMs in the mining tailings treated.[74] The addition of an organic fertilizer improved the phytoremediation. The results showed that phytostabilization was the main mechanism of phytoremediation and that this could be an alternative to conventional remediation methods.

In several studies amendments in combination with biochar has brought better removal efficiencies, e.g., ceramist, organic fertilizer, compost, ash, microbial inoculation and attapulgite[73,74]. Thus, the combined use of biochar, some amendments and a HM-tolerant plant species could be beneficial in the phytoremediation of HMs.

3.4 Application of biochar on advanced oxidation process

In environmental remediation, biochars have lately been used as catalysts for advanced oxidation processes (AOP). Several studies report degradation of pollutants, where biochar and biochar-based catalysts carry out advanced oxidation reactions.[75] Biochar with abundant surface functional groups, porous structures and high specific surface areas can donate, accept, or transfer electrons to the surrounding environment.[76] This renders biochar to be a

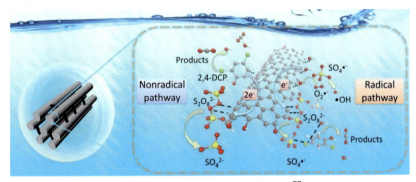

Fig. 5 Mechanism of persulfate activation in two pathways.[77]

good candidate as AOPs catalyst. According to He et al.[4] the biochar application in AOPs can be classified as (1) degradation of pollutants via biochar/H_2O_2 or biochar/PS system; (2) oxidation reaction without radicals.[77] Fig. 5 shows two degradation mechanisms of pollutants in biochar/PS system.

4. Microbes and biochar

The impact of biochar on soil microbes depends on the precursors of the biochar and soil type.[78] Biochars may act as material for electric conductivity. Co-cultures of *Geobacter metallireducens* with *Geobacter sulfurreducens* or *Methanosarcina barkeri* were attached to the biochar, but not in close contact. The authors suggested that electrons conduct through the biochar, rather than biologically. Activated carbon previously showed this type of electron transfer in materials due to their aromatic structures, which enhances the metabolism in microorganisms. Moreover, biochar has also been reported as habitat for immobilized bacteria to enhance the bioremediation of PAH-contaminated soil.[79] The immobilized bacteria could directly degrade the carrier-associated PAHs, phenanthrene and pyrene.

4.1 Microbial colonization of biochar

Field aged biochar was tested for microbial colonization with the idea of using biochar as a carrier for microbes.[80] The microbial colonization of field-aged biochar in this study was scant, with no clearly observed differences between the external and internal surfaces. After estimation of the total pore space in soil higher field application rate of 50 t ha-1 biochar displayed only 6.5% of the total soil pore space. In the top soil 7.3 + − 0.8% of the total soil surface area was of biochar origin (0–30 cm). They observed that 17.5% of the biochar pores were not readily habitable based on SEM analysis. This study[80] suggested that in <3 y perspective, biochar does not provide a substantial habitat for soil microbes. They reasoned that biochar is largely unavailable to soil microbes and the introduction of labile compounds into the 'charosphere' could change the soil microbial activity providing food for metabolism that may affect the community structure and soil-plant-microbe interactions. They suggested that in the long term, microbial colonization of biochar might be much greater when disintegration and microbial partial decomposition of the biochar will result in nutrient source and a habitat for microbes (Fig. 6).

Cadmium (Cd) contamination is a problem in several agricultural areas in several countries, e.g., China, Thailand.[81] When edible crop plants grow in contaminated soil, they can accumulate Cd, which through the food chain may reach the consumers.[82] In phytoremediaton, improvement Cd phytoextraction efficiency with inoculation of specific bacteria have widely been studied.[83,84] Many studies have used biochar for cell mobilization of organic pollutants, e.g., polycyclic aromatic hydrocarbons[79] and

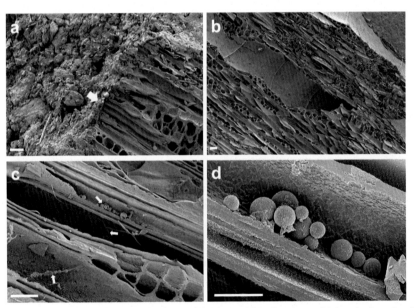

Fig. 6 Scanning electron micrographs of field-aged biochar buried in agricultural soil for 3 years. The biocharesoil interface on the outer surface, with arrow showing an example of pore blockage (A); spatial heterogeneity and sparsity of internal microbial colonization (B); internal colonization by hyphal and single-celled microbes (arrows) (C and D). Scale bar ¼ 20 mm (aec) and 5 mm (D).[80]

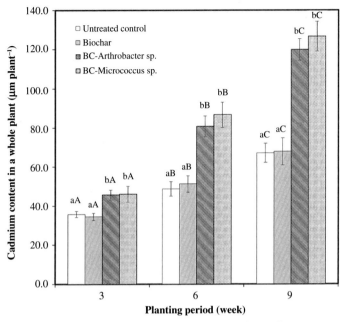

Fig. 7 Cadmium accumulation in a whole *C. laxum* planted in cadmium contaminated soil. The error bars are the SE (n = 3), and a different lowercase letter above a bar in the graph indicates a significant difference at $P < 0.05$ among different treatments at each planting period. A different uppercase letter indicates a significant difference at $P < 0.05$ among different planting periods according to Duncan's multiple range test.[86]

pesticides.[85] Chuaphasuk et al.[86] suggested that a low amount of biochar application in cadmium-contaminated soil does not change cadmium bioavailability in soil and it can serve as a bacterial habitat for cell immobilization (Fig. 7) Biochar application in the soil has positive effects on microbial groups and acts as a shelter for bacteria.[79,87] They concluded that it is suitable to use biochar as a natural carrier for bacterial immobilization.

4.2 Biochar as carrier of microbes

Bacterial fertilizers can improve the performance of plants in different climate conditions.[88,89] Plant growth promoting bacteria (PGPR) are suitable inoculants in the soil and the rhizosphere for beneficial symbiosis. For extending PGPR colonization ability various carrier substrates are needed. Suitable carrier materials include peat, alginate beads, methyl ethyl cellulose, corncobs, rice husk, sodium alginate, vegetable oils, coal perlite or clay.[90] Biochar is rich in nutrients and may provide a favorable habitat for bacterial proliferation and survival.[91] Biochar porosity and high surface area should favor proliferation of microbes. The inside pores may provide protection from desiccation, adverse pH, or toxic substances in soil *Pseudomonas libanensis* showed good survival in dynamotive, pyrovac and basque biochars used as carriers for bacterial preparations.[92] The *Bradyrhizobium japonicum* survived up to 9 months and pH together with pore area have been suggested as favorable in dynamotive and pyrovac biochar as carrier.[93] Plant growth promoting *Enterobacter cloacae* UW5 mixed with biochar (600 °C) produced from pinewood sustained higher population densities of bacteria compared to vermiculite.[94] Slow pyrolysis of agricultural wastes at 600 °C resulted in biochar formulation that augmented survival of *Burkholderia* sp. and plant growth, yield of tomato and soil activity. The biochar formulation even stimulated seed germination.[95]

In a study on biochar particle size and porosity[96] both small and large pores of biochar were clogged by soil particles after 56 d of incubation in soil as shown by scanning electron microscopy (SEM). Soil particles blocked smaller pores and attach to the exterior surfaces of the Simcoa biochars after 56 d of incubation. From the SEM studies, they concluded that all three woody biochars provided potential habitats for soil microorganisms due to their high porosity and large surface areas. The larger biochar pores showed after the 56-d incubation, hyphal colonization on biochar surfaces.

Pine wood biochar PBC-600 was produced at 600 °C to be used as bacterial carrier where inoculum treatments were compared with peat as a standard reference material.[97] After 1 month, the *Pseudomonas putida* UW4 population abundance in PBC-600 remained high around $4.9 \times 108\,CFU\,g-1$. After 5 months all tested PBC-600–related carriers had significantly higher cell numbers than that in peat. The authors suggested that the pore structure and internal surface area of biochar might provide better colonizable microsites for UW4 than what peat provided.

The utilization of biosorption processes emerge as an economical and eco-friendly alternative technology to remove pollutants present in the environment. One biosorption application is the immobilization of microbial cells for the removal of toxic pollutants in industrial wastewater.[98] Microbial cell carrier must be insoluble, non–biodegradable by the biomass, non-toxic, inexpensive, easy to handle and regenerate. Further, they must provide high cell mass loading capacity, high mechanical, biological and chemical stability in addition to the optimum diffusion of nutrients from flowing material to the center of carrier.[99,100]

The biochar surface contains several functional groups and elements and the pH determines the type of surface charge reported in the ever-increasing biochar literature.[3] It is though surprising that detailed studies of microbial interaction(s) on the biochar surfaces are lacking. One example of a detailed bacterial interaction study, an exception, is the study of Ye et al.[101] They hypothesized that specific bacterial communities would be recruited by the different mineral-enriched biochars (MEBs). These microbes carry out distinctive functions determined by specific aspects of the biochar surface. By sequencing of the bacterial 16S rRNA gene and metagenomics, they showed a substantial surface enrichment of specific soil bacteria with capacity for different kinds of chemolithotrophy based on genome-based predictions. They used a novel gold-label in situ hybridization (GISH) method and scanning electron microscopy to localize a dominant bacterium on the biochar particles. With scanning transmission electron microscopy, energy dispersive X-ray spectroscopy and electron energy loss spectroscopy, they were able to show that the dominant bacterium was involved in iron oxidation and assisting CO_2 fixation on the biochar surface. The metagenomes of these abundant surface bacteria were produced. The bacterium that was most abundant on all three tested biochar particles, OTU0001, represented a new gammaproteobacterial group of uncultured bacteria. These type of bacteria are often associated with carbonized materials that can deliver electrons to cells. The detected most abundant bacteria was a chemoautotroph that oxidizes Fe on the biochar surface. This novel group was only distantly related (about 10% 16S rRNA gene sequence divergence) to cultured organisms exemplified by *Acidiferrobacter thiooxydans*, which is an acidophilic iron-oxidizing bacterium. They concluded that deterministic processes govern the assembly of communities on biochar, which offers the potential to design MEBs that specifically enrich certain kind of bacteria.

Incorporation of organic pollutant-degrading bacteria into biochar to promote the biodegradation of organic contaminants by microorganisms and carbon material (CMs) immobilized complexes (McMICs) was the topic of very recent review.[102] Generally, McMICs-induced biodegradation of organic contaminants involves the initial adsorption on CMs through H-bonding, л-л stacking and hydrophobic effect and additionally the biodegradation by organic pollutant-degraders.[102] Bioreactors have extensively been used in bioremediation of organic pollutants by McMICs, since they are more efficient and reliable.[103]

Nonylphenol (NP) degrading bacteria were immobilized on bamboo biochar (BC) and wood biochar (WC) for short-term and long-term assays.[104] Results showed that cells immobilized on different biochar had different NP removal effects, and bacterial cells immobilized on bamboo biochar (I-BC) were more efficient. After eight rounds of long-term reuse, the cumulative removal rate and the degradation rate of NP in water by I-BC were 93.9% and 41.8%, respectively. That was significantly higher than with bacteria immobilized on wood biochar (69.6%, 22.7%) and free cells (64.7%, 19.4%) ($P < 0.01$). The long-term NP removal effect was more dependent on biodegradation. The amount of residual NP in I-BC still accounted for about 50%. SEM, quantitative PCR and 16SrRNA analysis showed promotion effect of biochar on microorganisms. *Pseudomonas, Achromobacter, Ochrobactrum* and *Stenotrophomonas* were predominant bacteria in NP degradation. The addition of biochar (especially bamboo charcoal) also effectively delayed the transformation of bacterial community structure.

References

1. Carson R. *Silent spring*. Houghton Mifflin Harcourt; 1962.
2. Winans K, Kendall A, Deng H. The history and current applications of the circular economy concept. *Renew Sustain Energy Rev* 2017;**68**:825–33.
3. Tisserant A, Cherubini F. Potentials, limitations, co-benefits, and trade-offs of biochar applications to soils for climate change mitigation. *Landarzt* 2019;**8**(12):179.
4. He X, Zheng N, Hu R, Hu Z, Yu JC. Hydrothermal and pyrolytic conversion of biomasses into catalysts for advanced oxidation treatments. *Adv Funct Mater* 2021; **31**(7):2006505.
5. Palansooriya KN, Wong JTF, Hashimoto Y, Huang L, Rinklebe J, Chang SX, et al. Response of microbial communities to biochar-amended soils: a critical review. *Biochar* 2019;**1**(1):3–22.
6. Xiao X, Chen B, Chen Z, Zhu L, Schnoor JL. Insight into multiple and multilevel structures of biochars and their potential environmental applications: a critical review. *Environ Sci Technol* 2018;**52**(9):5027–47.
7. Lehmann J, Joseph S. *Biochar for environmental management: science, technology and implementation*. Routledge; 2015.
8. Akolgo GA, Essandoh EO, Gyamfi S, Atta-Darkwa T, Kumi EN, de Freitas Maia CMB. The potential of a dual purpose improved cookstove for low income earners in Ghana—improved cooking methods and biochar production. *Renew Sustain Energy Rev* 2018;**82**:369–79.
9. Liu W-J, Jiang H, Yu H-Q. Development of biochar-based functional materials: toward a sustainable platform carbon material. *Chem Rev* 2015;**115**(22):12251–85.
10. Egamberdieva D, Hua M, Reckling M, Wirth S, Bellingrath-Kimura SD. Potential effects of biochar-based microbial inoculants in agriculture. *Environ Sustain* 2018; **1**(1):19–24.
11. Yuan J-H, Xu R-K, Zhang H. The forms of alkalis in the biochar produced from crop residues at different temperatures. *Bioresour Technol* 2011;**102**(3):3488–97.

12. Xu Y, Chen B. Investigation of thermodynamic parameters in the pyrolysis conversion of biomass and manure to biochars using thermogravimetric analysis. *Bioresour Technol* 2013;**146**:485–93.
13. Xiao X, Chen B, Zhu L. Transformation, morphology, and dissolution of silicon and carbon in Rice straw-derived biochars under different pyrolytic temperatures. *Environ Sci Technol* 2014;**48**(6):3411–9.
14. Qian L, Chen B, Hu D. Effective alleviation of aluminum Phytotoxicity by manure-derived biochar. *Environ Sci Technol* 2013;**47**(6):2737–45.
15. Gilli P, Gilli G. Hydrogen bond models and theories: the dual hydrogen bond model and its consequences. *J Mol Struct* 2010;**972**(1):2–10.
16. Zimmerman AR. Abiotic and microbial oxidation of laboratory-produced black carbon (biochar). *Environ Sci Technol* 2010;**44**(4):1295–301.
17. Yao Y, Gao B, Zhang M, Inyang M, Zimmerman AR. Effect of biochar amendment on sorption and leaching of nitrate, ammonium, and phosphate in a sandy soil. *Chemosphere* 2012;**89**(11):1467–71.
18. Taghizadeh-Toosi A, Clough TJ, Sherlock RR, Condron LM. Biochar adsorbed ammonia is bioavailable. *Plant and Soil* 2012;**350**(1):57–69.
19. Ngo P-T, Rumpel C, Ngo Q-A, Alexis M, Vargas GV, Gil MLM, et al. Biological and chemical reactivity and phosphorus forms of buffalo manure compost, vermicompost and their mixture with biochar. *Bioresour Technol* 2013;**148**:401–7.
20. Ni J, Pignatello JJ, Xing B. Adsorption of aromatic carboxylate ions to black carbon (biochar) is accompanied by proton exchange with water. *Environ Sci Technol* 2011;**45**(21):9240–8.
21. Bagreev A, Bandosz TJ, Locke DC. Pore structure and surface chemistry of adsorbents obtained by pyrolysis of sewage sludge-derived fertilizer. *Carbon* 2001;**39**(13):1971–9.
22. Li H, Dong X, da Silva EB, de Oliveira LM, Chen Y, Ma LQ. Mechanisms of metal sorption by biochars: biochar characteristics and modifications. *Chemosphere* 2017;**178**:466–78.
23. Chun Y, Sheng G, Chiou CT, Xing B. Compositions and Sorptive properties of crop residue-derived chars. *Environ Sci Technol* 2004;**38**(17):4649–55.
24. Nguyen BT, Lehmann J, Hockaday WC, Joseph S, Masiello CA. Temperature sensitivity of black carbon decomposition and oxidation. *Environ Sci Technol* 2010;**44**(9):3324–31.
25. Jin J, Li Y, Zhang J, Wu S, Cao Y, Liang P, et al. Influence of pyrolysis temperature on properties and environmental safety of heavy metals in biochars derived from municipal sewage sludge. *J Hazard Mater* 2016;**320**:417–26.
26. Wang X, Zhou W, Liang G, Song D, Zhang X. Characteristics of maize biochar with different pyrolysis temperatures and its effects on organic carbon, nitrogen and enzymatic activities after addition to fluvo-aquic soil. *Sci Total Environ* 2015;**538**:137–44.
27. Chen B, Zhou D, Zhu L. Transitional adsorption and partition of nonpolar and polar aromatic contaminants by biochars of pine needles with different pyrolytic temperatures. *Environ Sci Technol* 2008;**42**(14):5137–43.
28. Uchimiya M, Lima IM, Thomas Klasson K, Chang S, Wartelle LH, Rodgers JE. Immobilization of heavy metal ions (CuII, CdII, NiII, and PbII) by broiler litter-derived biochars in water and soil. *J Agric Food Chem* 2010;**58**(9):5538–44.
29. Chen T, Zhang Y, Wang H, Lu W, Zhou Z, Zhang Y, et al. Influence of pyrolysis temperature on characteristics and heavy metal adsorptive performance of biochar derived from municipal sewage sludge. *Bioresour Technol* 2014;**164**:47–54.
30. Subedi R, Taupe N, Pelissetti S, Petruzzelli L, Bertora C, Leahy JJ, et al. Greenhouse gas emissions and soil properties following amendment with manure-derived biochars: influence of pyrolysis temperature and feedstock type. *J Environ Manage* 2016;**166**:73–83.

31. Al-Wabel MI, Al-Omran A, El-Naggar AH, Nadeem M, Usman AR. Pyrolysis temperature induced changes in characteristics and chemical composition of biochar produced from conocarpus wastes. *Bioresour Technol* 2013;**131**:374–9.
32. Hossain MK, Strezov V, Chan KY, Ziolkowski A, Nelson PF. Influence of pyrolysis temperature on production and nutrient properties of wastewater sludge biochar. *J Environ Manage* 2011;**92**(1):223–8.
33. Zhang H, Voroney RP, Price GW. Effects of temperature and processing conditions on biochar chemical properties and their influence on soil C and N transformations. *Soil Biol Biochem* 2015;**83**:19–28.
34. Zhang J, Liu J, Liu R. Effects of pyrolysis temperature and heating time on biochar obtained from the pyrolysis of straw and lignosulfonate. *Bioresour Technol* 2015;**176**:288–91.
35. Qin L, Wu Y, Hou Z, Jiang E. Influence of biomass components, temperature and pressure on the pyrolysis behavior and biochar properties of pine nut shells. *Bioresour Technol* 2020;**313**:123682.
36. Li Y, Xing B, Ding Y, Han X, Wang S. A critical review of the production and advanced utilization of biochar via selective pyrolysis of lignocellulosic biomass. *Bioresour Technol* 2020; 123614.
37. Ahmad M, Lee SS, Lim JE, Lee S-E, Cho JS, Moon DH, et al. Speciation and phytoavailability of lead and antimony in a small arms range soil amended with mussel shell, cow bone and biochar: EXAFS spectroscopy and chemical extractions. *Chemosphere* 2014;**95**:433–41.
38. He L, Zhong H, Liu G, Dai Z, Brookes PC, Xu J. Remediation of heavy metal contaminated soils by biochar: mechanisms, potential risks and applications in China. *Environ Pollut* 2019;**252**:846–55.
39. Bashir S, Rizwan MS, Salam A, Fu Q, Zhu J, Shaaban M, et al. Cadmium immobilization potential of Rice straw-derived biochar, zeolite and rock phosphate: extraction techniques and adsorption mechanism. *Bull Environ Contam Toxicol* 2018;**100**(5):727–32.
40. Lahori AH, Guo Z, Zhang Z, Li R, Mahar A, Awasthi MK, et al. Use of biochar as an amendment for remediation of heavy metal-contaminated soils: prospects and challenges. *Pedosphere* 2017;**27**(6):991–1014.
41. Wang T, Sun H, Ren X, Li B, Mao H. Evaluation of biochars from different stock materials as carriers of bacterial strain for remediation of heavy metal-contaminated soil. *Sci Rep* 2017;**7**(1):12114.
42. Dehkhoda AM, Gyenge E, Ellis N. A novel method to tailor the porous structure of KOH-activated biochar and its application in capacitive deionization and energy storage. *Biomass Bioenergy* 2016;**87**:107–21.
43. Nowakowski DJ, Jones JM. Uncatalysed and potassium-catalysed pyrolysis of the cell-wall constituents of biomass and their model compounds. *J Anal Appl Pyrolysis* 2008;**83**(1):12–25.
44. Ahmad Z, Gao B, Mosa A, Yu H, Yin X, Bashir A, et al. Removal of cu(II), cd(II) and Pb(II) ions from aqueous solutions by biochars derived from potassium-rich biomass. *J Clean Prod* 2018;**180**:437–49.
45. El-Naggar A, Lee M-H, Hur J, Lee YH, Igalavithana AD, Shaheen SM, et al. Biochar-induced metal immobilization and soil biogeochemical process: an integrated mechanistic approach. *Sci Total Environ* 2020;**698**:134112.
46. Liu H, Xu F, Xie Y, Wang C, Zhang A, Li L, et al. Effect of modified coconut shell biochar on availability of heavy metals and biochemical characteristics of soil in multiple heavy metals contaminated soil. *Sci Total Environ* 2018;**645**:702–9.

47. Ramakrishnan M, Yrjälä K, Vinod KK, Sharma A, Cho J, Satheesh V, et al. Genetics and genomics of moso bamboo (Phyllostachys edulis): current status, future challenges, and biotechnological opportunities toward a sustainable bamboo industry. *Food Energy Secur* 2020;**9**(4):e229.
48. Qin P, Wang H, Yang X, He L, Müller K, Shaheen SM, et al. Bamboo- and pig-derived biochars reduce leaching losses of dibutyl phthalate, cadmium, and lead from co-contaminated soils. *Chemosphere* 2018;**198**:450–9.
49. Mukome FND, Buelow MC, Shang J, Peng J, Rodriguez M, Mackay DM, et al. Biochar amendment as a remediation strategy for surface soils impacted by crude oil. *Environ Pollut* 2020;**265**:115006.
50. Cao Y, Yang B, Song Z, Wang H, He F, Han X. Wheat straw biochar amendments on the removal of polycyclic aromatic hydrocarbons (PAHs) in contaminated soil. *Ecotoxicol Environ Saf* 2016;**130**:248–55.
51. Huang S, Wang T, Chen K, Mei M, Liu J, Li J. Engineered biochar derived from food waste digestate for activation of peroxymonosulfate to remove organic pollutants. *Waste Manag* 2020;**107**:211–8.
52. Chu J-H, Kang J-K, Park S-J, Lee C-G. Application of magnetic biochar derived from food waste in heterogeneous sono-Fenton-like process for removal of organic dyes from aqueous solution. *J Water Process Eng* 2020;**37**:101455.
53. Askeland M, Clarke BO, Cheema SA, Mendez A, Gasco G, Paz-Ferreiro J. Biochar sorption of PFOS, PFOA, PFHxS and PFHxA in two soils with contrasting texture. *Chemosphere* 2020;**249**:126072.
54. Silvani L, Cornelissen G, Botnen Smebye A, Zhang Y, Okkenhaug G, Zimmerman AR, et al. Can biochar and designer biochar be used to remediate per- and poly-fluorinated alkyl substances (PFAS) and lead and antimony contaminated soils? *Sci Total Environ* 2019;**694**:133693.
55. Bielská L, Škulcová L, Neuwirthová N, Cornelissen G, Hale SE. Sorption, bioavail-ability and ecotoxic effects of hydrophobic organic compounds in biochar amended soils. *Sci Total Environ* 2018;**624**:78–86.
56. Du J, Sun P, Feng Z, Zhang X, Zhao Y. The biosorption capacity of biochar for 4-bromodiphengl ether: study of its kinetics, mechanism, and use as a carrier for immobilized bacteria. *Environ Sci Pollut Res* 2016;**23**(4):3770–80.
57. Zhang C, Zhang N, Xiao Z, Li Z, Zhang D. Characterization of biochars derived from different materials and their effects on microbial dechlorination of pentachlorophenol in a consortium. *RSC Adv* 2019;**9**(2):917–23.
58. Frankel ML, Bhuiyan TI, Veksha A, Demeter MA, Layzell DB, Helleur RJ, et al. Removal and biodegradation of naphthenic acids by biochar and attached environmen-tal biofilms in the presence of co-contaminating metals. *Bioresour Technol* 2016;**216**:352–61.
59. Hagemann N, Schmidt H-P, Kägi R, Böhler M, Sigmund G, Maccagnan A, et al. Wood-based activated biochar to eliminate organic micropollutants from biologically treated wastewater. *Sci Total Environ* 2020;**730**:138417.
60. Hussain F, Hussain I, Khan AHA, Muhammad YS, Iqbal M, Soja G, et al. Combined application of biochar, compost, and bacterial consortia with Italian ryegrass enhanced phytoremediation of petroleum hydrocarbon contaminated soil. *Environ Exp Bot* 2018;**153**:80–8.
61. Denyes MJ, Langlois VS, Rutter A, Zeeb BA. The use of biochar to reduce soil PCB bioavailability to *Cucurbita pepo* and *Eisenia fetida*. *Sci Total Environ* 2012;**437**:76–82.
62. Yao Q, Liu J, Yu Z, Li Y, Jin J, Liu X, et al. Changes of bacterial community com-positions after three years of biochar application in a black soil of Northeast China. *Appl Soil Ecol* 2017;**113**:11–21.

63. Soja G. Interactions of biochar and biological degradation of aromatic hydrocarbons in contaminated soil. *Biochar Appl* 2016;247–67.
64. Song Y, Li Y, Zhang W, Wang F, Bian Y, Boughner LA, et al. Novel biochar-plant tandem approach for remediating Hexachlorobenzene contaminated soils: proof-of-concept and new insight into the rhizosphere. *J Agric Food Chem* 2016;**64**(27):5464–71.
65. Zhen M, Chen H, Liu Q, Song B, Wang Y, Tang J. Combination of rhamnolipid and biochar in assisting phytoremediation of petroleum hydrocarbon contaminated soil using *Spartina anglica*. *J Environ Sci* 2019;**85**:107–18.
66. Abbaspour A, Zohrabi F, Dorostkar V, Faz A, Acosta JA. Remediation of an oil-contaminated soil by two native plants treated with biochar and mycorrhizae. *J Environ Manage* 2020;**254**:109755.
67. Salt DE, Smith R, Raskin I. Phytoremediation. *Annu Rev Plant Biol* 1998; **49**(1):643–68.
68. Paz-Ferreiro J, Lu H, Fu S, Méndez A, Gascó G. Use of phytoremediation and biochar to remediate heavy metal polluted soils: a review. *Solid Earth* 2014;**5**(1):65–75.
69. Harindintwali JD, Zhou J, Yang W, Gu Q, Yu X. Biochar-bacteria-plant partnerships: eco-solutions for tackling heavy metal pollution. *Ecotoxicol Environ Saf* 2020;**204**: 111020.
70. Zhang M, Wang J, Bai SH, Zhang Y, Teng Y, Xu Z. Assisted phytoremediation of a co-contaminated soil with biochar amendment: contaminant removals and bacterial community properties. *Geoderma* 2019;**348**:115–23.
71. Gong X, Huang D, Liu Y, Zeng G, Chen S, Wang R, et al. Biochar facilitated the phytoremediation of cadmium contaminated sediments: metal behavior, plant toxicity, and microbial activity. *Sci Total Environ* 2019;**666**:1126–33.
72. Hamid Y, Tang L, Wang X, Hussain B, Yaseen M, Aziz MZ, et al. Immobilization of cadmium and lead in contaminated paddy field using inorganic and organic additives. *Sci Rep* 2018;**8**(1):17839.
73. Li X, Zhang X, Wang X, Cui Z. Phytoremediation of multi-metal contaminated mine tailings with *Solanum nigrum* L. and biochar/attapulgite amendments. *Ecotoxicol Environ Saf* 2019;**180**:517–25.
74. Li X, Wang X, Chen Y, Yang X, Cui Z. Optimization of combined phytoremediation for heavy metal contaminated mine tailings by a field-scale orthogonal experiment. *Ecotoxicol Environ Saf* 2019;**168**:1–8.
75. Fang G, Liu C, Gao J, Dionysiou DD, Zhou D. Manipulation of persistent free radicals in biochar to activate persulfate for contaminant degradation. *Environ Sci Technol* 2015;**49**(9):5645–53.
76. Yuan Y, Bolan N, Prévoteau A, Vithanage M, Biswas JK, Ok YS, et al. Applications of biochar in redox-mediated reactions. *Bioresour Technol* 2017;**246**:271–81.
77. Tang L, Liu Y, Wang J, Zeng G, Deng Y, Dong H, et al. Enhanced activation process of persulfate by mesoporous carbon for degradation of aqueous organic pollutants: Electron transfer mechanism. *Appl Catal Environ* 2018;**231**:1–10.
78. Li Y, Hu S, Chen J, Müller K, Li Y, Fu W, et al. Effects of biochar application in forest ecosystems on soil properties and greenhouse gas emissions: a review. *J Soil Sediment* 2018;**18**(2):546–63.
79. Chen B, Yuan M, Qian L. Enhanced bioremediation of PAH-contaminated soil by immobilized bacteria with plant residue and biochar as carriers. *J Soil Sediment* 2012;**12**(9):1350–9.
80. Quilliam RS, Glanville HC, Wade SC, Jones DL. Life in the 'charosphere' – does biochar in agricultural soil provide a significant habitat for microorganisms? *Soil Biol Biochem* 2013;**65**:287–93.
81. Liu X, Tian G, Jiang D, Zhang C, Kong L. Cadmium (cd) distribution and contamination in Chinese paddy soils on national scale. *Environ Sci Pollut Res* 2016; **23**(18):17941–52.

82. Rizwan M, Ali S, Hussain A, Ali Q, Shakoor MB, Zia-ur-Rehman M, et al. Effect of zinc-lysine on growth, yield and cadmium uptake in wheat (Triticum aestivum L.) and health risk assessment. *Chemosphere* 2017;**187**:35–42.

83. Liu Z, Lu Q, Wang C, Liu J, Liu G. Preparation of bamboo-shaped BiVO4 nanofibers by electrospinning method and the enhanced visible-light photocatalytic activity. *J Alloys Compd* 2015;**651**:29–33.

84. Li Y, Liu K, Wang Y, Zhou Z, Chen C, Ye P, et al. Improvement of cadmium phytoremediation by Centella asiatica L. after soil inoculation with cadmium-resistant Enterobacter sp. FM-1. *Chemosphere* 2018;**202**:280–8.

85. Liu J, Ding Y, Ma L, Gao G, Wang Y. Combination of biochar and immobilized bacteria in cypermethrin-contaminated soil remediation. *Int Biodeter Biodegr* 2017; **120**:15–20.

86. Chuaphasuk C, Prapagdee B. Effects of biochar-immobilized bacteria on phytoremediation of cadmium-polluted soil. *Environ Sci Pollut Res* 2019;**26**(23): 23679–88.

87. Yang X, Chen Z, Wu Q, Xu M. Enhanced phenanthrene degradation in river sediments using a combination of biochar and nitrate. *Science of The Total Environment* 2018;**619**–20:600–5.

88. Mendes R, Garbeva P, Raaijmakers JM. The rhizosphere microbiome: significance of plant beneficial, plant pathogenic, and human pathogenic microorganisms. *FEMS Microbiol Rev* 2013;**37**(5):634–63.

89. Hashem A, Abd Allah EF, Alqarawi AA, Al-Huqail AA, Wirth S. Egamberdieva D The interaction between arbuscular mycorrhizal fungi and endophytic bacteria enhances plant growth of *Acacia gerrardii* under salt stress. *Front Microbiol* 2016;**7**:1089.

90. Abd El-Fattah DA, Eweda WE, Zayed MS, Hassanein MK. Effect of carrier materials, sterilization method, and storage temperature on survival and biological activities of *Azotobacter chroococcum* inoculant. *Ann Agric Sci* 2013;**58**(2):111–8.

91. Pietikäinen J, Kiikkilä O, Fritze H. Charcoal as a habitat for microbes and its effect on the microbial community of the underlying humus. *Oikos* 2000;**89**(2):231–42.

92. Głodowska M, Husk B, Schwinghamer T, Smith D. Biochar is a growth-promoting alternative to peat moss for the inoculation of corn with a pseudomonad. *Agron Sustain Dev* 2016;**36**(1):21.

93. Glodowska M. *Biochar as a potential inoculant carrier for plant-beneficial bacteria.* LAP LAMBERT, Academic Publishing; 2015. ISBN 978-3-659-32047-7.

94. Hale L, Luth M, Crowley D. Biochar characteristics relate to its utility as an alternative soil inoculum carrier to peat and vermiculite. *Soil Biol Biochem* 2015; **81**:228–35.

95. Tripti, Kumar A, Usmani Z, Kumar V, Anshumali. Biochar and flyash inoculated with plant growth promoting rhizobacteria act as potential biofertilizer for luxuriant growth and yield of tomato plant. J Environ Manag. 2017;**190**:20–7.

96. Jaafar NM, Clode PL, Abbott LK. Soil microbial responses to biochars varying in particle size. *Surf Pore Properties Pedosphere* 2015;**25**(5):770–80.

97. Sun D, Hale L, Crowley D. Nutrient supplementation of pinewood biochar for use as a bacterial inoculum carrier. *Biol Fertil Soils* 2016;**52**(4):515–22.

98. Giese EC, Silva DDV, Costa AFM, Almeida SGC, Dussán KJ. Immobilized microbial nanoparticles for biosorption. *Crit Rev Biotechnol* 2020;**40**(5):653–66.

99. Bouabidi ZB, El-Naas MH, Zhang Z. Immobilization of microbial cells for the biotreatment of wastewater: a review. *Environ Chem Lett* 2019;**17**(1):241–57.

100. Qing W, Shanfeng Z. The application of immobilized microorganism technology in wastewater treatment. *Environ Sci Manag* 2008;**11**:020.

101. Ye J, Joseph SD, Ji M, Nielsen S, Mitchell DRG, Donne S, et al. Chemolithotrophic processes in the bacterial communities on the surface of mineral-enriched biochars. *ISME J* 2017;**11**(5):1087–101.

102. Wu P, Wang Z, Bhatnagar A, Jeyakumar P, Wang H, Wang Y, et al. Microorganisms-carbonaceous materials immobilized complexes: synthesis, adaptability and environmental applications. *J Hazard Mater* 2021; 125915.
103. Talha MA, Goswami M, Giri B, Sharma A, Rai B, Singh R. Bioremediation of Congo red dye in immobilized batch and continuous packed bed bioreactor by *Brevibacillus parabrevis* using coconut shell bio-char. *Bioresour Technol* 2018;**252**:37–43.
104. Lou L, Huang Q, Lou Y, Lu J, Hu B, Lin Q. Adsorption and degradation in the removal of nonylphenol from water by cells immobilized on biochar. *Chemosphere* 2019; **228**:676–84.
105. Wang Y, Li F, Rong X, Song H, Chen J. Remediation of petroleum-contaminated soil using bulrush straw powder, biochar and nutrients. *Bull Environ Contam Toxicol* 2017; **98**(5):690–7.
106. Hallin I, Douglas P, Doerr S, Matthews I, Bryant R, Charbonneau C. The potential of biochar to remove hydrophobic compounds from model sandy soils. *Geoderma* 2017;**285**:132–40.
107. Yavari S, Malakahmad A, Sapari NB, Yavari S. Synthesis optimization of oil palm empty fruit bunch and rice husk biochars for removal of imazapic and imazapyr herbicides. *J Environ Manage* 2017;**193**:201–10.
108. Ndoun MC, Elliott HA, Preisendanz HE, Williams CF, Knopf A, Watson JE. Adsorption of pharmaceuticals from aqueous solutions using biochar derived from cotton gin waste and guayule bagasse. *Biochar* 2021;**3**(1):89–104.
109. Ahmed MB, Zhou JL, Ngo HH, Guo W, Johir MAH, Sornalingam K. Single and competitive sorption properties and mechanism of functionalized biochar for removing sulfonamide antibiotics from water. *Chem Eng J* 2017;**311**:348–58.
110. Ahmed MB, Zhou JL, Ngo HH, Guo W, Johir MAH, Sornalingam K, et al. Chloramphenicol interaction with functionalized biochar in water: sorptive mechanism, molecular imprinting effect and repeatable application. *Sci Total Environ* 2017;**609**:885–95.
111. García-Delgado C, Eymar E, Camacho-Arévalo R, Petruccioli M, Crognale S, D'Annibale A. Degradation of tetracyclines and sulfonamides by stevensite-and biochar-immobilized laccase systems and impact on residual antibiotic activity. *J Chem Technol Biotechnol* 2018;**93**(12):3394–409.

CHAPTER ELEVEN

Influence of process parameters for production of biochar: A potential tool for an energy transition

Biswajit Samal[a], Kumar Raja Vanapalli[a], Brajesh Kumar Dubey[a,b,]*, and Jayanta Bhattacharya[a,c]

[a]School of Environmental Science and Engineering, Indian Institute of Technology, Kharagpur, West Bengal, India
[b]Department of Civil Engineering, Indian Institute of Technology, Kharagpur, West Bengal, India
[c]Department of Mining Engineering, Indian Institute of Technology, Kharagpur, West Bengal, India
*Corresponding author: e-mail address: bkdubey@civil.iitkgp.ac.in

Contents

1. Introduction	296
2. Source of biomass	298
3. Conversion techniques	299
3.1 Combustion	300
3.2 Gasification	300
3.3 Hydrothermal carbonization	301
3.4 Pyrolysis	301
4. Implications of process parameters on physicochemical and thermal properties of biochar	304
4.1 Implication of carrier gas and flow rate	304
4.2 Implication of heating rate	304
4.3 Implication of pressure	305
4.4 Implication of residence time	305
4.5 Implication of temperature	306
4.6 Implication of biomass composition	307
5. Application of biochar	307
6. Summary and future prospects	308
References	309

Abstract

Biomass refers to carbon-based materials derived from complexes of organic and inorganic materials or living matter either from the natural or anthropogenic source. The production of renewable energy options to replace traditional fossil fuels is necessitated by fossil fuel constraints, rising prices, and the advent of global environmental issues.

Thermochemical processes can turn biomass into biofuels and bioproducts, which has a lot of potential as a substitute source of energy. Pyrolysis is the most effective technique for converting biomass-based waste into value-added liquid, gaseous and solid products by heating in a deoxygenated atmosphere among all the available techniques. Biochar is a carbonaceous material with many applications in soil quality, water supply management, climate stability, energy generation and conservation. Pyrolysis conditions have a major impact on the concentrations and properties of this product. The organic elements of biomass, as well as pyrolytic process parameters such as time, temperature, heating intensity, carrier gas, carrier gas flow rate, and pressure influence the properties of biochar. Slow pyrolysis, low carrier gas flow rate, low heating rate, high pressure, longer residence time, low temperature, and feedstocks with high lignin content were found to be suitable conditions for biochar production. Biochar has desirable fuel properties and may be a reasonable way to solve some of the core problems with biomass as a fuel, such as high transportation costs and low grindability. Thermal conversion of biomass may be a feasible, renewable, and environmentally friendly solid fuel option in the future.

Keywords: Biomass, Biochar, Process parameters, Application of biochar, Solid fuel

1. Introduction

The foremost conventional energy supplies, such as gasoline, natural gas, and, coal are on the threshold of becoming extinct as energy demand continues to rise owing to fast industrialization and growth in population.[1] The current energy supplies are insufficient to meet the rising demands. Energy supplies in developing countries have doubled since 1973, but demand continues to rise.[2] reports that petrochemicals are set to account for nearly one third and half of the growth in oil demand by 2030 and 2050 respectively. The method of extracting energy from these sources pollutes the atmosphere, resulting in issues such as acid rain, global warming, and other issues. With the demand for energy and emissions concerns, a transition to a non-conventional source of energy such as biomass, wind, water, solar, and other renewables is unavoidable.

Biomass is an alternative source of energy that is renewable, potentially sustainable, and relatively environmentally friendly.[3] Biomass has been used as an energy source throughout history and is still a significant part of many countries' national energy sources today.[4] Biomass accounts for 43% of energy demand in developed countries and about one-seventh of overall global energy consumption.[5] Biomass is used to address a variety of energy needs, including power generation, vehicle fueling, and industrial heating. Biomass is one of the first energy source, mostly in rural areas where it is often the solitary available and reasonable source of energy. All of the living matter on the planet is made up of biomass and comes from growing plants such as trees, algae, and vegetables, as well as animal manure.[3] The organic matters in which solar energy is contained in chemical bonds are known as biomass resources. They are made up of nitrogen, oxygen, hydrogen, and carbon.[6]

Biomass is not concentrated in nature, so using it necessitates transportation, which raises the cost and decreases net energy output. The high moisture content, low bulk density, poor grindability and low energy density of biomass makes transportation and management tough and expensive.[7] Carbon monoxide, organic particulate matter, and other organic gases are generated when fuelwood is burned incompletely. In developing countries, where biomass is burned imperfectly in open fires for space heating and domestic cooking, the health effects of indoor air emissions are a major concern.[6] Biomass is unusual among renewable energy sources in that it directly retains energy from the sun and is the only renewable source of carbon that can be turned into convenient gaseous, liquid, solid fuels and chemical feedstock using thermochemical and biochemical technologies.[8–10] Most of the issues associated with using biomass directly can be addressed by thermochemical conversion of biomass to generate value-added products.

Due to its abundance, diversity, and flexible convertibility to fuels of solid, liquid, and gaseous phases, biomass has been the focus of research to promote it as an efficient source of energy.[11,12] Pyrolysis is a promising technique for bio-oil, syngas and bio-char among the thermochemical conversion processes (e.g., gasification, combustion and pyrolysis).[13,14] Syngas and bio-oil are two main intermediary products that can be used to make fuels that are more environmentally friendly than traditional fuels. Several studies concerning the upgrading and use of bio-oil and syngas for various applications have been conducted.[15,16] Biochar is a kind of carbonaceous material made by thermochemically converting biomass in an oxygen-deficient environment[17] has recently gained popularity for use in a variety of applications. Carbon is the most abundant component of biochar, followed by oxygen, ash, hydrogen and trace quantities of sulfur and nitrogen. Biochar's elemental composition varies depending on the raw material and condition of pyrolysis including temperature, heating rate, time, carrier gas flow rate, pressure, particle size, catalyst, and reactor bed height.[18–20]

Biochar has many advantages over traditional carbon materials including activated carbon and carbon black from coal, the feedstock is renewable, and the production procedure is easy. Soil amendment is the most widespread biochar application for reducing greenhouse gas emissions and improving soil quality. Since the char has a high calorific value, it can be used as activated charcoal or solid fuels for barbeques, and the gas component can fulfill the pyrolysis plant's overall energy needs.[21] Biochar is the strongest solid fossil fuel alternative for its low emissions and high calorific value.

Biochar applications are based on its characteristics, which are governed by pyrolysis process parameters. This chapter's main objective is to go further into the process parameters and understand how they affect the physicochemical and thermal characteristics of pyrolysis-produced biochar. For the enhancement of the fuel characteristics of biochar to use it as an energy source, the role of several process parameters including carrier gas and its flow rate, heating rate, pressure, residence time, temperature, and biomass composition was investigated.

2. Source of biomass

The sources and constituents of biomass can be used to classify it as shown in Fig. 1. Biomass can be anthropogenic (derived from the processing of natural biomass) or natural, and the polymeric composition of organic and inorganic molecules in its components can differ. Agricultural, forestry

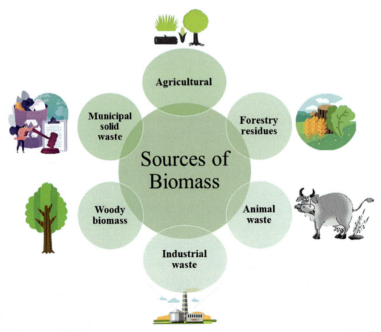

Fig. 1 Sources of biomass.

residues, human waste, wood processing residues, animal waste, algae, industrial waste and woody biomass are all types of biomass that can be categorized based on their sources. There are several ways to use agricultural resources on established lands without interfering with grain, feed, fiber, or forest product processing. Agricultural crop residues, such as stalks and leaves, are ample, varied, and generally available in the different parts of the country. Most countries around the world rely on woody and agricultural biomass for energy production.[22] Forest biomass feedstocks are classified as either forest residues leftover from harvesting (such as culled trees, tops, limbs, and tree components are unmarketable) or whole-tree biomass gathered specifically for biomass. On millions of acres of trees, there are still ways to use excessive biomass. Excessive woody biomass harvesting can help with forest regeneration, sustainability, vitality, and stability, as well as reduce the risk of fire and pests. This biomass could be harvested for bioenergy without compromising forest ecological structure and operation. Algae as bioenergy feedstocks apply to a large community of extremely active species such as microalgae, macroalgae (seaweed), and cyanobacteria (formerly known as "blue-green algae"). Many use sunlight and nutrients to produce biomass, which includes important components including proteins, carbohydrates, and lipids which can be transformed and promoted to biofuels and other items. Wood processing residues are byproducts and waste streams from the wood processing industry, and they have a huge amount of energy potential. Biofuels or bioproducts can be made from residues. These residues can be easy and relatively inexpensive sources of biomass for energy since they are already collected at the point of production.

Municipal solid waste (MSW) services include yard trimmings, paper and paperboard, bottles, rubber, leather, textiles, and food wastes, as well as mixed industrial and residential trash. Through diverting large volumes from landfills to the refinery, MSW for bioenergy also provides an incentive to eliminate residential and industrial waste. Non-food crops planted mainly for biomass on marginal land (land not suitable for conventional crops like soybeans and corn) are known as dedicated energy crops. When biomass is harvested, energy is not only produced and converted into usable energy, but it also takes care of the recycling and storage of waste materials to a large extent.

3. Conversion techniques

Biomass conversion can be accomplished by one of two methods: (i) biochemical conversion or (ii) thermochemical conversion (Fig. 2). To produce a value-added product, it is critical to choose an appropriate

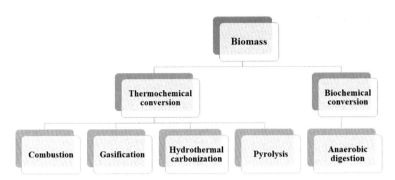

Fig. 2 Conversion technology of biomass.

technology for extracting energy from biomass. Anaerobic digestion for biogas processing and fermentation for alcohol production are examples of "biological conversion" techniques. The biochemical conversion processes provide a relatively small amount of value-added output. Nevertheless, this is a cost-effective and environmentally sustainable process that produces fewer greenhouse gases. Due to the above drawbacks, a thermochemical conversion is a common approach that is divided into four categories: combustion, pyrolysis, gasification and hydrothermal carbonization.

3.1 Combustion

One of the most viable and basic methods for biomass conversion is combustion.[23] When biomass is burnt at temperatures between 800 and 1000 °C under oxygenated conditions, a sequence of reactions occurs. The efficiency of heat generation is extremely high, and biomass heat is economically viable. However, in terms of environmental effects, there is a pressing need to boost biomass combustion. Steam loops are used in commercial electricity generation. Pretreatment strategies may increase productivity even more, but they may raise the overall expense of the operation. However, the increased cost can be offset by increased process efficiency.

3.2 Gasification

Gasification is a procedure that produces syngas as the main component, as well as certain char particles, ash, and tar, all of which may be used as fuel in the future.[24] However, since its first introduction in the 18th century, the implementation of such technology has met with mixed reception due to questions about the high cost and scarcity of other economic fuels, as well as an initial lack of environmental sensitivity.[24] Gasification is a process that entails heating biomass to high temperatures of up to 1000 °C, resulting in partial oxidation of fuel elements.[25] The gasification process is carried out in the presence of a vaporizing medium such as air, carbon dioxide (CO_2), steam (H_2O), or oxygen (O_2) inside a reactor called a gasifier, according to.[26] Set bed gasifiers, updraft gasifiers, fluidized bed gasifiers and downdraft gasifiers are among the various forms of gasifiers that have been stated to have evolved.[24,27] The yield of biochar produced during gasification is only 5–10% of raw biomass (dry basis), which is lower than that produced during fast pyrolysis.[24]

3.3 Hydrothermal carbonization

In comparison to other methods, hydrothermal carbonization is an effective hydrothermal conversion method for producing a specially engineered version of biochar, more commonly known as hydrochar. The hydrothermal treatment for converting biomass to biochar has many benefits, including the fact that it does not need an energy-intensive drying process, has a high conversion rate, is environmentally efficient, and requires a low operating temperature as compared to other thermal processes.[28] Biomass is heated to subcritical water temperature (180 to 350 °C) and autogenous pressure (>1 atmospheric pressure) within a reactor filled with steam in the hydrothermal conversion process. As compared to biochar obtained from slow pyrolysis, hydrochar has significantly different properties such as lower ash content, high carbon recovery, and more oxygen-containing functional groups on its surface.[29] In terms of cost and performance, hydrothermal carbonization is equivalent to slow pyrolysis for producing biochar with improved properties.

3.4 Pyrolysis

Pyrolysis can be defined as the thermal decomposition of biomass under oxygen-depleted conditions, yielding solid (i.e. biochar), liquid (i.e. oil) and gaseous products.[30] The absence of combustion of biomass is ensured by an oxygen-limited environment. Pyrolysis is one of the oldest techniques for producing biochar, but it encompasses a wide range of thermal decomposition, rendering it more difficult to distinguish from other methods. According to a study of the literature, scientists found pyrolysis as being close to carbonization, with the primary product becoming charcoal (solid).[30] Pyrolysis can be divided into two types based on the operating parameters: slow and fast pyrolysis.

3.4.1 Slow pyrolysis

Slow pyrolysis is the process of heating biomass for a long time in the absence of oxygen at temperatures ranging from 300 °C to 800 °C.[31,32] Low heating rate (5–7 °C min^{-1}) and long solid and vapor residence time characterize slow pyrolysis (usually more than 1 h).[33] Biochar is traditionally made by slow pyrolysis in batch kilns such as temporary pit kilns, mound kilns, stone kilns, metal kilns, and so on.[34] Furthermore, in the late 19th and early 20th century, commercial advances contributed to the establishment of industrial-scale pyrolysis setups.[35] Slow pyrolysis technology was updated for cleaner biochar processing during the late 20th century. Horizontal tubular kilns, agitated drum kilns, rotary kilns, and screw pyrolizers are some of the recorded examples of adapted setups for slow pyrolysis.[34,35] While biochar is the primary product of slow pyrolysis, the production of different quantities and forms of syngas as by-products is gaining growing interest for distillation-based capture, which is helping the process' commercial success.[35] The developments in pyrolysis technology have made biochar processing more controllable, simpler, economical, and environmentally friendly over the last 100 years.

3.4.2 Fast pyrolysis

Fast pyrolysis gained popularity after its first experiment in the late 1970s, and it has since been extensively used for the processing of bio-oil as the primary commodity, as well as some biochar.[36] To reduce secondary reactions, rapid pyrolysis uses a high heating rate (100 °C minute^{-1}) at a temperature range of 450–600 °C and a brief vapor residence period of around 2s to vaporize biomass and condensate it into a dark brown liquid.[37] This liquid has half the calorific value of natural oils or petroleum fuels. Fast pyrolysis is carried out in pressurized reactors, resulting in bio-oil yields of up to 80% on a dry feed basis, as well as the cogeneration of char and gas as by-products. So far, various types of reactors have been developed, including bubbling fluid beds, revolving fluid beds, spinning cone reactors, screw augur kilns, fixed bed reactors, and so on.[37]

Pyrolysis is considered one of the most suitable techniques among current technologies for converting biomass into value-added goods due to its broad applicability. This approach is unique in that it has intrinsic versatility and, most specifically, it deals with desired products.

Fast pyrolysis is favored for increased bio-oil yield whereas for increased biochar yield slow pyrolysis is recommended (Table 1). Through adjusting the operating conditions, the properties of pyrolysis products may be altered. Setting up larger pyrolysis plants may compensate for the higher cost of pyrolysis.

Table 1 Comparison of the different thermal conversion process of biomass.

Process	Temperature (°C)	Residence time (s/h/min/days)	Char yield (wt%)	Carbon content of char (wt%)	Carbon yield (Mass$_{carbon,product}$/ Mass$_{carbon,feedstock}$)	Reference
Slow pyrolysis	400–600	min to days	20–40	95	~0.58	38,39
Fast pyrolysis	500–1000	~1 s	12–26	74	0.2–0.26	40,41
Gasification	800–1000	5–20 s	~10	–	–	42
Hydrothermal carbonization	~180–250 °C	1–12 h	30–60	<70%	≈0.88	43
Torrefaction	~290 °C	10–60 min	61–84%	51–55%	0.67–0.85	44,45

4. Implications of process parameters on physicochemical and thermal properties of biochar

4.1 Implication of carrier gas and flow rate

The carrier gas (also known as purging gas or sweep gas) is used to remove volatiles from the pyrolysis system during the biomass pyrolysis process. Nitrogen, argon, hydrogen, carbon dioxide and are some of the purging agents available, However, nitrogen is the most often used because of its low cost, inert behavior, and ease of availability. The presence of H_2 during pyrolysis raises tar yield and coal fluidity significantly whereas N_2 increases the char yield.[46] As a consequence of the close relationship between char structure and thermoplastic properties, and the evolution of volatile matter, using hydrogen as a carrier gas can significantly alter the morphology of a char. By regulating the formation of vapors, the carrier gas flow rate influences the yield of products during pyrolysis. If these vapors aren't released, they will cause secondary reactions that change the structure and function of the pyrolysis components. The yield of biochar decreased with an increase in carrier gas (N_2) flow rate.[13,46] The yield of biochar from laurel (*Laurus nobilis* L.) extraction residues decreased from 24.4–22.6% to 28.48–27.21%, with increase in flow rate of nitrogen from 1.2–4.5 to 50–400 mL/min respectively.[13]

4.2 Implication of heating rate

The heating rate is considered to be an effective factor in pyrolysis product yields. The typical heating rates in slow pyrolysis are between 1 and 30 °C minute^{-1}.[47] Previous studies depict a positive relationship between the liquid oil yield and the heating rate of pyrolysis.[47,48] Secondary stages of pyrolysis and thermal cracking are absent at lower heating rates, preferring char production. The higher heating rates as practiced in fast pyrolysis usually resulted in increased liquid yield and reduces solid char yield.[48,49] This could have been possible due to higher conversion ratios of the thermally unstable biomass compounds to a liquid product due to reduced heat-mass transfers,- and resulting in faster depolymerization of solid biomass to primary volatiles that was responsible for the formation of char and coke.[50] Several reactions/processes take place concurrently during pyrolysis, including dehydration, decarbonization, volatilization, devolatilization, carbonization, and gasification/reforming. The high cellulose content of the biomass can explain the strong effect of the heating rate on the formation of char from biomass. In the past few authors have also opined that the production of char during pyrolysis was favored by a lower heating rate, with a longer residence time and within a moderate temperature range of 573–823 K.[48,51] However, most of the studies reported marginal to no significant effects of heating rate on any of the major thermal properties like proximate analysis results including fixed carbon and volatiles, H/C ratio, O/C ratio, high heat value (HHV), etc., were insignificant as compared to other major process parameters like pyrolysis temperature.[49,52,53] The HHV of the biochars produced from Safflower seeds increased from 28.15 to 30.06 MJ kg^{-1} for the heating rate of 10 °C min^{-1}, it increased from 28.51 to 30.17 MJ kg^{-1} for the heating rate of 30 °C min^{-1} and increased from 28.77 to 30.27 MJ kg^{-1} for the heating rate of 50 °C min^{-1} as the pyrolysis temperature was raised from 400 to 600 °C.[54]

4.3 Implication of pressure

Pyrolysis under increased pressure has also been shown to impact biochar yield. The pyrolysis reactions' heat demand is affected by both atmospheric pressure and particle size. When the pressure inside the reactor is higher than the atmospheric pressure, the yield of biochar increases. Increased pressure to 1 MPa increases char yield significantly compared to ambient pressure, with smaller improvements resulting from more rigorous conditions.[55] Pyrolysis at high pressure in a closed vessel produces high yields of biochar since the vapors are trapped and in contact with the stable products of pyrolysis. Water and tar vapors circulate at high pressure inside the reactor. If any of the feedstock in the reactor is comparatively cold, the water and tar vapors condense on it and transfer their heat of condensation to the feedstock. Condensing vapors are incredibly efficient at transferring heat. The feedstock's fixed-carbon yields (29.9%) are greatly improved as pressures are increased.[56] The industry will benefit greatly from a simple prediction of the fixed-carbon yield produced by pyrolysis at elevated pressure. Increasing pressure also increases carbon concentration, which improves the energy density of the biochar produced.[40]

4.4 Implication of residence time

Regardless of the raw materials, the pyrolysis residence time has a minor effect on the properties and yield of biochar as compared to the pyrolysis temperature.[57] The residence time that volatile matter produced during pyrolysis spent in contact with hot biochar affects biochar yield. For an increase in the duration of residence time, the yield of biochar marginally decreases. However, at a low temperature of 600 °C, residence time had no impact on biochar yield. This is because, under high-temperature conditions, increasing the residence time primarily affects the surface and internal structure of the biochar, while having a minor impact on biochar yield.[58] The ratio of carbon in biochar increases significantly as the residence time of slow pyrolysis increases, but differences in the proportions of nitrogen, hydrogen, and oxygen are not very precise.[59,60] The study by wang et al.[61] reported surface areas of the biochars produced by co-pyrolysis of sewage sludge and cotton stalks rose between 30 and 90 min, whereas values dropped between 90 and 150 min, which may be due to the gasification of the few well-developed micropore walls. Prolonged residence times raised the pH and ash content of the biochars, but they lowered the C, N, and H content.[61]

4.5 Implication of temperature

The temperature of pyrolysis has a significant impact on the properties of biochar.[62,63] The yield of biochar is negatively correlated with pyrolytic temperature.[57] As the temperature rises, the thermal cracking of high-molecular-weight hydrocarbons in biomass increases, increasing the production of liquid and gaseous products while decreasing the yield of biochar. Due to the partial decomposition of biomass, the yield of biochar was observed to be high at lower temperatures. The yield of biochar from agricultural waste was 51.1% at 250 °C that reduced to 33.5% and 24.3% at 350 and 450 °C.[64]. According to studies, the carbon percentage in biochar increases as the slow pyrolysis temperature rises, but its nitrogen, oxygen, hydrogen and sulfur content declines.[59,65] The lower nitrogen and sulfur content of biochars suggests that, like raw biomass combustion, biochar combustion will have comparable environmental benefits. Biochar have slightly lower O/C and H/C ratios than parent biomass, which decrease with rising pyrolysis temperature.[7] With the rise in pyrolysis temperature, the volatile matter content of biochar decreases while the fixed carbon content rises.[28] As a result, the fuel ratios (the proportion of fixed carbon to volatile carbon) of biochar increase as the temperature rises. According to ASTM 388, a fuel ratio is a characteristic value that represents a solid fuel's property and is used to distinguish coal rank. When comparing raw biomass combustion to biochars combustion, the increased fuel ratios mean higher combustion efficiencies and lower pollutant emissions. The mass energy density of biochars increases significantly as a result of pyrolysis.

This can be explained by the fact that carbon-carbon bonds provide more energy than carbon-hydrogen and carbon-oxygen bonds. Due to extreme pyrolysis reactions at elevated temperatures, the mass energy density of biochar increases with pyrolysis temperature. Despite improved HHVs, the energy yields decline as the temperature rises due to the biochar's considerably lower mass yields.[28] Biochar's ash content and specific surface area increase with the rise in temperature[66] whereas, the volatile matter content and cation exchange capacity in the biochar decreases.[57,66] A low temperature and a long residence time are ideal for producing an extremely stable biochar. The ideal treatment temperature for making biochar is below 300 °C, since this generates the maximum biochar yield (when compared to yields at higher temperatures) and the best biochar properties.[67]

4.6 Implication of biomass composition

Biomass constituents affect the quality and structure of pyrolytic products. Biomass is made up of lignin, hemicellulose and cellulose which affect the existence of thermochemical process items. Hemicellulose is the first to decompose during pyrolysis, beginning at 220 °C and completing at 315 °C; while cellulose is converted to non-condensable gas, condensable organic vapors, and aerosols once 400 °C is attained.[68] Lignin-rich biomass, on the other hand, is used to make solid products like biochar.[69] Feedstock with high fixed carbon contents yields large quantities of char, whereas those with higher volatile contents generate a better quality of liquid and gaseous products. On the other hand, feedstock with high moisture contents produce large amounts of liquid byproducts in contrast to gaseous or solid byproducts, biochar production is favored by biomass with a low moisture content.[70] At higher pyrolysis temperatures, biochar made from animal litter and solid waste feedstocks has lower surface areas, carbon content, volatile matter, and CEC than biochar made from crop residue and wood biomass. The reason for this difference is because the composition of lignin and cellulose, as well as the moisture content of biomass, varies greatly.[69]

5. Application of biochar

Biomass is a renewable fossil fuel source and intermediate that can be turned into a variety of useful materials. Due to its inherent properties, biochar is regarded as the most essential of these products. Biochar is used in a variety of areas, including soil, air, water, and energy. Biochar is more desirable for direct carbon fuel (DCFC) use for energy production.[71]

Biochar-based catalysts are a potential alternative to conventional catalysts, which are either costly or non-renewable. Biofuel processing using biochar-based catalysts reflects a more sustainable and optimized biorefinery scheme.[72] Because of its well-developed synthesis techniques and organic feedstock biochar processing is inexpensive and convenient. Second, different techniques were built to adapt biochar's physicochemical properties to its intended use. Supercapacitors are used in portable electric devices and hybrid cars to store electricity. Secondary batteries are said to have lower strength, lower energy capacity, a shorter life cycle and lower chemical stability than supercapacitors.[73] Since carbon materials have a large pseudo capacitance and a high electric conductivity, they have recently been used as supercapacitor electrodes. Biochar is a promising electrode for supercapacitors and lithium-ion batteries because of its low cost, electric conductivity, pore structure, surface functional groups, and specific surface area.[74] Biochar can also be used as an energy source because of its high calorific value and fixed carbon, replacing the non-renewable fossil fuel as it has O/C and H/C ratios lower or equal to coal.[7] Biochar could be used for the co-combustion of fire coal for large boilers because blended combustion of biochar and coal revealed that biochar could efficiently reduce the ignition and burnout temperatures of fuel mixtures while increasing their conversion efficiency and combustion characteristics. Small households and large-scale central heating systems can also benefit from biochar, which is environmentally friendly and has a high combustion efficiency. Biochar's use in energy recovery technology, as well as its ability for CO_2 capture and catalytic conversion of CO_2 to fuels and energy, would require further study in the future. To fully understand the uses of biochar, process optimization must be carried out with reduced costs associated and energy inputs with the by-product's strong applicability on a higher scale.

6. Summary and future prospects

Biomass is a unique resource since it is renewable and can be turned into a variety of energy fuel sources and chemical. Even after the invention of coal and oil, the use of charcoal as a superior energy source is quite common. Pyrolysis is a thermochemical process that turns biomass into usable byproducts like biochar. Biochar is an environmentally sustainable substitute for activated carbon and other carbon materials created through the thermochemical decomposition of biomass. It not only declines the amount of carbon absorbed into the environment but also lowers the amount of carbon released into the atmosphere. Biochar, among the different biomass products, will be the most desirable for its unique chemical and physical properties. The carrier gas and flow rate, time, pressure, heating rate, temperature and type of feedstock all play a role in the production of biochar. Biochar's physical and chemical properties, as well as its combustion characteristics, are primarily determined by the biomass feedstocks and carbonization processes used. Slow pyrolysis, a low heating rate, high pressure, low carrier gas flow rate, extended residence time and low temperature were found to be suitable conditions for biochar processing. Biochar is continuously being used in a variety of areas. Biochar has a wide range of traditional and emerging uses and can be both a cheap fuel and a costly carbon material. With the severity of global warming predicted to rise, biochar's value is expected to grow evenmore. In this respect, increasing biochar's uses and improving its productivity would lead to the improvement of global carbon cycle sustainability.

References

1. Demiral İ, Şensöz S. The effects of different catalysts on the pyrolysis of industrial wastes (olive and hazelnut bagasse). *Bioresour Technol* 2008;**99**(17):8002–7. https://doi.org/10.1016/j.biortech.2008.03.053.
2. *The future of petrochemicals*. International Energy Agency; 2018. p. 11.
3. Saxena RC, Adhikari DK, Goyal HB. Biomass-based energy fuel through biochemical routes: a review. *Renew Sust Energ Rev* 2009;**13**(1):167–78. https://doi.org/10.1016/j.rser.2007.07.011.
4. Gerçel HF. Production and characterization of pyrolysis liquids from sunflower-pressed bagasse. *Bioresour Technol* 2002;**85**(2):113–7. https://doi.org/10.1016/S0960-8524(02)00101-3.
5. Hall DO, Barnard GW, Moss PA. *Biomass for energy in the developing countries: current role, potential, problems, prospects*. Elsevier; 2013.
6. Demirbaş A. Biomass resource facilities and biomass conversion processing for fuels and chemicals. *Energy Convers Manag* 2001;**42**(11):1357–78. https://doi.org/10.1016/S0196-8904(00)00137-0.
7. Abdullah H, Wu H. Biochar as a fuel: 1. Properties and grindability of biochars produced from the pyrolysis of Mallee wood under slow-heating conditions. *Energy Fuel* 2009;**23**(8):4174–81. https://doi.org/10.1021/ef900494t.
8. Ozbay N, Pütün AE, Uzun BB, Pütün E. Biocrude from biomass: pyrolysis of cottonseed cake. *Renew Energy* 2001;**24**(3):615–25. https://doi.org/10.1016/S0960-1481(01)00048-9.
9. Qian Y, Zuo C, Tan J, He J. Structural analysis of bio-oils from sub-and supercritical water liquefaction of Woody biomass. *Energy* 2007;**32**(3):196–202. https://doi.org/10.1016/j.energy.2006.03.027.
10. Sensoz S. Slow pyrolysis of wood barks from Pinus Brutia ten. And product compositions. *Bioresour Dent Tech* 2003;**89**(3):307–11. https://doi.org/10.1016/S0960-8524(03)00059-2.
11. Abnisa F, Wan Daud WMA. A review on co-pyrolysis of biomass: an optional technique to obtain a high-grade pyrolysis oil. *Energy Convers Manag* 2014;**87**:71–85. https://doi.org/10.1016/j.enconman.2014.07.007.

12. Dewangan A, Pradhan D, Singh RK. Co-pyrolysis of sugarcane bagasse and low-density polyethylene: influence of plastic on pyrolysis product yield. *Fuel* 2016;**185**:508–16. https://doi.org/10.1016/j.fuel.2016.08.011.
13. Ertas M, Hakkı Alma M. Pyrolysis of Laurel (Laurus Nobilis L.) extraction residues in a fixed-bed reactor: characterization of bio-oil and bio-char. *J Anal Appl Pyrolysis* 2010;**88**(1):22–9. https://doi.org/10.1016/j.jaap.2010.02.006.
14. Islam MN, Islam MN, Beg MRA. The fuel properties of pyrolysis liquid derived from urban solid wastes in Bangladesh. *Bioresour Technol* 2004;**92**(2):181–6. https://doi.org/10.1016/j.biortech.2003.08.009.
15. Mortensen PM, Grunwaldt J-D, Jensen PA, Knudsen KG, Jensen AD. A review of catalytic upgrading of bio-oil to engine fuels. *Appl Catal A Gen* 2011;**407**(1):1–19. https://doi.org/10.1016/j.apcata.2011.08.046.
16. Swain PK, Das LM, Naik SN. Biomass to liquid: a prospective challenge to research and development in 21st century. *Renew Sust Energ Rev* 2011;**15**(9):4917–33. https://doi.org/10.1016/j.rser.2011.07.061.
17. Qian K, Kumar A, Zhang H, Bellmer D, Huhnke R. Recent advances in utilization of biochar. *Renew Sust Energ Rev* 2015;**42**:1055–64. https://doi.org/10.1016/j.rser.2014.10.074.
18. Kambo HS, Dutta A. Strength, storage, and combustion characteristics of densified lignocellulosic biomass produced via torrefaction and hydrothermal carbonization. *Appl Energy* 2014;**135**:182–91. https://doi.org/10.1016/j.apenergy.2014.08.094.
19. Yuan H, Lu T, Wang Y, Huang H, Chen Y. Influence of pyrolysis temperature and holding time on properties of biochar derived from medicinal herb (Radix Isatidis) residue and its effect on soil CO2 emission. *J Anal Appl Pyrolysis* 2014;**110**:277–84. https://doi.org/10.1016/j.jaap.2014.09.016.
20. Zhao C, Jiang E, Chen A. Volatile production from pyrolysis of cellulose, hemicellulose and lignin. *J Energy Inst* 2017;**90**(6):902–13. https://doi.org/10.1016/j.joei.2016.08.004.
21. Horne PA, Williams PT. Premium quality fuels and chemicals from the fluidised bed pyrolysis of biomass with zeolite catalyst upgrading. *Renew Energy* 1994;**5**(5):810–2. https://doi.org/10.1016/0960-1481(94)90093-0.
22. Hall DO. Biomass energy in industrialised countries—a view of the future. *For Ecol Manag* 1997;**91**(1):17–45. https://doi.org/10.1016/S0378-1127(96)03883-2.
23. Nussbaumer T. Combustion and co-combustion of biomass: fundamentals, technologies, and primary measures for emission reduction. *Energy Fuel* 2003;**17**(6):1510–21. https://doi.org/10.1021/ef030031q.
24. Couto N, Rouboa A, Silva V, Monteiro E, Bouziane K. Influence of the biomass gasification processes on the final composition of syngas. *Energy Procedia* 2013;**36**:596–606. https://doi.org/10.1016/j.egypro.2013.07.068.
25. Arafat HA, Jijakli K. Modeling and comparative assessment of municipal solid waste gasification for energy production. *Waste Manag* 2013;**33**(8):1704–13. https://doi.org/10.1016/j.wasman.2013.04.008.
26. Sikarwar VS, Zhao M, Clough P, Yao J, Zhong X, Memon MZ, et al. An overview of advances in biomass gasification. *Energy Environ Sci* 2016;**9**(10):2939–77. https://doi.org/10.1039/C6EE00935B.
27. Bulkowska K, Gusiatin ZM, Klimiuk E, Pawlowski A, Pokoj T. *Biomass for biofuels.* CRC Press; 2016.
28. Liu Z, Quek A, Kent Hoekman S, Balasubramanian R. Production of solid biochar fuel from waste biomass by hydrothermal carbonization. *Fuel* 2013;**103**:943–9. https://doi.org/10.1016/j.fuel.2012.07.069.
29. Kang S, Li X, Fan J, Chang J IE. Characterization of hydrochars produced by hydrothermal carbonization of lignin, cellulose, d-xylose, and wood meal. *Ind Eng Chem Res* 2012;**51**:9023–31. https://doi.org/10.1021/ie300565d.

30. Basu P. *Biomass gasification, pyrolysis and torrefaction—2nd Edition.* https://www.elsevier.com/books/biomass-gasification-pyrolysis-and-torrefaction/basu/978-0-12-396488-5.
31. Laird D, Fleming P, Wang B, Horton R, Karlen D. Biochar impact on nutrient leaching from a Midwestern agricultural soil. *Geoderma* 2010;**158**(3–4):436–42. https://doi.org/10.1016/j.geoderma.2010.05.012.
32. Liu Z, Han G. Production of solid fuel biochar from waste biomass by low temperature pyrolysis. *Fuel* 2015;**158**:159–65. https://doi.org/10.1016/j.fuel.2015.05.032.
33. Liu W-J, Jiang H, Yu H-Q. Development of biochar-based functional materials: toward a sustainable platform carbon material. *Chem Rev* 2015;**115**(22):12251–85. https://doi.org/10.1021/acs.chemrev.5b00195.
34. Brown R. *Biochar production technology.* Routledge; 2012. p. 159–78. https://doi.org/10.4324/9781849770552-15.
35. Brownsort PA. *Biomass pyrolysis processes: performance parameters and their influence on biochar system benefits.* MASc thesis from the University of Edinburg; 2009.
36. Bridgwater AV, Meier D, Radlein D. An overview of fast pyrolysis of biomass. *Org Geochem* 1999;**30**(12):1479–93. https://doi.org/10.1016/S0146-6380(99)00120-5.
37. Bridgwater AV. Review of fast pyrolysis of biomass and product upgrading. *Biomass Bioenergy* 2012;**38**:68–94. https://doi.org/10.1016/j.biombioe.2011.01.048.
38. González JF, Román S, Encinar JM, Martínez G. Pyrolysis of various biomass residues and char utilization for the production of activated carbons. *J Anal Appl Pyrolysis* 2009;**85**(1):134–41. https://doi.org/10.1016/j.jaap.2008.11.035.
39. Uchimiya M, Wartelle LH, Klasson KT, Fortier CA, Lima IM. Influence of pyrolysis temperature on biochar property and function as a heavy metal sorbent in soil. *J Agric Food Chem* 2011;**59**(6):2501–10. https://doi.org/10.1021/jf104206c.
40. Antal MJ, Grønli M. The art, science, and technology of charcoal production. *Ind Eng Chem Res* 2003;**42**(8):1619–40. https://doi.org/10.1021/ie0207919.
41. DeSisto WJ, Hill N, Beis SH, Mukkamala S, Joseph J, Baker C, et al. Fast pyrolysis of pine sawdust in a fluidized-bed reactor. *Energy Fuel* 2010;**24**(4):2642–51. https://doi.org/10.1021/ef901120h.
42. Meyer S, Glaser B, Quicker P. Technical, economical, and climate-related aspects of biochar production technologies: a literature review. *Environ Sci Technol* 2011;**45**(22):9473–83. https://doi.org/10.1021/es201792c.
43. Libra JA, Ro KS, Kammann C, Funke A, Berge ND, Neubauer Y, et al. Hydrothermal carbonization of biomass residuals: a comparative review of the chemistry, processes and applications of wet and dry pyrolysis. *Biofuels* 2011;**2**(1):71–106. https://doi.org/10.4155/bfs.10.81.
44. Bridgwater A. v. the production of biofuels and renewable chemicals by fast pyrolysis of biomass. *Int J Glob Energy Issues* 2007;**27**(2):160–203. https://doi.org/10.1504/IJGEI.2007.013654.
45. Yan W, Acharjee TC, Coronella CJ, Vásquez VR. Thermal pretreatment of lignocellulosic biomass. *Environ Prog Sustain Energy* 2009;**28**(3):435–40. https://doi.org/10.1002/ep.10385.
46. Yan Q, Toghiani H, Yu F, Cai Z, Zhang J. Effects of pyrolysis conditions on yield of bio-chars from pine chips. *For Prod J* 2011;**61**(5):367–71. https://doi.org/10.13073/0015-7473-61.5.367.
47. Lua AC, Yang T, Guo J. Effects of pyrolysis conditions on the properties of activated carbons prepared from pistachio-nut shells. *J Anal Appl Pyrolysis* 2004;**72**(2):279–87. https://doi.org/10.1016/j.jaap.2004.08.001.
48. Uzoejinwa BB, He X, Wang S, El-Fatah Abomohra A, Hu Y, Wang Q. Co-pyrolysis of biomass and waste plastics as a thermochemical conversion technology for high-grade bio-fuel production: recent progress and future directions elsewhere worldwide. *Energy Convers Manag* 2018;**163**:468–92. https://doi.org/10.1016/j.enconman.2018.02.004.

49. Zhao B, O'Connor D, Zhang J, Peng T, Shen Z, Tsang DCW, et al. Effect of pyrolysis temperature, heating rate, and residence time on rapeseed stem derived biochar. *J Clean Prod* 2018;**174**:977–87. https://doi.org/10.1016/j.jclepro.2017.11.013.

50. Uzun BB, Apaydin-Varol E, Ateş F, Özbay N, Pütün AE. Synthetic fuel production from tea waste: characterisation of bio-oil and bio-char. *Fuel* 2010;**89**(1):176–84. https://doi.org/10.1016/j.fuel.2009.08.040.

51. Anuar Sharuddin SD, Abnisa F, Wan Daud WMA, Aroua MK. Energy recovery from pyrolysis of plastic waste: study on non-recycled plastics (NRP) data as the real measure of plastic waste. *Energy Convers Manag* 2017;**148**:925–34. https://doi.org/10.1016/j.enconman.2017.06.046.

52. Anuar Sharuddin SD, Abnisa F, Wan Daud WMA, Aroua MK. A review on pyrolysis of plastic wastes. *Energy Convers Manag* 2016;**115**:308–26. https://doi.org/10.1016/j.enconman.2016.02.037.

53. Li A, Liu H, Wang H, Xu HB, Jin LF, Liu JL, et al. Effects of temperature and heating rate on the characteristics of molded bio-char. *BioResources* 2016;**11**(2):3259–74. https://bioresources.cnr.ncsu.edu/.

54. Angın D. Effect of pyrolysis temperature and heating rate on biochar obtained from pyrolysis of safflower seed press cake. *Bioresour Technol* 2013;**128**:593–7. https://doi.org/10.1016/j.biortech.2012.10.150.

55. Itoh T, Iwabuchi K, Maemoku N, Chen S, Taniguro K. Role of ambient pressure in self-heating torrefaction of dairy cattle manure. *PLoS One* 2020;**15**(5):e0233027. https://doi.org/10.1371/journal.pone.0233027.

56. Antal MJ, Allen SG, Dai X, Shimizu B, Tam MS, Grønli M. Attainment of the theoretical yield of carbon from biomass. *Ind Eng Chem Res* 2000;**39**(11):4024–31. https://doi.org/10.1021/ie000511u.

57. Sun J, He F, Pan Y, Zhang Z. Effects of pyrolysis temperature and residence time on physicochemical properties of different biochar types. *Acta Agric Scand Sect B Soil Plant Sci* 2017;**67**(1):12–22. https://doi.org/10.1080/09064710.2016.1214745.

58. Braadbaart F, Poole I. Morphological, chemical and physical changes during charcoalification of wood and its relevance to archaeological contexts. *J Archaeol Sci* 2008;**35**(9):2434–45. https://doi.org/10.1016/j.jas.2008.03.016.

59. Crombie K, Mašek O, Sohi SP, Brownsort P, Cross A. The effect of pyrolysis conditions on biochar stability as determined by three methods. *GCB Bioenergy* 2013;**5**(2):122–31. https://doi.org/10.1111/gcbb.12030.

60. Suman S, Gautam S. Effect of pyrolysis time and temperature on the characterization of biochars derived from biomass. *Energy Sources, Part A* 2017;**39**(9):933–40. https://doi.org/10.1080/15567036.2016.1276650.

61. Wang Z, Liu K, Xie L, Zhu H, Ji S, Shu X. Effects of residence time on characteristics of biochars prepared via co-pyrolysis of sewage sludge and cotton stalks. *J Anal Appl Pyrolysis* 2019;**142**:104659. https://doi.org/10.1016/j.jaap.2019.104659.

62. Chandra S, Bhattacharya J. Influence of temperature and duration of pyrolysis on the property heterogeneity of rice straw biochar and optimization of pyrolysis conditions for its application in soils. *J Clean Prod* 2019;**215**:1123–39. https://doi.org/10.1016/j.jclepro.2019.01.079.

63. Wu W, Yang M, Feng Q, McGrouther K, Wang H, Lu H, et al. Chemical characterization of Rice straw-derived biochar for soil amendment. *Biomass Bioenergy* 2012;**47**:268–76. https://doi.org/10.1016/j.biombioe.2012.09.034.

64. Waqas M, Aburiazaiza AS, Miandad R, Rehan M, Barakat MA, Nizami AS. Development of biochar as fuel and catalyst in energy recovery technologies. *J Clean Prod* 2018;**188**:477–88. https://doi.org/10.1016/j.jclepro.2018.04.017.

65. Hossain MK, Strezov V, Chan KY, Ziolkowski A, Nelson PF. Influence of pyrolysis temperature on production and nutrient properties of wastewater sludge biochar. *J Environ Manag* 2011;**92**(1):223–8. https://doi.org/10.1016/j.jenvman.2010.09.008.

66. Suliman W, Harsh JB, Abu-Lail NI, Fortuna A-M, Dallmeyer I, Garcia-Perez M. Influence of feedstock source and pyrolysis temperature on biochar bulk and surface properties. *Biomass Bioenergy* 2016;**84**:37–48. https://doi.org/10.1016/j.biombioe.2015.11.010.

67. Jouhara H, Ahmad D, van den Boogaert I, Katsou E, Simons S, Spencer N. Pyrolysis of domestic based feedstock at temperatures up to 300 °C. *Therm Sci Eng Prog* 2018;**5**:117–43. https://doi.org/10.1016/j.tsep.2017.11.007.

68. Ozsin G, Putun AE. A comparative study on co-pyrolysis of lignocellulosic biomass with polyethylene terephthalate, polystyrene, and polyvinyl chloride: synergistic effects and product characteristics. *J Clean Prod* 2018;**205**:1127–38. https://doi.org/10.1016/j.jclepro.2018.09.134.

69. Tomczyk A, Sokołowska Z, Boguta P. Biochar physicochemical properties: pyrolysis temperature and feedstock kind effects. *Rev Environ Sci Biotechnol* 2020;**19**(1):191–215. https://doi.org/10.1007/s11157-020-09523-3.

70. Demirbas A. Effect of initial moisture content on the yields of oily products from pyrolysis of biomass. *J Anal Appl Pyrolysis* 2004;**71**(2):803–15. https://doi.org/10.1016/j.jaap.2003.10.008.

71. Palniandy LK, Yoon LW, Wong WY, Yong S-T, Pang MM. Application of biochar derived from different types of biomass and treatment methods as a fuel source for direct carbon fuel cells. *Energies* 2019;**12**(13):2477. https://doi.org/10.3390/en12132477.

72. Cheng F, Li X. Preparation and application of biochar-based catalysts for biofuel production. *Catalysts* 2018;**8**(9):346. https://doi.org/10.3390/catal8090346.

73. Inal IIG, Holmes SM, Banford A, Aktas Z. The performance of supercapacitor electrodes developed from chemically activated carbon produced from waste tea. *Appl Surf Sci* 2015;**357**:696–703. https://doi.org/10.1016/j.apsusc.2015.09.067.

74. Xiu S, Shahbazi A, Li R. Characterization, modification and application of biochar for energy storage and catalysis: a review. *Trends Renew Energy* 2017;**3**(1):86–101. https://doi.org/10.17737/tre.2017.3.1.0033.

Printed in the United States
by Baker & Taylor Publisher Services